Principles of Horticulture

Principles of Horticulture

ERVIN L. DENISEN Ph.D.

Professor, Department of Horticulture, Iowa State University / **Second Edition**

MACMILLAN PUBLISHING CO., INC.

New York

Collier Macmillan Publishers

London

Earlier edition © 1958 by Macmillan Publishing Co., Inc.

Macmillan Publishing Co., Inc.
866 Third Avenue, New York, New York 10022

Collier Macmillan Canada, Ltd.

Library of Congress Cataloging in Publication Data

Denisen, Ervin L
  Principles of horticulture.

  Includes bibliographies and index.
  1. Horticulture.   I. Title.
SB418.D4   1979   635   78–5687
ISBN 0–02–328380–7

PRINTING   56789   YEAR   89

ISBN   0-02-328380-7

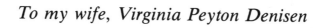

To my wife, Virginia Peyton Denisen

Preface to the Second Edition

When the first edition was being written, one goal was to produce a book that would not need revision, at least not very soon, for it emphasized principles, and principles "don't change." Basically this has proved to be nearly correct with a few exceptions: pictures can become dated in spite of definite efforts to avoid fads in design and model changes in automobiles, tractors, and equipment. Also new developments and discoveries can affect principles in such instances as mechanical harvesting technology, discovery of phytochrome and its role in the influence of light on plants, and emphasis on horticultural hobbies and their role in horticulture therapy. It even seems that a date on a book can cause it to be read less and used less, although most teachers will recognize that the most recent copyright does not guarantee the best textbook. Adaptability is a characteristic that cannot be overlooked. Since no book is perfect for all teachers, it leaves instructors with an opportunity to show creativity, current knowledge in the field, and the stature of an authority relating to students.

The interest of students and the general public in horticulture has increased tremendously since the first edition was written. It is this interest that has encouraged me to revise this book. When the publishers approached me for a revision after about ten years, I was deeply involved in department administration and knew I shouldn't take on another "iron-in-the-fire." However, now that I am no longer department chairman, I view things differently. In addition to being able to concentrate my time and energies on a favorite topic, the broad field of horticulture, I am again fascinated with the challenge of bringing the messages of horticulture to interested people.

Experience has shown that certain rearrangements in organization or sequence are desirable. Thus, "Hobbies in Horticulture" is removed as the last chapter and has been revised and reinserted in Chapter 1 as a part of "Horticulture in Everyday Life." Chapters on "Soil Management" and "Water Management" have been combined and placed earlier in Part II. Also, the chapters "Protection from Heat and Cold" and "Controlling Pests" have been placed earlier in Part II with the objective of having a more logical sequence. A chapter on "Improvement of Horticulture Plants by Breeding and Selection" has been added because of popular request. There is no change in chapter sequence for Part III, but of course there are both added information and drastically up-dated illustrations through line drawings and photographs.

It has been a tremendous undertaking to thoroughly revise this book, but the challenge was such that I found no alternative. Again it has been an exciting endeavor.

ACKNOWLEDGMENTS

My first "thank you" goes to Mr. Jim Smith, College Editor of Macmillan, for if we had not signed a contract this Second Edition would not have materialized.

I want to thank the special reviewers of the First Edition who made very excellent constructive criticisms. They are Professors C. C. Zych of the University of Illinois, Paul Prasher of South Dakota State University, and Gary McDaniel of the University of Tennessee.

Coworkers at Iowa State University who were helpful in developing the revised edition include Karl Nilsen who advised me on the *influence of light on plants* and in his work has emphasized teaching techniques with students and the total program in curriculum development. Melvin Garber supplied some pictures of class experiments and philosophized with me on laboratories and their use in the learning process. Pictures were also supplied by Paul Domoto, William Knoop, William Summers, Henry Taber, Charles Sherwood, and Ed Cott. Other staff members who were helpful in Horticulture topics include C. V. Hall, R. J. Bauske, G. J. Buck, J. A. Cook, D. G. Grove, C. F. Hodges, J. D. Kelley, H. E. Nichols, L. E. Peterson, D. F. Sopp, J. L. Weigle, B. F. Vance, and E. C. Volz. Several other members of the department who have been very helpful include Herb Taylor, greenhouse foreman; Allen Kemp, station foreman; Gordon Rolf, laboratory technician; and also Betsy Thomas, my secretary, who typed the manuscript at her home during her evenings. I am also grateful for the precise work of Inge Herigstad of the ISU Photo Service who produced excellent prints of publication quality.

My daughter Kathy, who is employed as a professional photographer and journalist by the Metropolitan Waste Control Commission of the Twin Cities area as their Public Information Officer, used her vacation taking hundreds of new pictures to be used for this edition. She had the assistance on that vacation trip of her sister Mary Clare who is an art graduate and illustrator. My wife, Virginia, has been professional adviser, editor, and organizational director of a task which seemed sometimes almost insurmountable.

To Virginia Yelland, a very close friend of the family, I again extend my grateful appreciation for the superb detailed line drawings, artwork, and technical and professional innovations she suggested.

Not to be ignored in my acknowledgments are the hundreds of students from my classes, other friends, neighbors, staff members at ISU, coworkers at other universities, my constituents in horticulture in the state of Iowa and elsewhere, and coworkers and friends in other countries.

Ervin L. Denisen

Preface to the First Edition

The primary objective of this book is to provide a balanced general text emphasizing the *principles* of *horticulture* for the college student. A further objective is to enlighten the reader in the far-reaching aspects of horticulture and to show how it enters into his daily life.

In Part I the student is oriented in the area of study and is provided with a foundation on which to build his knowledge of horticulture. This part of the text is presented briefly and concisely in order that the student may proceed promptly into the applied science of horticulture. It has been the author's observation that many writers and teachers of horticulture become so engrossed in basic details that they actually discuss horticulture itself only when time or space is running out. The supporting and basic sciences are important to horticulture but should not become a "tangent" that deviates from the objective of "general" horticulture. Too much time spent on remotely related subjects has been known to quell the student's enthusiasm instead of preparing him for exciting adventures in horticulture.

The skills and practices of modern horticulture and the bases for their use are covered thoroughly, with numerous examples, in the chapters of Part II. These skills and practices and the bases for them make horticulture an art and a science. They are the principles for the commercial grower, the processor, the retailer, the salesman, the home owner, or the hobbyist—any one of whom may be a horticulturist in his own right.

Most students of general horticulture are not preparing themselves for a life's work in horticulture. It is, non-the-less, an important part of their educational background, especially in home development and in the complex of social relationships. Part III of this book relates the principles of horticulture to a familiar area, the home grounds. Here the student can visualize the values of home horticulture for persons in all walks of life. He sees opportunities for developing a program to enhance the beauty, joy, and comfort of his own home. He learns of hobbies or avocations for those who like outdoor life and enjoy growing their own plants, creating new varieties, or doing both usual and unusual things with plants.

ACKNOWLEDGMENTS

A book is the product of many people. I am grateful to the thousands of students who have been in my classes in General Horticulture during the past twelve years. Many of these former students have given comments, criticisms, and suggestions on course content and teaching. Their ideas and reactions were a guiding factor in organizing and developing this book. The horticultural problems of friends, neighbors, and the general public also helped spur its development, especially the part dealing with horticulture in the home. To these many friends, I also wish to express my appreciation for photogenic subject matter, on which I drew freely to capture on film those scenes and objects that showed, described, or narrated the points being discussed. Dr. L. C. Snyder, Head of the Department of Horticulture at the University of Minnesota, made many helpful suggestions and criticisms in his thorough review of the manuscript.

I am particularly grateful to Iowa State College and, more specifically, its Division of Agriculture; to Dr. Floyd Andre, Dean of Agriculture, and Dr. Roy Kottman, Associate Dean of Agriculture, for their administrative policy of stimulating effective instruction and encouraging authorship; to Dr. E. S. Haber, Head of the Department of Horticulture, for his suggestion that I write a book on general horticulture and his help and encouragement; to my coworkers in horticulture and other fields who helped in many and various ways; and to Miss Lucille Pihl who typed the manuscript.

My wife, Virginia, is "Co-author incognito" because of her constant vigilance in editing, reviewing, and proofreading the manuscript and illustrations. My three daughters, Kathy, Peggy, and Mary Clare, were cooperative and understanding beyond their years. I extend sincere thanks to my wife's parents, Dr. and Mrs. William T. Peyton, for their encouragement and the generous loan of their Hasselblad camera during the entire sequence of my quest for photographs. To my friend Miss Virginia Yelland I offer my grateful appreciation for the excellent and detailed line drawings.

Ervin L. Denisen

Ames, Iowa

Contents

1 Horticulture in Everyday Life

WHAT IS HORTICULTURE?

Modern *horticulture* may be defined as an agricultural science which treats of the production, utilization, and improvement of fruits, vegetables, and ornamental plants. The term is derived from the Latin *hortus*, garden, and *cultura*, cultivation. Today's horticulture encompasses much more than garden cultivation. It is a tremendous industry composed of numerous commercial enterprises and even more numerous home gardens, orchards, lawns, and ornamental plantings. Millions of people are engaged in horticulture on a full-time, part-time, leisure-time, or amateur basis. It is a field that affects and influences all people. We live daily with horticulture. It provides a large portion of our food supply. It is a bounteous source of beauty in our homes, cities, rural landscapes, parks, campuses, gardens, conservatories, greenhouses, and areas of the great outdoors. It furnishes the setting for many recreational events, from picnics in the outdoor living area of a home to the tough turf of a football field or the "carpet like" putting green of a golf course. All these things are horticultural.

The role of horticulture in daily life will vary greatly among individuals. It may be a profession as with research workers, plant breeders, and teachers. Those in production phases of horticulture consider it an occupation or a vocation. For the merchandiser and salesman it is strictly a business. Horticulture may be a pastime, an adventure, or a means of exercise and health. It is a hobby for many people.

Various aspects of the science and art of horticulture may be the favorite subject for the cartoonist, the author, the poet, the humorist, the artist, the composer, the photographer, and the connoisseur of food, the gourmet. It is a fertile topic of conversation among neighbors and home owners, at service club luncheons and women's meetings, and at social gatherings in general. It even approaches the weather in popularity as a conversation piece.

The Boundaries of Horticulture   The art or practice of horticulture is an applied plant science as distinguished from the pure plant science of botany. In addition to horticulture the applied plant sciences include agronomy and forestry. These applied sciences are distinguished from each other by one or more of the following factors: (1) intensiveness of production, (2) purpose for which a crop is grown, and (3) custom. Intensive production is characteristic of most horticultural crops. This implies a greater

FIGURE 1-1 **Horticulture includes the cultivation of fruits, vegetables, and ornamental plants. (Lower and upper right photographs by K. A. Denisen.)**

expenditure of capital and labor per unit of land to obtain high production. When an equivalent amount of labor and capital is extended over more units of land we have more extensive agriculture. The contrast is apparent between a strawberry enterprise which may require several hundred man-hours of labor per acre and a wheat enterprise which may

FIGURE 1-2   Conservatories and parks provide horticultural grandeur. (Photographs by
                  K. A. Denisen.)

use well under fifty man-hours per acre. Capital costs are likewise greater
per acre for the strawberries. Gross and net returns per acre have to be
greater from strawberries than from wheat to justify the high per-acre
investment. Strawberries are a horticultural crop with intensive production. Wheat is an agronomic crop with extensive production.

Since horticulture involves the cultivation of fruits, vegetables, and
ornamental plants, the purpose for which horticultural crops are grown
is twofold: for human food or for aesthetic value. Thus, bluegrass in a
lawn is horticultural, but in a pasture or meadow, bluegrass is

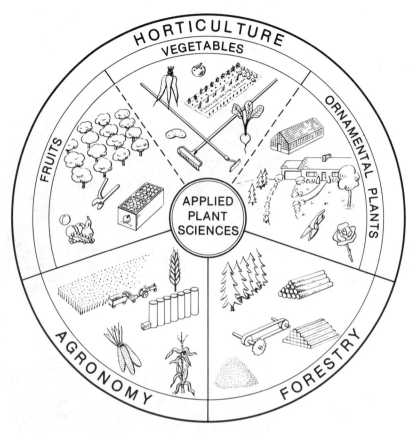

FIGURE 1-3   The applied plant sciences.

agronomic. Walnut trees grown for nut production or for shade are horticultural, but walnut trees grown for the furniture industry are a forestry enterprise.

Custom often determines our actions and viewpoints. Even the intensiveness of production and the purpose for which a crop is grown are sometimes overruled by custom in establishing the boundaries of horticulture. Tobacco is a crop grown and handled in an extremely intensive manner, yet, by custom, if often is considered an agronomic crop. Field corn and sweet corn do not differ greatly in methods of production. Sweet corn, probably because it is present in many gardens, is a product of horticulture, and field corn is a product of agronomy. The boundaries of horticulture cannot be specifically established by definition. However, the borderline examples help to emphasize the relationship of horticulture to the other plant sciences of agriculture.

The Divisions of Horticulture

One of the great horticulturists in America, the late Liberty Hyde Bailey, said, "There are very few true horticulturists." Bailey meant that most workers in horticulture select a division or phase of the field and become specialists in that division or even with a single crop. Since many diverse species of plants are included in horticulture, it is only natural that there be a classification, a division of interests and specialization among those engaged in the field. Many of the divisions of horticulture are further subdivided where certain crops or groups of crops are of great commercial significance. The general divisions of horticulture follow.

Pomology or **Fruit Science** includes the culture of apples and pears, stone fruits, citrus fruits, nuts, grapes, the various berry crops and miscellaneous temperate, sub-tropical and tropical fruits. The tree fruits represent a long term investment because of the time required to grow trees to bearing age. Orchardists must be keenly aware of the ravages of insects, diseases, and inclement weather and take the necessary precautions for producing high quality crops. Grape and berry crops are

FIGURE 1-4   Research workers are "specialists" in the various divisions of horticulture. Growth chambers are among the facilities used for controlled environments. (Photograph by K. A. Denisen.)

shorter term investments. They are adaptable for both fresh market and processing. Improved shipping, storage, and refrigeration methods have given impetus to production for distant markets. Both tree fruits and small fruits are adapted to home and farm fruit production.

Olericulture or **Vegetable Science** is a large commercial industry and a very popular home and farm activity. Certain crops in areas of favorable soil and climate have been developed into large commercial enterprises. Large scale production with good shipping and refrigeration facilities have brought the produce to distant markets in good condition. Market gardeners in local production near metropolitan areas have felt the competition of the more distant production areas. They have been forced to increase their efficiency and in many cases specialize in fewer crops. Vegetable growing is a pastime for many people.

Ornamental horticulture includes the culture and use of many hundreds of species of plants. Subdivisions are based on types of culture in the various groupings. *Floriculture* is the art of growing, selling, designing, and arranging flowers and foliage plants. Flowers are in demand throughout the year. The local florist may grow flowers himself or have them shipped from other areas. There are many hobbyists in floriculture. *Trees and shrubs* are important ornamentals for homes, farmsteads, parks, public buildings, avenues, and recreation areas. Sales of

FIGURE 1-5   **A young pear orchard. The tree fruits represent a long-term investment. (Courtesy National Rain Bird Sales and Engineering Corp., Azusa, California.)**

Horticulture in Everyday Life

FIGURE 1-6   **Commercial production of chrysanthemums and the retailing of potted plants are phases of floriculture. (Photographs by K. A. Denisen.)**

trees and shrubs are handled mainly by local and mail order nurseries. Local tree services are established in many communities, but many home owners prefer to grow and maintain their own trees and shrubs. *Turf grasses* receive considerable attention as ornamental plantings. The vast numbers of golf courses, football fields, baseball lots, parks and playgrounds, boulevards, highway plantings, cemeteries, and even more numerous home lawns have created a big demand for good turf. Commercial enterprises specializing in turf culture have developed throughout America. The average home owner is very concerned with this crop.

Landscaping is receiving increased attention from both the home owner and home builder. This phase of horticulture consists of planning and arrangement of home grounds and farmsteads, public areas, highways, and business establishments. It involves not only the use and placement of horticultural plants but also placement of buildings, walks, drives, fences, service areas, recreation areas, and other parts of the landscape. It is sometimes included as a subdivision of ornamental horticulture.

Nursery production provides a commercial source of plants used in the home, orchard, and garden. Wholesale nurseries produce

FIGURE 1-7   A golf course is an important turf grass enterprise. (Courtesy Iowa State University, Ames, Iowa.)

tremendous quantities of trees, shrubs, flowering plants, fruit plants, or vegetables. Mail order and local retail nurseries and garden centers make these plants available to the public. The propagation of plants is an important technique of the nursery industry. Because of the interests of many home owners and gardeners in producing their own plants, they too have become plant propagators of many species of plants.

Seed production is a vital part of vegetable and flower growing. Some retail and mail order nurseries are also in the seed business. However, their seed is usually produced by others. The seed business is highly specialized and production areas are centered in those parts of the country where conditions are most favorable to seed production of specific crops. Home production of seed is not widely practiced except insofar as certain growers are interested in maintaining their own special strains of vegetables and flowers.

Processing and storage of horticultural crops have effected tremendous changes in the eating habits of people throughout the world within the period of a single generation. The canning and frozen food industries have preserved much of the quality and freshness of fruits and vegetables. The potato chip industry has shown rapid development due primarily to the increasing popularity of its product. Research and engineering have combined efforts to improve storage conditions of

FIGURE 1-8 Seed production in an onion breeding program.

FIGURE 1-9 The local retail nursery has plants available to the public. (Photograph by K. A. Denisen.)

FIGURE 1-10   A large-scale tomato enterprise. Tomatoes are an important fresh vegetable and canning crop. (Courtesy John Bean Division, Food Machinery and Chemical Corporation, Lansing, Michigan.)

horticultural crops. As a result, many fruits and vegetables can be stored for several months with maintenance of high quality, equal or superior to the quality at harvest time. With the increased use of home freezers and cold-storage locker plants, many home gardeners are storing their own produce by freezing. Improved canning equipment and techniques for use in the home have also developed interest among many gardeners and homemakers to process their production in excess of immediate needs.

Each of these divisions of horticulture can be a science, an art, a business, or an industry in itself. They all have characteristics which fit into the meaning of horticulture. The field of horticulture is complex, often highly specialized, and yet it is within the realm of every citizen to participate in one or more of the horticultural activities so omnipresent in everyday life.

IMPORTANCE OF HORTICULTURE TO DAILY LIVING

In the economy of the world and within each nation, horticulture is a basic industry. It is an important source of the food supply of the world.

Approximately 33 percent of the foods consumed in the United States are of horticultural origin. This compares to 25 percent for dairy products, 18 percent for grain and sugar crops, 18 percent for meats, fish, and eggs, 4 percent for fats and oils, and 2 percent for miscellaneous sources. Production, processing, and marketing of horticultural crops provide gainful occupations for many citizens of the world. Horticultural crops enter interstate commerce and world trade as fesh fruits, vegetables, and ornamental plants. Even greater quantities of fruits and vegetables are shipped in a processed form.

In the United States, approximately 3 percent of the cropland is in commercial production of horticultural crops. This compares with 83 percent for field crops, 5 percent for cotton, and 9 percent for all other crops. This does not account for the vast numbers of home gardens and home orchards. With intensive production the low percentage of cropland in horticulture provides a large portion of the food supply. This feature of horticultural crops, high production per unit of land, offers promise for the future for expanding populations. Arid and semi-arid areas, which have been developed for irrigation, have become important horticulturally. Irrigation is usually costly to install and operate, thus most operators turn to intensive production which is characteristic of horticultural crops.

From the nutrition standpoint, horticulture is extremely important to our daily living. Nutritionists have discovered basic facts concerning the relationship of our health and the foods we eat. Today we are a much

FIGURE 1-11   **The "basic four" food groups. Note the role of horticulture.**

better fed people than were our ancestors. The knowledge available to us about kinds of foods, their content of food nutrients, and human requirements enable us to have the best balanced diets known to the world. Fruits and vegetables play a vital role in satisfying the nutritive requirements of the human body. Among those foods high in carbohydrates are several garden vegetables, including white potatoes, sweet potatoes, beans, peas, and sweet corn which are commonly called energy foods. Wheat and rice are the only other energy food items which rank with these vegetables in quantities consumed. The banana and date are among the few fruits high in carbohydrates.

Although minerals are not needed in large quantities by the human body, they are vital to our health. The leaves of most plants are good sources of minerals which is the reason for leafy salad vegetables and greens in our diet.

Most vegetables and fruits are sources of one or more of the vitamins or vitamin complexes needed for good health. In many instances of vitamin deficiency, fruits and the green leafy vegetables have not been part of the diet. If people eat adequate amounts and variety of fruits and vegetables there is rarely need for vitamin pills as a supplement to the regular diet.

Other food nutrients include proteins and fats. Horticultural foods are not generally considered high in either protein or fat. However, the nut crops as a group are relatively high in fat content and, to a lesser extent, are a source of protein.

Many people have beome aware of the vitamin value of fresh fruits and vegetables and tend to shift their eating habits accordingly. This increase in per capita consumption must thus reflect a decrease in other foods. Those affected have been the high energy foods such as potatoes and sweet potatoes. In an age of technology where much labor is done by machine, the "human machine" needs less of the fuel-supplying foods. Thus with an increased consumption of salad crops we have a decreased use of high carbohydrate, high energy vegetables on a per capita basis. As more of our work is done by machine, as leisure time increases, and as the nature of our work changes, people become more calorie conscious. It is reflected in eating habits. The problem of obesity can often be corrected by judicious use of fruits and vegetables.

HORTICULTURE AS PART OF OUR HOMES

The early Greeks and Romans placed great emphasis on beauty and attractiveness as evidenced by those phases of their culture which exist today. Throughout history beauty has played an important role in the development of cities, homes, countrysides, architecture, the fine arts, music, personalities, ad infinitum. We have it with us today in ever increasing importance. In its broad aspects beauty can be considered one of the major industries of the world. Exclusive of any man-developed versions, beauty occurs abundantly in nature. The horticulturist combines these beauties of nature with artistic beauty to enhance the overall beauty of our homes to make them more pleasant places in

FIGURE 1-12   **The horticulturist combines the beauties of nature with artistic beauty.**

which to live. Plants of diverse types are used to provide a setting for the home. They are further used to blend the man-made structures with their surroundings.

Much of our planning is for convenience. In our work, in our play, in our various routines, we like to be efficient. Efficiency goes hand-in-hand with convenience. Planning direct routes between service areas in a yard can save countless steps in performing routine jobs. Convenience for maintenance and upkeep of lawns, hedges, fruit trees, drives, walks, buildings, and so forth is an important part of home development. Convenience to the outdoor living area is essential for frequent use, as is access to the charcoal burner, picnic table, patio, and vegetable garden.

The harvest of fresh fruits and vegetables is the reward of effort in the production phase of home horticulture. Yield in aesthetic value of flowers, trees, and shrubs is more difficult to measure but is non the less a stimulating incentive for horticulture in the home.

General health and happiness of the family are major objectives and results of a good horticultural program for the home. In addition to better nutrition from fresh fruits and vegetables, the home grounds also provide exercise through work and play, all of which may be conducive to good health. Many ideas for hobbies or avocations occur to those who take an active interest in home gardening, whether it be on a farmstead, a city lot, a window sill, or an artificially lighted room. The home grounds provide an ideal setting for family projects. The vegetable garden may be a joint endeavor. Combined efforts of all members of the family to build a terrace, to develop a lawn and picnic area, or to plan a badminton court increase family ties and strengthen home unity.

Horticulture is also an important part of the home when monetary values are considered. Home grounds that are well planned, attractive, and functional are appealing to the eye. Good horticultural practice increases the real estate value of homes and farms. Many people's first impressions are frequently based on the appearance of the home grounds.

FIGURE 1-13   **Fresh fruit is an attraction at the market. (Photograph by K. A. Denisen.)**

Energy for fuel consumption can be controlled by environmental considerations in planting of trees and shrubs both for winter heating and summer cooling.

Horticulture as a Hobby

A hobby or a vocation is a favorite endeavor of an individual to be doing something constructive, creative, or interesting solely because of that person's interest or desire. Gardening offers opportunity for unlimited pleasures while discovering the beauties of nature, the mysteries of growth, and the role of man in improving plant production. It offers relief from pressures of business life, from the fears and anxieties of an atomic age. The serenity of a garden has a relaxing influence in the fast pace of modern society. It can be excellent therapy following illness. When a diversion is desired, the pent-up energies and emotions can be released in some of the mechanical skills.

African violets (*Saintpaulia* spp.) are often considered difficult to grow under average treatment and conditions. It has been a challenge to many plant lovers to develop a "green thumb" and master the difficult. Previously, growing hybrid tea roses in areas of severe winters was considered practically impossible; now, the rose fancier who has the real desire to grow roses can modify cultural practices to harden the tissues and use special techniques of mulching with considerable success. It makes a great difference how much desire the individual has; how persistently he meets the challenge; how soon and how successfully he can conquer the obstacle.

Even pest control may be a challenge to hobbyists. Insect and disease-free fruits and vegetables are indeed much to be desired in any garden. New techniques of pesticide application in the home and new physical control methods are always welcome. They may only need the ingenuity of a hobbyist to bring them to light and prove their usefulness.

Specialities and Techniques

A bare wall on one side of the house can be a challenge to create an attractive or subdued effect forgotten in the original plan. Espaliered trees or shrubs may be the answer. If so, it then becomes a hobby for developing skills involving the training and pruning of espaliers.

Training and pruning of any type of horticultural plant can be an interesting and useful hobby skill. Keeping plants within bounds and yet maintaining attractiveness and usefulness shows definite skill in pruning. Resultant healing, shifting of apical dominance, stimulation of certain buds or shoots, increased bloom, larger fruits, and other interesting results make it a fascinating skill. One of the best uses of a clipped hedge is for the hobbyist who takes pride in his skill at sculpturing the foliage and twigs of plants to beautiful and symmetric perfection. The skilled pruner may adapt the practice of topiary in certain locations of the home grounds. Topiary is the art or skill of training and pruning plants to resemble certain objects, animals, or forms. Plants easily adapted to clipped hedges, like yew or privet, are most commonly used in topiary. The products of topiary are best suited to the more private, less conspicuous areas of the home grounds lest they become a curiosity or

FIGURE 1-14   Wisteria vine trained to a post and twisted and looped to form a convenient hanger.

distraction and an unintentional central feature of the landscape. Espaliered trees and shrubs or the manipulation of vines provides interesting results to the hobbyist of skills. Summer pruning, tying, and mechanical pressure are used in accomplishing the desired effects.

Propagation of plants used in the home grounds is a skill which appeals to many home owners. They take considerable pride in having produced a hedge of yews, privet, or honeysuckle by rooting their own cuttings. House plants give twofold satisfaction when home propagated. An apple tree with five or more cultivars of fruit is much more proudly displayed when its owner did the topworking than is a tree purchased from the nursery with the completed grafts. Proficiency at a hobby is good for the ego. Skills in horticultural practices are readily learned when the desire is provided by the hobbyist. Often the acquiring of the skill is as soul satisfying as the attainment of perfection.

Special treatment and attention is often required for the successful growing and production of some horticultural plants. It is relatively easy to grow tulips, hyacinths, and other ornamental bulbs out-of-doors. It takes extra effort and careful handling to "force" them, that is, grow them indoors and under artificial conditions. Forcing of plants is very adaptable for an interesting hobby. It consists of planting bulbs in pots containing rich, well-aerated soil. These are given a cold treatment at temperatures just above freezing in a heavily mulched cold frame. During this period they form a network of roots. When the roots reach the outer part of the soil ball they are ready for placing in a warmer

FIGURE 1-15   **Succulents and varied-textured plants form the Liberty Bell and ascending stairs.**

Horticulture in Everyday Life

room. Top growth then proceeds at a rapid rate. Lilies, tulips, daffodils, and other bulbs are produced out-of-season. The forcing of cucumbers, strawberries, and blueberries can be done if adequate space, effort, and some precautions are taken. Strawberries do not yield well the year they are planted because of root loss in transplanting. If they are allowed to establish their new runner plants, each in a pot of soil sunk in the garden, the roots will be unmolested when the pot is brought indoors. They make attractive conversation pieces and are extremely interesting to observe as they flower, set fruit, and grow fruit to maturity.

Growing plants under artificial light has become an interesting study. It is a speciality of some people who have poor natural lighting, inadequate space by windows, or who have the desire to grow plants under artificial light. Such factors as light intensity, distance from light source to plants, quality (color) of light, and photoperiod must be studied and adjusted to meet plant requirements. It requires fairly precise calculations, an appreciation of the use of electricity, and an alertness to plant response.

Soil-less culture, or growing plants in nutrient solution, is a specialty requiring tanks rather than benches. An extremely important factor for consideration is root aeration. Air must be pumped into the solution periodically or the level of water must be lowered and raised at intervals during the day. A support of wires, gravel, crushed brick, or other inert material holds the plants in a rooting medium and an upright position. The nutrient solution is prepared by dissolving fertilizers and other chemicals in water. Relatively small quantities of nutrients are needed. The micro elements are an essential part of the solution but are supplied in very minute quantities. The pH of the solution must be controlled. Soil-less culture is not new but may be an intriguing topic for an indoor garden.

Luther Burbank made a great impact on history as one of the early plant breeders. He made major contributions to modern horticulture by producing many new and better cultivars of fruits, vegetables, and ornamentals. He was able to do much of this even before the science of genetics became a tool for plant breeders. Observation, familiarity with his plant materials, and recognition of superior qualities were basic to his successful production of good cultivars. Gregor Mendel, an Austrian monk, through his experiments and complete records, laid the foundation for the science of genetics by discovering basic principles of inheritance. Plant breeding in modern horticulture utilizes genetic principles, selection, and adaptability for the production of new cultivars and hybrids. Plant breeding is readily adaptable for a horticultural hobby, and like most hobbies, does not always require a broad technical background for its development. Many of the techniques and procedures are learned by doing (see Chapter 14).

Arts and Skills Since horticulture is an art as well as a science, it can be further developed as an art through hobbies involving painting, sketching, drawing, and photography. Literature and music have been inspired by horticulture topics, including flowers, trees, and beauty in general.

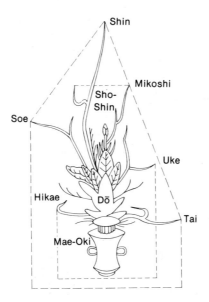

FIGURE 1-16   **The art of flower arranging is a fertile area for creativity in design.**

The arts in horticulture are especially well expressed in flower arranging and floral design. In addition to being a vital area in floriculture, flower arrangement has become one of the most popular hobbies in horticulture. The Japanese ceremonial art of flower arranging has been introduced to the West; its simplicity and its relationship with nature make it adaptable to modern study and use. The art of flower arranging is a fertile area for creativity in design, color relations, and adaptability of unusual plant materials. It is one of the very popular hobbies in the home.

That food preparation is an art, any experienced cook will agree. It is widely used as a hobby. With the great increase in outdoor cookery, more men have become interested in culinary arts and have actively adopted it as a hobby. Since nearly half the food consumed is of horticultural origin, it is only natural that fruits and vegetables form a basis for enjoyment of the culinary arts.

Processing for preservation has its variable techniques. Pickling of both fruits and vegetables has numerous adaptations. Some hobbyists have developed their own processes for making wine from grapes and other products. Jams and jellies vary within the season for similar raw products merely because of differing techniques. Much experimental cookery of hobbyists has resulted in significant advances in food preparation and increased palatability. Desserts as specialities are often based on fruits and fruit products.

Exhibition can be an interesting horticultural hobby. Shows are widespread customs. Fruits, vegetables, and flowers that are grown, selected, and exhibited in a highly creditable manner are rewarding to the grower and attractive to the patrons of the show. Special precautions in growing for exhibition are often the difference between championships and lesser awards. Fruits on a tree can be selected for an exhibition plate long before being picked at maturity. They may be bagged with cellophane or kept continually coated with insecticides and fungicides that are noninjurious to epidermal tissues. Flowers may also receive special consideration by exhibitors.

Bonsai, the Japanese art of growing miniature trees and shrubs, is a fascinating hobby. It involves techniques of extreme dwarfing. Selections of hardened woody plants that have been subjected to adverse conditions is a good starting point. Some pruning of both tops and roots is usually done at planting time. Trunk and branches may be bent, forced, and tied by coiling them with heavy wire. This wire is removed after several months when training to shape has been accomplished. The bonsai plant is fed sparingly of soil nutrients and given only minimum

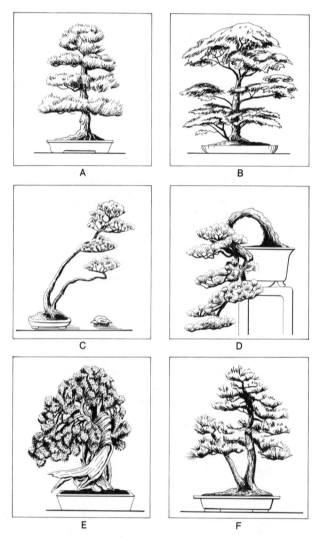

FIGURE 1-17   Bonsai, the Japanese art of growing miniature trees and shrubs, is a fascinating hobby: *A*, Formal Upright; *B*, Informal; *C*, Slanting (Literati); *D*, Cascade; *E*, Slanting Trunk (windswept); *F*, Twin Trunk.

requirements of moisture. At periodic intervals, usually annually, the plant is either repotted or lifted from the pot, root pruned, and reset. It is possible to grow plants for many years with little or no increase in size. Some Japanese azaleas and conifers are hundreds of years of age and yet may be only 12 to 18 in. (30 to 45 cm) in height. Plants adaptable to extreme dwarfing in this manner include juniper, holly, yew, elm, maple, azalea, pine, arborvitae, cypress, cryptomeria, and euonymus. Miniature landscapes can also be developed using the miniature trees and small, low-growing grasses, flowers, and moss.

Terrariums are small gardens encased in glass bowls, aquariums, and bottles. They differ from the bonsai miniatures in that they contain only herbaceous plants and have an environment that is conditioned for them. The soil of the terrarium is high in organic matter content and thus has high water-holding capacity. Plants best adapted to this closed environment are those most resistant to common leaf fungi, the molds and mildews. The humidity in the terrarium is usually high because of the organic content of the soil. Succulents can also be grown in terrariums, but the soil must then be lower in organic matter content and higher in sand. Humidity is lower, which results in waxier leaves and stems. Growing plants in small-necked bottles often raises speculation as to how such plants were ever placed inside. The hobbyist, of course, placed them there while they were still very small. A series of long-handled instruments can be adapted for placement of materials in a terrarium.

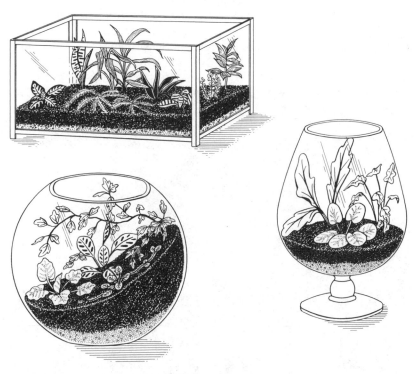

FIGURE 1-18   **Terrariums provide a controlled environment for miniature herbaceous gardens.**

Preparation for retirement should predate retirement, so it is not a shock or cause for panic when the time arrives for cessation of regular duties. It is well to develop an attitude of anticipation for retirement. The development of a hobby program that increases in intensity as retirement approaches contributes to a smoother transition from an active to a less strenuous mode of life. A hobby program in preparation for retirement can be as important to the psychological welfare of the individual as financial preparation through annuities, endowments, social security, and savings. The hobby program should extend over a considerable period of time.

Long-range programs are actually begun when hobbies are carried through life. Development of a hobby program may lead to anticipation of full-time devotion to the hobby as retirement nears. Hobbies and retirement have more appeal if the spirit of service prevails in their planning and accomplishment. Time and effort given to projects that tend to make the world a better place in which to live give greater satisfactions than those purely for entertaining the whims of the individual. Too often retirement is a withdrawal from service to others when it could be a doorway to greater adventures in promoting beauty, cheerfulness, ideas, fun, relaxation, new creations, and appreciation of Nature. Hobbies in horticulture provide the raw materials for all of these.

Therapy is correlated with the duration of the hobby program, its consoling and distracting influence, and its broadening of interests. For physical incapacitation, it may be desirable to select a hobby that can be easily picked up and easily dropped, such as flower arranging, wood sculpturing, photography, or sketching. For mental therapy, a long-term program of propagation, pruning, vegetable and flower gardening, and

FIGURE 1-19   **Hobbies in horticulture can give a lifetime of adventure.**

home grounds development may hasten the return to a brighter outlook on an abundant, generous, beautiful, and soul-satisfying future through Nature.

STUDY QUESTIONS

1. What are your interests in the field of horticulture? Are they compatible with your career objectives? Be specific.
2. How does your daily food intake relate specifically to horticulture?
3. How does horticulture affect your daily living pattern?
4. What is your favorite hobby interest? Is it part of a long-term program? Explain.
5. What is "Horticulture Therapy"? Why does it become an increasingly important application to the well-being of the general public?
6. How does horticulture become important to you from a travel and geographical view?

SELECTED REFERENCES

1. Abraham, George, and Kathy Abraham. 1977. *The Green Thumb Garden Handbook*. Prentice-Hall, Englewood Cliffs, N.J.
2. American Society for Horticultural Science. *Horticulture, A Satisfying Profession* (updated annually). A.S.H.S. Headquarters, Mt. Vernon, Va.
3. Bailey Hortorium Staff. 1977. *Hortus III*. Macmillan, New York. The most current classification and descriptions of ornamental horticultural plants.
4. Bauske, R. J. 1976. *Home Horticulture*. West Publishing Co., St. Paul, Minn.
5. Brooklyn Botanic Garden Handbooks, Brooklyn Botanic Garden, Brooklyn, New York.
   1977. No. 13, *Dwarfed Potted Trees, The Bonsai of Japan.*
   1976. No. 62, *Gardening Under Artificial Light.*
   1976. No. 79, *Gardening Guide.*
   1976. No. 80, *Designing With Flowers.*
   1977. No. 82, *The Environment and the Home Gardener.*
   These and numerous other handbooks are available for a nominal charge, usually $1.75 each.
6. Denisen, E. L. 1958. *Principles of Horticulture*, 1st ed., Chapter 20: Hobbies in horticulture. Macmillan, New York.
7. Director, Japan Bonsai Association. 1975. *Bonsai, Masters Book.* Kodansha International, Ltd., Tokyo.
8. Edmond, J. B., T. L. Senn, F. S. Andrews, and R. G. Halfacre. 1974. *Fundamentals of Horticulture*, 4th ed. McGraw-Hill, New York.
9. Janick, Jules. 1974. *Horticultural Science*. W. H. Freeman Co., San Francisco, Calif.

10. Sato, Sho. 1969. *Art of Arranging Flowers.* Harry Abrams Publishing, New York.
11. U.S. Census (Agriculture). 1976. U.S. Government Printing Office, Washington, D.C.
12. Yearbook of Agriculture. 1959. *Food.* U.S. Department of Agriculture, Washington, D.C.
13. Yearbook of Agriculture. 1972. *Landscape for Living.* U.S. Department of Agriculture, Washington, D.C.
14. Yearbook of Agriculture. 1977. *Gardening for Food and Fun.* U.S. Department of Agriculture, Washington, D.C.

Fundamentals of Horticulture

I

2 Classification, Anatomy and Growth of Horticultural Plants

To study and explore the principles and practices in horticulture, a certain basic knowledge of the classification, structure, functions, and characteristics of plants is essential. In the broad field of plant life, there is great diversity of plant kinds and types. Along with these many differences, plants also have numerous similarities. These criteria are the basis for our discussions on classification, anatomy and the phenomenon of growth in plants.

CLASSIFICATION OF HORTICULTURAL PLANTS

Differences and similarities are important criteria for describing and distinguishing among people, animals, nations, civilizations, eras, and a host of other tangible and intangible entities. This is also true of plants. Horticulture deals with fruits, vegetables, and ornamentals. We distinguish between them and between their kinds, types, and cultivars by comparing characteristics of form, habit of growth, and longevity. With a few rather commonly used terms, we can express characteristics of, and thus become acquainted with, many horticultural plants.

Terms Describing Horticultural Plants

Plants in horticulture are either *herbaceous* or *woody*. Herbaceous stems are softer, more succulent, and less fibrous than are woody stems. Lilies and tomatoes are herbaceous; lilacs and pines are woody.

Some plants complete their life cycle, seed→plant→seed, during a single season. These are *annuals*. Other plants require two years to complete their life cycle and are called *biennials*. They grow during one season, store their manufactured food, and produce seeds the following year. The long-lived plants, those living more than two years, are *perennials*. Annuals and biennials are usually herbaceous. Perennials may be woody or herbaceous.

Woody plants may be either *evergreen* or *deciduous*. The evergreen retains living leaves at all times. It loses its leaves gradually, never all at one time, and forms new leaves before the old ones are lost. A deciduous plant is without leaves during the winter in temperate climates or during the dry season in tropical climates.

Some horticultural terms are difficult to define but are so commonplace they require no definition. These include such terms as tree, shrub, and vine. Other terms have slightly different meanings from the botanical description or definition, and so require delineation.

FIGURE 2-1   **The banana is produced as an upright herbaceous plant in tropical climates.**

Fruits   To the botanist a fruit is simply a ripened ovary. To the horticulturist, from common usage of the term, a fruit is the fleshy, edible product of a woody or perennial plant which in its development is closely associated with a flower. Thus we can readily see how the range from the apple or orange to the strawberry or pineapple is all inclusive.

By first dividing the fruits into those borne on woody plants and those borne on herbaceous plants a means of classification is begun. The woody plants may be trees, shrubs, or vines. They may be evergreen or deciduous. The woody bush and vine fruits together with the herbaceous plants of the strawberry are also called small fruits. The woody deciduous tree fruits are also divided between the drupe (stone) fruits like the peach, plum, and cherry and the pome fruits such as apple, pear, and quince which have an embedded ovary, commonly called the core. An outline of fruits with some typical examples will clarify relationships.

I. Fruits borne on woody plants
  A. Tree Fruits
    1. Deciduous
      a. Pome
        Apple, pear, quince
      b. Drupe (stone fruits)
        Peach, plum, apricot, cherry
    2. Evergreen
      a. Citrus
        Orange, grapefruit, tangerine, lemon, lime
      b. Palm
        Date, coconut
      c. Avocado
      d. Papaya
  B. Small Fruits
    1. Grape
    2. Brambles
      Raspberry, blackberry, boysenberry
    3. Blueberry and cranberry
    4. Currant and gooseberry
II. Fruits borne on herbaceous perennial plants
  A. Prostrate growth
    Strawberry (also considered a small fruit)
  B. Upright growth
    Banana, pineapple

Fruits may also be grouped according to climatic adaptability as temperate, sub-tropical, and tropical.

Vegetables   The group of plants known as vegetables includes many diverse types, and the lines of delineation are not clear cut. In general, a vegetable is the edible product of a herbaceous garden plant. Thus the cereals are not vegetables since they are not considered garden plants. The strawberry, while a product of a herbaceous plant, is a fruit by virtue of custom and its production of a berry resulting from a flower. Asparagus, conversely, is not a fruit, even though it is a herbaceous perennial, since the edible part is not associated with a flower. The tomato, pepper, watermelon, etc. are considered vegetables even though their edible products are botanical fruits (ripened ovaries). They are annual garden plants and do not qualify by definition or custom as horticultural fruits.

Overlap or lack of a clear-cut delineation between fruits and vegetables is similar to combinations or dual characteristics in other pursuits. For instance, in the automotive industry there is no fine line of distinction between a car and a truck. The extremes are easily recognized, but it becomes a matter of custom or definition whether a station wagon, van, panel-delivery, pick-up or a carry-all is a car or a truck. It is seemingly unimportant, but it can be extremely important when considering license fees, chauffeur's licenses, and other factors. Similarly with fruits, vegetables, and cereals, it can be vital information to a grower, merchandiser, or even a home gardener as to which type of plant he grows or handles. Such legal terminology as is involved in tariffs, tax laws, support prices, soil bank specifications, etc. must clarify meanings or establish lines of demarcation. Thus our concern over proper classification. A typical example is the merchandiser who imported canned sweet corn and tried to have it declared simply as corn. As such it would have been subject to import duty of only a few cents per bushel. However, as a canned vegetable he had to pay duty on a per can basis which resulted in a much higher total tax.

Vegetables can be grouped according to length of life cycle as annuals, biennials, and perennials. They can be classed as cool or warm season types. Sometimes vegetables are described by edible portion of the plant. They are frequently placed in categories of similar characters because of the similarity of culture within the group. Such a classification follows.

Root crops
    Carrot, beet, rutabaga, turnip, radish, parsnip, sweet potato
Bulb crops
    Onion, garlic, shallot
Tuber crops
    White potato, Jerusalem artichoke

FIGURE 2-2   **Longitudinal section of red cabbage shows the head consists of distinct leaves arising from the stem of this cole crop.**

Vine crops
Cucumber, pumpkin, squash, cantaloupe, watermelon
Cole crops
Cabbage, broccoli, cauliflower, Brussels sprouts, kale, kohlrabi
Greens
Spinach, chard, chicory, mustard
Salad crops
Lettuce, celery, cress, endive, Chinese cabbage
Perennial vegetables
Asparagus, rhubarb, horse-radish
Solanaceous fruits
Tomato, pepper, eggplant, husk tomato
Pod crops
Bean, pea, okra
Corn
Sweet corn, popcorn

Ornamentals   By the very wide usage of the term ornamental plants one would expect a great many kinds and types. A fairly complete nursery catalog shows this to be true. If the nursery catalog is well illustrated, it is a good means of becoming acquainted with ornamentals. A visit to an arboretum is also effective in learning to know more ornamental plants. A brief outline with a few of the numerous examples suggests the vast numbers of kinds and diverse types of ornamentals.

I. Woody Ornamental Plants
  A. Trees
    1. Deciduous
       Elm, oak, linden, ash, beech, birch, willow, poplar, larch
    2. Evergreen
       a. Narrowleafed
          Pine, fir, hemlock, cedar, spruce
       b. Broadleafed
          Citrus, holly, magnolia, live oak
  B. Shrubs
    1. Deciduous
       Spirea, lilac, honeysuckle, mock orange, barberry, privet
    2. Evergreen
       a. Narrowleafed
          Spreading junipers, arbor vitae, Mugho pine
       b. Broadleafed
          Boxwood, rhododendron, gardenia
  C. Vines
    1. Deciduous
       Clematis, bittersweet, Virginia creeper
    2. Evergreen
       Wintercreeper, Baltic English ivy

FIGURE 2-3   **A bed of ornamental plants accenting the statuary. (Photograph by K. A. Denisen.)**

        II. Herbaceous Ornamental Plants
           A. Perennials
             1. For flowers
                Peony, lily, tulip, delphinium, phlox, geranium
             2. For foliage
                Coleus, peperomia, sanseveria, the lawn grasses
           B. Biennials
             Hollyhock, foxglove, Canterbury bells
           C. Annuals
             Marigold, zinnia, petunia, alyssum, nasturtium

Ornamentals are also classed according to use such as bedding, border, shade, screening, specimen, etc. Other criteria for grouping include hardiness, response to day length, and light intensity requirements.

Botanical Classification of Horticultural Plants   The Swedish botanist, Carolus Linnaeus is considered the founder of the current widely used binomial nomenclature in plant classification. The terms *genus* and *species* are used to designate the identity of each plant. There are no duplications of genera and species according to the rules established by the International Code of Botanical Nomenclature.

Three kinds of plants are here classified as examples:

The Apple:
  Kingdom—*Plantae*
    Division—*Anthophyta*
      Class—*Dicotyledonae*
        Order—*Rosales*
          Family—*Rosaceae*
            Genus—*Pyrus*
              Species—*malus*
                Cultivar—'Delicious'

The Bean:
  Kingdom—*Plantae*
    Division—*Anthophyta*
      Class—*Dicotyledonae*
        Family—*Leguminosae*
          Genus—*Phaseolus*
            Species—*vulgaris*
              Cultivar—'Stringless Green Pod'

The Hollyhock:
  Kingdom—*Plantae*
    Division—*Anthophyta*
      Class—*Dicotyledonae*
        Order—*Malvales*
          Family—*Malvaceae*
            Genus—*Altheae*
              Species—*rosea*
                Cultivar—'Newport Pink'

The term *cultivar* is a contraction of "*culti*vated *variety*" to distinguish from botanical variety, a term used in botanical literature.

ANATOMY OF PLANTS

The principal parts of horticultural plants are stems, leaves, buds, flowers, fruits, and roots. They assume important roles in plant production, propagation, food manufacture, and in providing balance for the cycles of nature.

The Woody Dicot Stem   Woody stems differ from herbaceous stems in that they show greater maturation and hardiness of stem tissue. Principal parts of the woody stem are the xylem, cambium, and phloem. These tissues are quite distinct when viewing a cross-section cut of the trunk or branch of a tree. The xylem is the wood, the phloem is a major portion of the bark, and the cambium is a thin layer between the wood and the bark. The xylem, phloem, and cambium are all involved in such horticultural practices as grafting, pruning, and wound treatment. They play vital roles in growth, development, and production of horticultural plants.

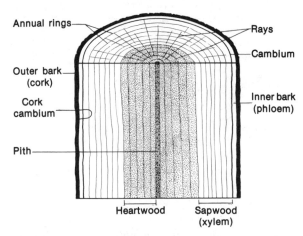

Annual rings
Rays
Camblum
Outer bark (cork)
Inner bark (phloem)
Cork cambium
Pith
Heartwood
Sapwood (xylem)

FIGURE 2-4   **A diagrammatic cross-sectional and longitudinal view of a woody dicot stem.**

The Xylem

Conducts H₂O & food from Roots To Leaves.

The wood, or xylem, is the tissue in greatest quantity in a tree. It is the central cylinder which conducts water and dissolved nutrients from the root to upper parts of the plant. The xylem consists of countless tiny tubes or vessels extending from the roots to the leaves. As the tree or branch develops and grows these vessels are formed from chains of long cells located end-to-end. The end walls or partitions disintegrate forming, in effect, long pipes for carrying water and nutrients. The xylem also contains extremely thick-walled cells called wood fibers which give further structural strength to woody stems. In young woody stems the center of the xylem area consists of *pith*, a soft storage tissue. The pith plays a prominent role in young woody stems but loses its significance as the woody plant increases in diameter. It ceases growth and is no longer produced after the cambium becomes completely functional.

The Cambium

function is to form New Cells

each year a New Layer

Outside the xylem and surrounding the central cylinder of wood is the layer of meristematic (dividing) cells known as the cambium. In cross-section this tissue appears, to the naked eye, as a very fine line between the wood and the bark. Its function is to form new cells. Those located inside the cambium layer become xylem cells and those outside the cambium become phloem cells. Cell division occurs in the cambium throughout the growing season. Those xylem cells formed early in the season enjoy very favorable growth conditions and become large. Cells formed later in the season are smaller. The alternate groups of large and smaller cells around the periphery of the xylem give the effect of rings in the cross-section of a tree. These are called *annual rings* since each ring is usually the demarcation of a single year's growth. This ring effect is also present in the phloem tissue but is not as distinct.

When grafting it is the callus tissue produced from the cambia of the two grafted pieces which must unite to form a successful union. It is the cambium which produces wound tissue that surrounds and envelops a scar or cut on a woody stem. The cambium is the regenerating tissue, the healing tissue, and the tissue-maker of the woody stem.

FIGURE 2-5   A scar showing wound tissue formed by cambial activity. (Photograph by K. A. Denisen.)

The Phloem   Outside the cambium, and forming the inner part of bark tissue, is the phloem. This tissue has the function of transporting manufactured food (dissolved in water) both upward and downward in the plant. Since the phloem is actually a part of the bark, we can readily see that the bark contains a vital tissue for plant growth in addition to its role as a protective tissue in the woody stem. If a tree is girdled, that is, the bark removed from around its entire periphery, the flow of manufactured food is stopped. Unless the severed phloem tissues are reunited by cambial growth, the roots will starve for lack of food and the plant will eventually die. Rabbit injury to young trees and shrubs can be very serious for this reason. Precautionary or remedial measures (e.g., bridge grafting) are necessary to save these plants when rabbits or other rodents feed on the bark.

Outside the phloem is corky tissue of the bark which prevents drying of the inner tissues and otherwise serves in a protective capacity. Longitudinal cracks or ridges in the bark are a result of expansion of the circumference of a tree due to cambial activity.

The Herbaceous Dicot Stem

The stem of a herbaceous plant is similar in anatomy to the shoot, or current season's growth, of a woody plant before hardening or maturation of tissues occurs. It has the components, xylem, cambium, and phloem but they are not as distinct as in the woody stem. The cambium again separates the xylem and phloem but may not be a continuous ring near the outer periphery of the stem.

Herbaceous stems, like those of geranium, are more succulent, less fibrous, more pithy, and less tough than woody stems. They, characteristically, are not as strong, not as hardy, and not as rigid.

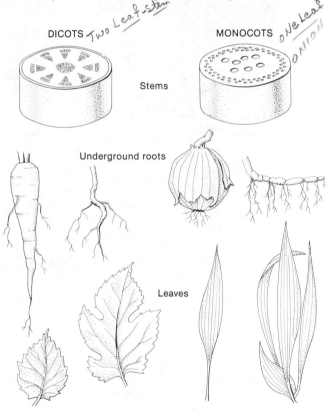

FIGURE 2-6   Dicot and monocot comparisons of stems, roots, and leaves.

The Monocot Stem   The *monocots*, or monocotyledonous plants, differ in stem characteristics from the dicots in that they generally do not have a cambial ring between the xylem and ploem tissues. Such plants as sweet corn, lilies, onions, and grasses are herbaceous monocots. The palm, bamboo, and screw pine are woody monocots, which have an intercalary meristem for increased diameter. Xylem and phloem tissues have adjacent locations in structures called *vascular bundles*, which are scattered throughout the pith tissue of the stem. Cell division in monocot stems occurs principally in the *apical* or tip region and in intercalary meristems of woody types.

Modified stems   In many horticultural plants, modifications of stem tissues play an important role in their usefulness (see Chapter 7). The *tuber* of the potato is an enlarged underground stem which is adapted for food storage and is also capable of reproducing the plant. The *rhizomes* of iris and lily-of-the-valley are horizontal underground stems capable of producing both roots and shoots at the nodes, thus forming new plants. A *corm* is a globe-shaped, fleshy stem used for food storage through the dormant season in the gladiolus and crocus. *Bulbs* are similar to corms in shape, but consist of concentric layers of fleshy leaves that arise from a *stem-plate* at the base. Examples include onions, lilies, and tulips. *Thorns* are also modified stems useful to plants for protection and in some cases, as with certain plants, for support. *Tendrils* are twining branches or stems which give support to vines.

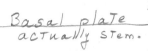
Basal plate actually stem.

FIGURE 2-7   The tuber of the potato is an enlarged underground stem. (Photograph by W. L. Summers, Iowa State University.)

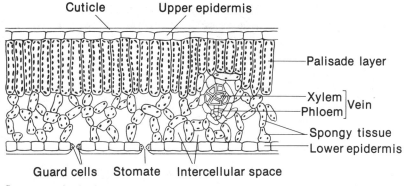

Cuticle   Upper epidermis

Palisade layer

Xylem ⎤
Phloem ⎦ Vein

Spongy tissue

Lower epidermis

Guard cells   Stomate   Intercellular space

FIGURE 2-8   **Structure of a leaf.**

The Leaf   The primary function of the leaf is to synthesize food for the plant by the complex process of photosynthesis. The structure of the leaf is designed to facilitate this important manufacturing activity. The *petiole* (leaf stem), *mid-rib*, and *veins* of the leaf supply water and minerals to the leaf cells through their basal connection with the xylem of the stem. Manufactured food is also transported in the veins to the phloem of the stem.

The tiny pores, or *stomates*, on the under side of the leaf admit carbon dioxide needed as a raw material for photosynthesis. Stomates are the gateways for the escape of water vapor in transpiration and for the release of oxygen as a by-product of photosynthesis. *Chlorophyll*, the green coloring material of plants, is found in great abundance in leaf cells. The loose structure of the cells (spongy tissues) in the lower region of the leaf blade readily permits gaseous exchange between the leaf cells and the atmosphere.

Prior to leaf fall of deciduous plants an *abscission layer* is formed at the base of the petiole, adjacent to the stem. This layer of cells undergoes maturation so that, as the time of leaf fall arrives, the leaves separate readily from the tree without leaving an open wound. This is one of Nature's techniques of protection against drying of tissue or entrance of disease organisms.

Many horticultural plants have leaves with an obviously waxy surface. The waxy material on the outside of leaves is *cutin*. The layer of cutin on leaves is called the *cuticle*, and it is present, to some degree, on all leaves. It is a precaution against excess water loss from leaves. Under dry weather conditions the cuticle is thicker than under conditions of high humidity.

The Bud   A bud is an undeveloped shoot, or flower, or shoot and flower. By position, a bud is either *terminal* or *lateral* on a twig or shoot. By activity, a bud is either resting, dormant, or latent. A *resting* bud is undergoing internal change and development, and will be ready to resume growth when the rest period is over, provided other conditions are favorable. A *dormant* bud is not growing because the weather is not favorable. A *latent* bud is inactive and may remain so for the life of the plant unless stimulated in some way. Latent buds are the source of *water sprouts* which sometimes suddenly appear from old branches and trunks of trees because of very heavy pruning, over-fertilization, and other variable factors.

Buds are located at *nodes* on shoots and twigs. Nodes are the place of origin of leaves, and buds develop in the axils of the leaves. The stem sections between nodes are called *internodes*.

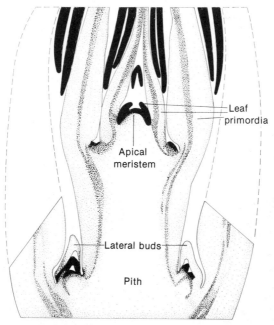

FIGURE 2-9   **Structure of a lilac (*Syringa vulgaris*) bud.**

FIGURE 2-10   **Dormant buds of lilac (*Syringa vulgaris*).**

Classification, Anatomy and Growth of Horticultural Plants   **39**

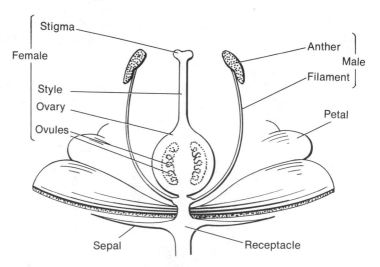

FIGURE 2-11   **A diagrammatic longitudinal section of a typical angiospermous complete flower.**

FIGURE 2-12   **Photograph of parts of a flower can be compared with diagram of Fig. 2-11. (Photograph by K. A. Denisen.)**

The Flower   The flower precedes, and is essential for, fruit development. It is a reproductive organ of the plant, since, in complete fulfillment of its function, seed is produced.

Horticultural plants vary widely in flower type, but the essential parts of the flower are found in all of them. The essential parts are those which constitute the sex organs of the flower and are needed for pollination and fertilization.

A flower which has both sexes present is called a *perfect* (hermaphroditic) flower. A typical perfect flower contains the pistil (female) and the stamen (male). The pistil is "bottle-shaped" and consists of the *stigma*, which has a sticky surface; the *style*, or "neck of the bottle"; and the *ovary*, the globe shaped structure which contains the *ovules* that develop into seeds following fertilization. The stamen consists of a

Pistil—(female)
Stigma
Style
ovary

STAMEN—(male)
filement
anther

FIGURE 2-13   **Sweet corn is a monoecious plant. The tassel is the male flower, located terminally. The ear is the female flower, located laterally. (Photograph by K. A. Denisen.)**

FIGURE 2-14   Squash flowers. *Left*, female flower. *Right*, male flower. (Photograph by K. A. Denisen.)

filament or stalk which supports the *anther*, the *pollen* producing organ of the flower.

The petals and sepals are considered accessory parts of the flower, yet in many floricultural specimens they have been developed into extremely large and showy flower parts. The petals as a group form the *corolla*. The sepals constitute the *calyx*, which encased the flower before it opened.

Imperfect flowers are those with one of the sexes absent. These are of two types, *monoecious* and *dioecious.* Monoecious plants are those which have male and female flowers on the same plant: e.g., sweet corn, pumpkin, squash. Dioecious plants are those with flowers of only one sex on each plant; that is, male plants and female plants: e.g., asparagus, cottonwood, spinach.

Pollination occurs when pollen is transferred from the anther to the stigma. The pollen grain germinates on the sticky surface of the stigma, a pollen tube grows down through the style and into the ovule of the ovary. When the nuclei of the pollen and egg cell fuse to form a new organism, *fertilization* has taken place.

The Fruit   Pollination and fertilization provide a stimulus to fruit set and seed development. The fruit, whether horticultural or botanical, is the object in growing tree and small fruits, many vegetables, and some ornamentals. Each specific kind of plant has its own flowering and *fruiting habits.* Some arise from the current season's growth and others are borne directly on woody tissue. The fruiting habits of horticultural plants are the basis for many of the cultural practices used in getting high yields, large size, and good quality. Fruit of the grape is borne laterally, near

FIGURE 2-15   **Fruits of the white potato. Potatoes do not usually set fruit or produce seed, but will under long days and with hand pollination.**

the base, of current season's growth. Fruit of the apple is mostly borne terminally on spurs, one year old or older, sometimes laterally on twigs and occasionaly terminally on twigs, but always from woody tissue. About 90 percent of the previous season's growth is removed in annual pruning of the grape; nearer 10 percent is removed from the apple. Fruiting habits also influence other cultural practices. The relationships of fruiting habits and the skills and practices in horticulture are discussed more completely in Part II.

Some plants develop fruit without fertilization of the flower and thus are seedless or *parthenocarpic*. This may be due to a natural condition of the plant, e.g., the banana. It may be induced by removing a ring of phloem (bark) from the stem of some woody plants like the grape to produce seedlessness. Certain growth-regulating chemicals have been used to stimulate ovary growth with or without fertilization, the object being to increase fruit set and not necessarily parthenocarpy.

The Root   There are many similarities between the internal structure of the stem and the root, especially within the same plant. The root of the woody dicot has xylem, cambium, and phloem tissues which join the counterparts of the stem. This applies also to herbaceous plants and to the monocots.

Roots do not have nodes. The branch roots arise at irregular intervals and locations instead of from buds. This is a determining feature which

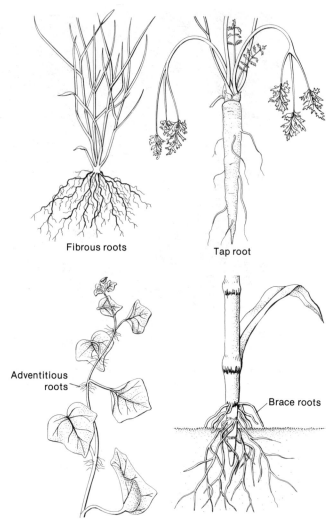

Fibrous roots

Tap root

Adventitious roots

Brace roots

FIGURE 2-16   **Four types of roots.**

classes the sweet potato as a root and the Irish potato as a tuber (modified stem).

The chief function of roots is absorption of moisture and minerals from the soil. Absorption occurs through the *root hairs* which are outgrowths or protrusions from the epidermal cells of the *feeder roots* of the plant. The root hairs dry out readily upon exposure to a dry atmosphere. This is an important point to remember in transplanting, especially with succulent plants. Roots may also be a place of storage, a means of propagation, and a method of anchorage and support for plants.

PHENOMENON OF GROWTH IN PLANTS

Growth in plants, as in animals and humans, is often taken for granted. It is a complex process of progressive development resulting in

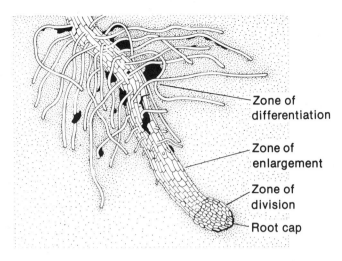

Zone of differentiation

Zone of enlargement

Zone of division

Root cap

FIGURE 2-17   **Zones of cell growth in a root.**

greater size, larger yield, increased maturity, more cells, or increased specialization of plant parts. In nature, plants grow without the help of man. However, man can guide the destinies of plants by influencing their growth. An examination of the inner parts and inner workings of the plant will introduce us to the peculiar phenomenon called growth.

Phases of
Cell Growth

The cell is the basic unit for growth in plants. As the cells increase in number and size and weight, the plant becomes larger.

Cell *division* is the first phase of growth. Some areas of plants, such as the root and stem tips and the cambium, consist almost entirely of dividing or meristematic cells. The chief function of these areas is to produce more cells. But having more cells does not in itself increase plant size. Certain daughter cells are pushed away from the zone of division and proceed to the next phase of growth, cell *enlargement*. These cells are supplied with large quantities of water and food and may become several times their former size. However, growth by cell enlargement does not represent a substantial increase in plant production merely because an abundant supply of water and some food reserve entered and helped to stretch the cell walls. Cell enlargement must be followed by another phase, cell *differentiation*, in order to represent significant growth.

Cell differentiation involves specialization of cells, tissue development, and maturation. The enlarged cell is supplied with food which may be differentiated in various ways. Inside the cambium of a woody plant it may become woody xylem tissue or strengthening fiber. It may serve as food conducting tissue in the phloem. It may accumulate food reserves and become storage tissue as in the tuber of a potato or the root of a carrot. It may differentiate into floral initials and become part of the flower. It may give rise to the sex cells and be instrumental in transmitting hereditary factors to the next generation through the seed. Each

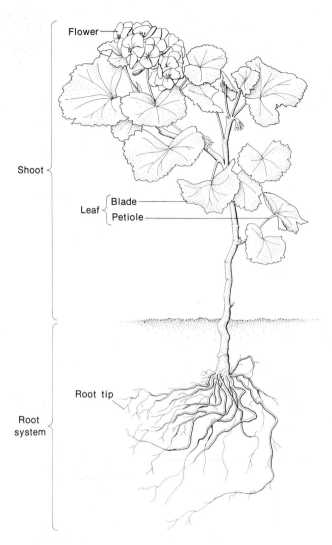

FIGURE 2-18   **The complete plant structure.**

specialized cell of a plant originated through cell division, underwent a degree of enlargement, and then became differentiated into its specific shape, size, and function. The cells of an actively growing root tip under the microscope can be readily grouped into zones of division, enlargement, and differentiation.

Processes of Growth   Basic processes which plants depend on for growth are *photosynthesis*, for food supply; *respiration*, for energy release; *assimilation*, for food utilization; *absorption*, for raw materials; *transpiration*, for upward flow of water; and *translocation*, for movement of manufactured food. In order for horticultural plants to grow properly and be of maximum usefulness, these processes must be integrated to operate at optimum efficiency.

Photosynthesis, the process of food manufacture by plants, is often described as the most important chemical reaction in the world. The implication of its universal importance is apparent when we consider that green plants alone have the ability to manufacture sugars from carbon dioxide and water. These raw materials of photosynthesis, carbon dioxide and water, are actually the waste products of oxidation, combustion, and respiration. Oxygen, which is needed for all life, is released as a by-product of photosynthesis. Thus, the complex process of photosynthesis is of utmost importance in maintaining the balance of nature.

Only plants containing the green material *chlorophyll* can carry on photosynthesis. Light, as the source of energy, is also an essential factor in carbohydrate manufacture. Plants are a means whereby the energy of the sun can be stored for future use. Wood, coal, and petroleum are fuels produced by plants; they contain stored energy of the sun from the complex process of photosynthesis. Heating and lighting of our homes; propulsion of cars, trains, and airplanes; and other advances in technology are possible because energy was stored in the form of carbohydrates and carbon compounds in plants.

The main area of photosynthetic activity is the leaf. In cross-section (Fig. 2-8) we find cells in a unique arrangement functioning as the machines of food manufacture. Xylem cells supply water from the leaf veins, and guard cells regulate the opening of stomates for admitting carbon dioxide. Chlorophyll is present to facilitate the reaction when light strikes the leaf. Energy is absorbed, carbon dioxide and water are reduced to elemental forms, a complex series of intermediate reactions take place, and glucose, a simple sugar, is produced. Oxygen is released as a by-product. The general equation is

$$\underset{\text{water}}{12H_2O} + \underset{\text{carbon dioxide}}{6CO_2} \xrightarrow[\text{light energy}]{\text{chlorophyll}} \underset{\text{glucose}}{C_6H_{12}O_6} + \underset{\text{oxygen}}{6O_2} + \underset{\text{water}}{6H_2O}$$

The simple carbohydrate, glucose, is a basic material for synthesis of other foods by plants or animals. It is a building block for other carbohydrates, proteins, fats, vitamins, pigments, enzymes, hormones, organic acids, alcohols, and other organic materials. Thus, photosynthesis is really fundamental to life itself. Nature has entrusted green plants with the role of utilizing the energy of the sun to manufacture food for the world from carbon dioxide and water and to release oxygen to the atmosphere.

Respiration is an oxidation process found in all living organisms. Food is utilized as a source of energy and carbon dioxide and water are released as end products. In effect, respiration is the opposite chemical reaction from photosynthesis.

$$\underset{\text{glucose}}{C_6H_{12}O_6} + \underset{\text{oxygen}}{6O_2} \longrightarrow \underset{\text{carbon dioxide}}{6CO_2} + \underset{\text{water}}{6H_2O} + \text{energy}$$

Thus, our source of energy from the sun, which has been stored in the food manufactured by green plants, is released during the metabolic process of respiration.

Both plants and animals undergo respiration. A growing green plant under favorable conditions has a photosynthesis rate which far exceeds the respiration rate. Yet a plant needs energy during growth to provide for the mechanics of absorption, translocation, assimilation, and other basic processes.

Oxygen in the atmosphere is replenished by plants, and it is essential for energy release through respiration. When present in abundant quantity and oxidation is complete, *aerobic* respiration results. When available oxygen is limited, respiration is *anaerobic*, and the reaction stops short of complete oxidation. Intermediate compounds, such as alcohols and organic acids are produced by anaerobic respiration. *Fermentation*, a form of anaerobic respiration, results in incomplete oxidation of sugars and is used in the production of wines and vinegar.

Assimilation is the process of utilizing food for growth. *Digestion*, the converting of foods into simpler compounds, is the first step in assimilation. *Enzymes*, biological catalysts, play an important role in digestion or food conversion in plants. For example, starch is converted to sugar in the presence of the enzyme, amylase. The reaction will proceed in the opposite direction under different temperature and enzyme conditions. Certain minerals are essential as components of enzymes or their associated groups. Thus, a deficiency of one of these minerals can result in abnormal enzyme action and abnormal plant metabolism.

As foods are digested, the simpler forms which result, are used as "building blocks" for more complex compounds. Assimilation in its later stages is the integration of the numerous simple and complex compounds into protoplasm, the living substance of all cells. Thus, assimilation has the vital role in growth of taking food manufactured by the plant and transforming it into living tissue.

Absorption is the means whereby minerals and water enter the plant. Through the physical phenomena of *osmosis* and *diffusion*, water and dissolved minerals enter the plant through the root hairs which are in intimate contact with the soil. Leaves are also capable of absorption. Carbon dioxide enters the food manufacturing cells of the leaf by absorption in its role as a raw material of photosynthesis. Certain foliage sprays such as growth regulators and liquid fertilizers enter the plant by absorption through the leaves.

Transpiration is the loss of water by evaporation from the leaf surface of plants. Tremendous quantities of water enter and leave a plant during the course of a single season. A mature apple tree will transpire several hundred gallons of water annually. It has long been a mystery why plants lose so much water by transpiration. Scientists have calculated that plants can bring nutrients from the soil with much less water than is used in transportation. A very low percentage of the water intake of a plant is used as a raw material for photosynthesis. Even the cooling effect of evaporation has been ruled insignificant as a role of transpiration. Yet transpiration does occur and plants need lots of water. Transpiration is regulated to a degree by the plant since the size of the stomates is controlled by guard cells on each side of the openings. Under conditions of excessive heat, guard cells lose turgidity and the stomates close. When a plant wilts, the stomates close. At night, when carbon dioxide is not needed for photosynthesis, the stomates close. As we shall see, transpiration is greatly influenced by environmental factors.

Translocation is the process of movement of synthesized and absorbed food and nutrients within the plant. Movement of sugars following photosynthesis is primarily through phloem tissue. To be moved through the tubes and vessels of a plant, the materials must be in water solution. Thus, the need for sugar to starch and starch to sugar conversions in the plant becomes apparent. Sugars are moved to storage organs where they are generally converted to starch which is insoluble in water. Starches, as reserve carbohydrates, remain in various storage organs until needed in other areas. Enzymes convert the starch back to sugars and they are translocated via either xylem or phloem conduits. The sugars of the sugar maple move upward through the xylem when growth is about to begin in the spring. The starch of the seed is converted to sugar at germination, and it moves to the growing point of the stem and the root tip where it provides food for energy release during growth. Starch is stored in the potato tuber until growth begins anew. Following enzyme action the resulting sugars move through conductive tissue to the growing points.

Not all translocation is upward and downward, but may be horizontal as well. Rays are tissues which radiate from the center to the periphery of stems and roots and are present in both phloem and xylem.

The complete intergration of the basic processes of growth is the goal of horticultural production. When they operate at maximum coordination and efficiency, the plant produces its maximum. If one process is

FIGURE 2-19   **Synthesized foods are translocated from leaves to storage areas; in this case, to the blossom of broccoli. (Photograph by K. A. Denisen.)**

inefficient, the whole system is less productive just as "a chain is no stronger than its weakest link."

Stages of Growth in Horticultural Plants

Whether a plant be annual, biennial, or perennial, it has certain stages or periods of growth. During these stages all horticultural plants seem to follow an established pattern of development even as the cells and tissues follow a plan in their development. The stages of growth in horticultural plants are termed *juvenile, transition*, and *productive*. These are descriptive terms which enable us to visualize horticultural plant development as a parallel of our own childhood, youth, and adult periods in life.

JUVENILE STAGE. This period is characterized by rapid *vegetative* growth. It is the time during which a fruit tree is attaining size, framework, and strength for its ultimate role of fruit production. It is the period of early growth of a tomato plant in preparation for blossoming and fruiting later in the season. It is the stage during which shrubs or other ornamentals are becoming established to adequately fulfill their destined goals as producers of beauty and usefulness.

During the juvenile stage, increase in size is rapid due to the utilization of manufactured food for further shoot, leaf, and root development. Carbohydrates are used in great quantities for cell enlargement and for differentiation of xylem and phloem tissues. The absorbing surfaces of the roots and food manufacturing areas in the leaves operate at maximum capacity to produce more food for assimilation and continued growth. Then, as with all types of growth among all species of life, increase in size tapers off and a gradual change or transition in growth occurs. This shift may be created by various sets of conditions. It may be due to limiting factors of the environment, such as shortage of moisture, inadequate light, lack of sufficient nutrients, unfavorable temperatures, etc. Competition between plants may cause these factors to be limiting and hasten the end of juvenility. However, within species and kinds of plants the approximate end of the juvenile stage is generally predetermined by inherited characteristics. We can modify plant size to a certain degree by varying its environment, applying growth regulators, and pruning, but plants are still subject to hereditary influences.

TRANSITION STAGE. A plant in its "youth" is undergoing a change from utilization of carbohydrates for growth to accumulation of carbohydrates in storage organs. The transition stage may last for several years in some fruit and nut trees to as little as a week or less in some annual plants. In either case the change from juvenile to productive is not abrupt, but represents a gradual changeover beginning in the cells. The naturally occurring plant hormones play an important role in stimulating differentiation of cells to form reproductive and storage organs. The stimulus of the cells destined to give rise to these organs of storage and reproduction is actually the beginning of the transition stage. The production of these growth substances by the plant could be called an even earlier beginning of transition. The end of transition, like

FIGURE 2-20   **Two-year-old wood of apple. *Left*, late transition stage showing fruit buds on spurs. *Right*, productive stage. Spur has fruited previously, will grow during current season, then continue in its cycle of production.**

the beginning, is not clear cut. Some vegetative growth continues to occur even during the most productive stages of plant growth. However, a balance is reached during which carbohydrate accumulation is favored over carbohydrate utilization and the plant is considered to be in its productive stage of growth.

PRODUCTIVE STAGE. Depending upon the kind of horticultural plant, the productive stage results in an accumulation of food reserves in either vegetative storage organs, reproductive parts or fruits. Examples of accumulation in vegetative storage organs include root crops, such as beets, carrots, and radishes; bulbs, including onions, tulips, and garlic; petioles, like rhubarb and celery; and stems, e.g., asparagus and potato (tuber). The principal area of food storage in reproductive parts is in the seed. In beans, peas, sweet corn, and nut crops, sufficient quantities of carbohydrates are stored in the seed to render them important sources of human food. Most seeds have sufficient carbohydrates to carry them safely through the period of gemination until seedlings become established. Cauliflower and broccoli are examples of floral parts used as storage sites. The horticultural fruits, borne on perennial or woody plants, are an important area of study for the productive stage of plant growth. Such practices as prunning, dwarfing, fertilizing, and chemical treatment may have a pronounced effect on hastening or delaying the onset of the fruiting or productive stage. Many ornamentals are considered to be in the productive stage when they bear flowers. Others are preferred for their juvenile habits such as the attractive twig growth of certain shrubs.

SEASONAL PHASES OF THE PRODUCTIVE STAGE. Most woody or perennial fruits have seasonal fluctuations between carbohydrate accumulation and carbohydrate utilization. Deciduous fruit trees, which are in the productive stage, bloom early in the growing season, set fruit, and begin vegetative development of shoots. During shoot growth the fruits do not increase appreciably in size. However, after the initial burst of shoot growth, the rate of shoot enlargement subsides and stem tissues harden. This is followed by a high rate of photosynthesis in the enlarged leaf area and the phase of carbohydrate accumulation has begun. Fruit buds for the succeeding year are differentiated while the current season's fruit are still very small. These buds then remain in a resting condition while the current crop is being grown and matured.

STUDY QUESTIONS

1. Why do we concern ourselves about the classification of plants? What problems might we have if we didn't concern ourselves about classification?
2. The term cultivar is of quite recent vintage. Why the introduction of this new term as a replacement for variety?
3. Why do we concern ourselves with the anatomy of plants? Does it help in growing, harvesting, or consuming plants?
4. What is the chief function of the cambium? What role does it have in grafting?
5. Give several differences between monocotyledonous and dicotyledonous stems.
6. What process occurs in green leaves which makes them extremely important organs of the plant?
7. Why is the bud such an important structure in a woody plant? What are the different kinds of buds?
8. What kinds of imperfect flowers are there? How can these differences affect a breeding program?
9. Are parthenocarpic fruits desirable? What causes fruits to be seedless?
10. What is the role of the root in plant development?
11. Give the three phases of cell growth and what occurs in each phase.
12. Follow the steps (or processes) in plant growth from photosynthesis to assimilation. How do absorption and transpiration fit into the scheme of plant growth?
13. What is an enzyme? What role does it play in germination?
14. Distinguish between juvenile, transition, and productive stages of growth in an apple tree.

SELECTED REFERENCES

1. Butler, W. L., and R. J. Downs. 1960. *Light and Plant Development. Sci. Amer. 203*:56–63.
2. Curtiss, Helena. 1972. *Invitation to Biology*, Worth Publishers, New York.
3. Esau, K. 1960. *Anatomy of Seed Plants.* Wiley, New York.
4. Galston, A. W. 1968. *The Green Plant.* Prentice-Hall, Englewood Cliffs, N. J.
5. Troughton, J., and L. A. Donaldson. 1972. *Probing Plant Structure.* McGraw-Hill, New York.

3 Plants and Their Environment

The environmental factors, temperature, water, light, and nutrients, are essential for the growth of green plants. If one becomes short in supply, it is called a *limiting* factor, since normal growth and development are limited by its scarcity. The horticulturist strives for maximum production of his plants, but in order to attain it, he needs the cooperation of the factors which might limit plant growth. Some of these factors are under the control of man, but others cannot be supplied or augmented in a practical manner. Irrigation is practiced in many regions, but before such a program can be adopted a good source of water must be available. Temperatures can be modified only to the extent of the economic feasibility of greenhouse production, artificial heat, and other costly procedures. Since we must live with our various environments, the most effective and practical method of dealing with the various factors which may become limiting is to adapt our production methods to the environment. Thus, we may grow hardy plants in cold areas, apply irrigation in dry regions, grow long-day plants during long-day seasons, and apply fertilizers where nutrients are needed. We can either grow plants adapted to the environment or change the environment so plants can prosper.

TEMPERATURE

A change of a relatively few degrees in temperature can make us feel warm or cold. We respond by modifying our clothing, turning up the heat, or using fans and air conditioners. Plants, when subject to temperature changes, respond quite differently. Since they cannot modify the temperature directly they can only react to it through their basic processes.

Optimum Conditions   Each species or variety of horticultural plant has its own temperature requirements. At the optimum range of conditions, the various processes of photosynthesis, respiration, assimilation, etc. proceed at "peak" capacity. Based on the optimum temperature ranges, plants may be classed as cool-season or warm-season types.

Cool-season plants are those which thrive best under relative cool conditions, usually under 70°F (20°C) mean daily temperature. They include many of the early spring vegetables such as onions, peas, radishes, and lettuce which germinate at relatively low temperatures and

FIGURE 3-1   *Left*, hollyhocks (*Althea rosea*) in a favorable environment. *Right*, hollyhock in an unfavorable environment. This plant is six inches (15 cm.) in height and is growing in a lawn under severe competition.

will tolerate a light frost. Apples, pears, plums, and American type grapes are examples of cool-season fruits. They will tolerate very cold winter conditions, need cold weather for subsequent fruiting, but will produce good crops under warm summer conditions. Many herbaceous perennials, such as tulips, crocuses, daffodils, and violets are cool-season types. The early spring blooming shrubs forsythia, lilac, spirea, and honeysuckle are cool-season woody ornamentals.

Warm-season plants are those which need mean daily temperatures of 70°F (20°C) or above during much or all of the growing season for best performance. Vegetables in this group are extremely frost sensitive, e.g., tomatoes, peppers, beans, sweet corn, melons, and pumpkins. Fruits of the temperate zone have less chilling requirements, withstand higher summer temperatures, and are less hardy to winter cold. They include peaches, sweet cherries, European grape, and figs. Ornamentals in this category include holly, gardenias, hybrid tea roses, many lilies, and numerous house plants. Tropical and sub-tropical plants are usually warm season types unless native to high altitudes in otherwise tropical areas.

Adverse Conditions   Winter-killing, frost injury, and sunburn are examples of plant injury that may be due to extremes in temperature. However, temperature conditions which are not that extreme may be adverse to good growth and development. Lettuce is a cool-season crop and during periods of above-optimum temperature, it becomes virtually inedible because of the bitter flavor which develops during hot weather. Radishes are stronger and more pithy during hot weather. Celery and cauliflower may form a seedstalk (called bolting) as a result of a dip in temperature below optimum during early growth. Adverse temperatures either above or below optimum can result in abnormal development or merely stunted underdevelopment.

Distribution and Use   The occurrence of optimum and adverse temperature conditions has a great effect on the distribution and use of many horticultural plants. Specialized production areas have developed in regions that have temperatures which favor a certain crop or crops. Oranges are produced abundantly in southern California and Florida because they need warm temperatures and mild winters. The Mediterranean area is likewise an important citrus area. Potatoes are a cool-season crop and produce high yields in northern United States, parts of Canada and Alaska, and throughout northern Europe. Bluegrass is the principal turf grass in cool climates of the world and Bermuda grass is more common in lawns of the milder regions. Large-flowered everblooming roses do best where winters are mild, but the limitations of their adaptability are extended by various methods of cold protection. Plants are often grown in climates outside their adapted areas but only if given special treatment or if the grower is satisfied with less growth.

Periods of No Visible Growth   There are times when plants or plant parts are seemingly in periods of non-growth. These are the dormant period and the rest period. The two terms are sometimes used interchangeably, yet the two conditions are distinctly different in both cause and effect.

The *dormant period* in biennial and perennial plants is due to unfavorable environmental conditions for growth. In temperate climates this occurs during the winter months. The lowering of temperatures during the autumn prepares the plant for its period of dormancy. As winter sets in, temperatures are too low for deciduous and herbaceous plants to carry on their basic processes of growth and they enter the dormant period. Woody deciduous plants lose their leaves and herbaceous perennial plants die back to the ground. They have ceased growth and are dormant because of unfavorable environmental conditions. Some tropical plants also have dormant periods which are not due to low temperatures but rather occur with the dry season.

The *rest period* in biennial and perennial plants is found only in certain plant parts and is controlled by internal conditions of the plant. The bud is the principal plant structure which has a rest period. Some seeds also have a rest period. Other plant structures such as roots are not affected by rest periods.

Buds of woody plants which are formed during shoot growth continue to develop and enlarge until they reach a certain size for the species. They then "rest" for the remainder of the season or until stimulated to resume growth. The onset of the rest period is gradual and not specific as to time or environment. It is controlled by physiological and chemical conditions within the plant. The rest period is broken or interrupted and the bud "awakens" to resume growth when both internal and external conditions of the plant again favor growth activities. Various conditions can bring about the cessation of rest. Exposure to cold temperatures is the usual manner in which the rest period is broken in woody plants of temperate climates. Thus the rest period is broken or shortened by cold temperatures in contrast to the dormant period, which is lengthened by cold temperatures.

Observation of the various phases and activities of a bud will show the contrast, relationship, and overlap of the rest and dormant periods. We shall start with a developing bud in the leaf axil of a shoot. The bud is in a growing state and its internal conditions favor enlargement. It continues to grow until it reaches full size for the species. Growth in size gradually ceases at this point even though the environment of the plant is favorable for growth. This is the beginning of the rest period. While at rest the bud carries on the process of respiration at a low rate, it remains the same size, its enzyme and hormone systems may change up to the point of activation for growth. As cold weather approaches, growth of the plant is slowed, tissues harden, and abscission layers form at the bases of leaf petioles. Then comes leaf-fall and the plant is dormant for the winter. Two forces are now preventing growth of the bud, the rest period and the dormant period. This is the period of overlap and both are governing the activities of the bud. This seems to be a precaution or protection of Nature. If mild weather (Indian summer) prevailed following leaf-fall, the bud would no longer be prevented from growth by unfavorable environment. However, since the rest period is still in full effect, the bud is prevented from growing and producing new tender growth that would be subject to the winter temperatures which follow. As the plant accumulates sufficient hours and days of temperatures below a certain thermometer reading, the buds complete their rest period. For example, about 1400 hours of temperature below 45°F (8°C) are required to break the rest period of the Latham raspberry.

After the breaking of the rest period, buds do not usually start growth immediately because the dormant period is still in effect as it may still be mid-winter. However, the period of overlap is past. When temperatures go up and remain at a point favorable for growth, the dormant period will also be ended. A January thaw which occurs after the rest period is over may cause winter killing if the start of growth activities is followed by a severe dip in temperature. Such environmental occurrences limit the adaptation of certain so-called hardy plants in areas of milder climates. The Latham raspberry has withstood temperatures of −50°F (−45°C), yet has winter-killed in areas with minimum readings of 0°F (−18°C). Wide fluctuations in temperature after the rest period is over may cause winter killing in many plants.

Other methods of breaking the rest period include pruning during the growing season and treating resting buds with specific chemicals. In the case of pruning during the growing season, the buds are not usually fully developed at that time and may not have reached the resting stage. This

FIGURE 3-2   **Apple shoots in mid-summer.** *Left,* **buds are in their rest period.** *Right,* **several buds have "broken" as a result of abundant food for growth. This was "forced" by severe pruning.**

could actually be called by-passing the rest period. Very severe pruning before growth starts in the spring may also force buds to by-pass the rest period.

Herbaceous biennial and perennial plants may also be subject to rest and dormant periods. The Irish potato has a rest period during which the "eyes" or buds of the tuber will not sprout. It can be broken by heat rather than cold. Storage at 80°F (26°C) will break the rest period of newly harvested Irish potatoes in three to four weeks instead of the usual six to nine weeks. Some chemicals, e.g., ethylene chlorohydrin will break the rest period in a few hours. The tulip goes into its rest period shortly after bloom in the spring. Its foliage dries up and the bulb remains in a non-growing state until fall. Cold weather breaks the rest period and growth starts after the dormant period is over.

Hardiness   Some plants can tolerate extremely low temperatures and other adverse conditions. These plants are called *hardy*. Hardiness is a complex of several factors: (1) It involves the ability of plant cells to hold water against freezing. (2) It requires adequate maturation of tissues before cold weather begins. (3) It is greatest when the rest period is of long duration or not easily broken by low temperatures. (4) It is decreased by fluctuating winter temperatures. (5) It is seriously reduced by drouth conditions both during the growing season and the dormant season. (6) It is characteristic of native plants for they have survived the rigors of their environment through the years and have reproduced their kind.

WATER

One needs but to observe the effects of drouth, drive through an arid region, or just forget to water the house plants to realize the important role of water in plant life. The amount of rainfall or the quantity of water available for irrigation are often the limiting factors which regulate the growing of certain kinds of horticultural plants in specific areas. In some localities, where water is at a premium, there may be drastic competition between the needs of plants, industrial needs, and needs of the human population. Water thus plays an important economic role in addition to its specific functions in plant and animal life.

Fresh fruits and vegetables are very high in water content. The potato is nearly 80 percent water, carrots 88 percent, pears about 83 percent and melons about 94 percent. Even dry beans, peas, and most seeds have a moisture content of 10 to 15 percent.

Why Is Water Needed in Such Large Quantities

Calculations show that an apple tree uses approximately 10 barrels (50 gal. per bbl.) of water in producing one bushel of apples. Similar high ratios of water intake to food produced are found with other crops. It will be recalled that the processes of growth previously discussed all utilize water. Absorption and translocation are entirely dependent on water for movement of materials into the root and for distribution within the plant. Cell enlargement is primarily dependent on water to expand the cell wall and cell membranes. After cell size has been increased by this expanding effect of water, the size is maintained by turgor pressure of water, while the solids content of the cell is built up in the differentiation process.

Photosynthesis could not proceed without considerable water for several reasons. Water is a raw material of food manufacture by plants. Photosynthesis proceeds at the highest rate and efficiency in those cells adequately filled with moisture. The various elements which must be present for photosynthesis to proceed normally are present through the medium of water transport. Manufactured food is conveyed from the areas of photosynthesis by water. Thus, photosynthesis continues without being hampered by a stockpile of manufactured food in the leaf, the manufacturing site.

The tremendous loss of water from plants through transpiration accounts for the greatest amount of the water which a plant absorbs. Perhaps this could be termed luxury use of water, for seemingly more water passes through the transpiration stream of plants than is necessary for the various growth processes. However, plants do play a vital role in maintaining a balance in Nature, and replenishing the atmosphere with moisture is undoubtedly an important phase in the water cycle. More water enters the atmosphere through transpiration from leaf surfaces of plants for a given area than by evaporation of an equivalent area of open water. The transpiration rate of plants is increased by such environmental factors as wind, low humidity, light, and rising temperatures. Some plants have structural adaptations which tend to reduce water loss by transpiration. These include a reduction of leaf surface in the needles of conifers, heavily cutinized leaves of the jade plant, pubescent or hairy

FIGURE 3-3   The jade plant has heavily cutinized leaves for reduced water loss. This 15-year-old plant has prospered under a low watering regime. (Photograph by K. A. Denisen.)

FIGURE 3-4   **Cacti and cactus-like plants have modified stems and leaves for water storage and reduced transpiration.**

leaves on the African violet, or leaves reduced to spines, with photosynthesis occurring in the stem, found in the cactus group.

Why Is Water Needed in the Soil   The soil is the reservoir, the supplier, the storehouse for moisture. Since minerals for plant nutrition are also stored in the soil, water just be present as a medium of transport for these minerals so they may enter the plant. All the plant processes require that the soil furnish water for optimum growth and development. Root growth is basic for plant growth since roots determine the extent of the plant's ability to absorb water and nutrients. In a dry soil root growth is greatly retarded. In an extremely wet soil, root growth is also retarded because of restricted aeration. Consequently, an optimum supply of water in the soil is a requisite for maximum root development.

Water is also needed in the soil to encourage fixation of nutrients. Many of the minerals in the soil are not readily available to the plant as nutrients. They must undergo chemical changes which require the presence of moisture. Thus, plants not only have great water requirements in themselves, but also require that water be present to perform specific functions in the soil.

Effects of Abnormal Amounts of Water   Water in deficiency or excess can create problems for the horticulturist. A shortage of water during the growing season may have diverse effects on various crops but the end effect is reduced growth or production. Fruits will not be as large. Edible parts of vegetables will be smaller and may develop coarse and fibrous tissues. Grass in lawns may cease growth and become dormant. Trees and shrubs may defoliate as a precaution against water loss. In general, differentiation is hastened, tissues become hardened and mature, and sugars accumulate in storage organs instead of being utilized in growth. A shortage of moisture during the dormant season may increase the amount of winter killing or winter injury. Cold drying winds will increase evaporation from dormant twigs and canes. A deficit in the soil under these conditions may lead to winter killing as a result of desiccation, or drying out.

An excess of water throughout the growing season exerts its principal adverse effects by restricting aeration of the roots. This results in reduced root growth. A less extensive root system means a smaller absorbing area for mineral nutrients and thus becomes a limiting factor in plant growth. Stunted, dwarfed, and unthrifty plants are the result.

Special situations for horticultural plants make their appearance with sudden or drastic changes in the water supply. If a period of excess precipitation and saturated soil is followed by a period deficient in rainfall, serious consequences may result especially in mid-summer. Root systems of annual plants have become dwarfed during the period of excess moisture. Suddenly the root system is inadequate to absorb moisture. The result, wilting, and under severe conditions, death of the plant due to "blasting" or drying up of the leaf and stem tissues. If a period deficient in adequate precipitation is followed by a period of excess moisture, difficulties may also arise. The sudden intake of water may result in cracked fruits in apples, peaches, cherries, etc., and in such vegetables as tomatoes. The dry period has caused the cells of these fruits to become somewhat mature and they lose much of their ability to expand. Consequently, the increase in the water supply bursts the cells. This can also cause cracks in potato tubers or result in secondary growths or "knobby" tubers. Cabbage heads often burst under similar conditions. Root pruning can prevent bursting of cabbage heads.

LIGHT

The essential role of light with regard to the complex process of photosynthesis has been discussed. Light is the source of energy for plants. The response of plants to light is dependent upon its amount or *intensity*, its kind or *quality*, and its daily duration or *photoperiod*.

Intensity   Intensity or quantity of light affects the rate of photosynthesis. For a given species the amount of light a plant receives can be a limiting factor in its growth. Some plants require nearly full sunlight for maximum growth and others prefer shade or partial shade. Light intensity not only influences the amount of photosynthesis which occurs but also has an effect on leaf structure. In full light there are often two or three layers of palisade cells which contain considerable chlorophyll. Since these cells constitute the principal region for photosynthesis, the food manufacturing function can proceed at a high level. In the leaves of shade grown plants, there are fewer palisade cells and less chlorophyll. Also the leaf

FIGURE 3-5   **Geranium plant grown in full light.**

structure is more open, the spongy mesophyll has larger intercellular spaces and the veins and interveinal areas are more succulent. For this reason, some vegetables such as lettuce, celery, and other salad crops are often of higher quality when grown under conditions of cloudy weather.

Within a species the transpiration rate is lower in shade or partial shade than in full light; thus the shading of young growing plants that have been transplanted. Part of their root systems has been lost, resulting in decreased water absorption. Shading reduces transpiration and tends to equalize water loss with water intake.

Extremely high light intensity may cause sunburning of fruits and may result in fading of deep colors in flowers especially during hot weather.

Quality   Quality or kind of light refers to its color or location in the spectrum as seen in the array of color in a rainbow or through a prism. The total blend of colors in visible light is called white light. Photosynthesis proceeds through all the components of white light from violet through blue, green, yellow, orange, and red. Artificial light can supply energy for photosynthesis if the intensity is sufficiently great. Better growth is obtained with white light than if certain portions are screened out. Ultraviolet, which is also known as black light, can be harmful to plants when present in excess. It does play an important role in color development in some fruits and in autumn coloration of leaves. The pigment, anthocyanin, develops in greater abundance in fruits growing in a clear, dry atmosphere where the ultraviolet rays are not screened out. Ordinary greenhouse glass and water vapor and dust particles of the

FIGURE 3-6   **Geranium plant grown under low light intensity in a poorly lighted basement window.**

atmosphere screen out much of the ultraviolet coming from the sun. The autumn color of leaves is more intense when there are frequent sunny days and clear skies.

Photoperiod

Photoperiod is the term given the daily duration of light. Some plants are very specific in their response to the number of hours of light during a 24-hour period. The response of plants to photoperiod is generally noted in regard to vegetativeness or flowering and fruiting. A *short-day* plant is one which will form flower buds and bloom as a result of exposure to relatively short days, usually 12 hours or less. Under long days these plants are vegetative. *Long-day* plants, conversely, are stimulated to flower on exposure to relatively long days, usually 14 hours or more. Under short days these plants are vegetative. Not all plants respond to short or long days in this manner and are called *day-neutral* or indeterminate day length plants.

Numerous investigations have been conducted in determining the photoperiods of various species. A short-day plant such as the cosmos can be covered with black cloth for several hours daily during long summer days and will soon form flower buds and begin to flower. Plants not so treated will remain vegetative. The upper part of a cosmos can be subjected to a 10-hour day and the lower part to the full light of early summer. The upper portion will bloom and the basal portion will remain vegetative. If the treatment is reversed on another plant, the lower part

FIGURE 3-7   **The poinsettia is a short-day plant, thus its wide use as a Christmas flower in the northern hemisphere.**

FIGURE 3-8   **Chrysanthemums are forced to bloom during long days by covering daily to shorten the photoperiod. (Photograph by K. A. Denisen.)**

will bloom and the top portion will remain vegetative. Many chrysanthemums grow vegetatively during the long days of summer and bloom during the short days of autumn. Florists are successful in forcing bloom out-of-season by altering the day length.

Long-day plants, such as sweet corn and peas, will flower during the relatively long days of summer. If grown in the greenhouse during the short days of winter, they need supplemental light during morning or evening in order to bloom. The tomato is a day-neutral plant, for it will flower and fruit under either long or short days. Its production will be greatest during long days, however, since there are more hours of light for carrying on photosynthesis.

NUTRIENTS

The elemental nutrients form an important "link" in the "chain" of optimum environmental conditions. Numerous elements have been shown to be essential for plant growth. Some of these are needed in greater abundance than others, but this does not mean that some are more important than others. All of the essential elements must be available to the plant or deficiency disorders of various types will result.

Non-mineral Nutrients

Non-mineral nutrients are carbon, hydrogen, and oxygen. They are the ingredients of carbohydrates and as such must be present for both photosynthesis and respiration to occur. The *carbon* utilized by plants is obtained from the carbon dioxide of the atmosphere, one of the end products of respiration and oxidation of organic materials. The atmosphere consists of about .03 percent carbon dioxide. Decaying vegetative matter supplies considerable carbon dioxide to the atmosphere and growing plants.

Hydrogen is obtained from water. Thus, we see water not only as a source of moisture for its physical properties but also are aware of its actual chemical participation as a raw material of photosynthesis. Water is the end product of combustion of carbohydrates. Thus both carbon and hydrogen are part of the end products of respiration and combustion, and conversely these same end products become raw materials in the manufacturing process of photosynthesis.

Oxygen utilized by plants in photosynthesis is derived from carbon dioxide (CO_2). Water (H_2O) also carries oxygen; it is the source of free oxygen liberated during photosynthesis. Oxygen for respiration in plants as for animals is obtained from the atmosphere. The oxygen content of the air we breathe is about 20 percent and is maintained principally by plants through photosynthesis.

Mineral Nutrients

Mineral nutrients are absorbed by plants principally through the root hairs. The essential mineral elements occur naturally in most soils. However, the mineral nutrient content of many soils has become greatly depleted due to such factors as continued cropping, leaching, and erosion. Some nutrients are present in abundant quantities but may be unavailable to plants. In such cases additional quantities in available form are sometimes added, or the unavailable form is so treated that it will release nutrients to the soil solution. This is often accomplished by adjusting the soil reaction or pH. The following group of minerals are considered essential to normal plant growth and development. The first three listed (nitrogen, phosphorus, and potassium) are needed in greatest quantities and are often called the macro or primary mineral nutrients.

Nitrogen is used in large quantities by plants. It is responsible for the deep green color of healthy leaves. It is present in the chlorophyll molecule and is a component of all proteins. When abundant in the soil, plants respond with heavy vegetative growth. Nitrogen is present in most commercial fertilizers and is found in animal manures and plant remains. Deficiency of nitrogen is manifested by yellow or yellowish-green leaves and a general dwarfing of all plant parts.

Phosphorus is important for early maturity and for seed and fruit production. It functions as part of the enzyme system having a vital role in the synthesis of other foods from carbohydrates. Phosphorus is a constituent of nuclear proteins. Deficiency symptoms include purpling of the leaf veins, stunted growth, shrunken seeds, and late maturity.

Potassium is important to plant growth. However, its specific functions are not thoroughly understood. It appears to be necessary for carbohydrate manufacture, and its presence seems to make plants more resistant to disease. Deficiency symptoms include abnormal leaf color, weak stems, and underdeveloped roots.

The second group of mineral nutrients has been given various terms such as secondary, trace, minor, and micro nutrients. They are needed in smaller quantities than nitrogen, phosphorus, and potassium. However, calcium, sulfur, and magnesium are used in greater than trace quantities. *Calcium* is an important constituent of the cell wall. *Sulfur* is present in many fats and some proteins. *Magnesium* is in the center of the chlorophyll molecule and a deficiency results in chlorosis, a lack of sufficient chlorophyll in the leaves.

Other essential nutrients include manganese, iron, copper, zinc, boron, and molybdenum. These are needed in very small quantities and are appropriately called trace or micro elements. Excess amounts of some trace elements such as copper and zinc may be toxic to plants. Where deficient, the micro elements may be applied to the soil or foliage.

INTERRELATIONSHIPS

Of the environmental factors essential to plant growth and production, each has its own individual role. Each can be considered a limiting factor to plant growth. Unless these factors are present in optimum quantity or quality, plants will not show maximum growth response. In addition to the vital role which each plays, the essential conditions or factors of the environment are interrelated. It is often a combination of factors that exerts a particular effect.

Photoperiod influences the time of flowering of long-day and short-day plants. Yet temperature often interacts with photoperiod and modifies the result. Lettuce will form a seedstalk during the long days of summer but the appearance of the flower may be delayed or even prevented if the temperature is cool.

FIGURE 3-9   Exacting environmental requirements are especially evident in the African violet. It likes cool temperatures, partial shade and high organic matter content in the soil. It is quite subject to injury from over-watering.

As light intensity increases, the rate of photosynthesis increases, and nutrients and moisture are needed in greater quantities. As a result, high light intensity may indirectly cause nutrients or water to be limiting factors to maximum growth. Conversely, before light intensity was at a high level, it could have been limiting. Even within the light factor, photoperiod, intensity, and quality have interacting influences.

During spring weather of below normal temperatures, growth is usually retarded. In this case, nitrogen may be a limiting factor, since warm temperatures favor bacterial activity which is instrumental to nitrogen availability. Here we see the interrelationship of temperature and nutrients.

STUDY QUESTIONS

1. What are the limiting factors to growth and well-being of horticultural plants?
2. Can just one of the limiting factors have a "veto" power over growth of the plant? Give an example.
3. We hear the term, "Air-conditioned greenhouses." Do we actually mean air-conditioning? Is it economically practical or feasible for plants?

Plants and Their Environment   **69**

4. Give examples of cool-season and warm-season types of fruits, vegetables, and ornamentals.
5. Distinguish between *rest* period and *dormant* period for biennial and perennial plants. Set up an outline with "Rest" period on one side and "Dormant" period on the other, and list those characteristics which fit under each category or how they respond to the environment, both internal and external.
6. Why is water needed in such great quantities by growing plants? What happens during drouth?
7. What adaptations to shortages of water are found in cacti?
8. How do plants respond to light intensity, light quality, and photoperiod?
9. How do they respond to different wavelengths of light?
10. List the essential non-mineral elements for growth. Where do they come from?
11. Which are the essential mineral elements for plant growth? What is their source?

SELECTED REFERENCES

1. Borthwick, H. A., S. B. Hendricks, and M. W. Parker. 1952. The reaction controlling floral initiation. *Proc. Nat. Acad. Sci. 38*:929–934.
2. Curtiss, Helena. 1972. *Invitation to Biology*, Chapter 5–4: Biological clocks: Circadian rhythms and photoperiodism. Worth Publishers, New York.
3. Stulte, C. A. (editor). 1977. *Plant Growth Regulators: Chemical Activity, Plant Responses, and Economic Potential. Adv. Chem. Series No. 159.* American Chemical Society, Washington, D.C. and American Society for Horticultural Science, Mount Vernon, Va.
4. U.S. Department of Agriculture Yearbook, Washington, D.C.:
   1957. *Soils.*
   1955. *Water.*
   1941. *Climate and Man.*

Skills and Practices in Horticulture

4 Soil and Water Management

The two environmental factors, soil and water, are closely allied in nature, each very dependent on the other in furnishing favorable conditions for plant growth and production. Our soil is the product of glaciers, winds, rivers, oceans, plants, rocks, and centuries of exposure to weathering. Both flood and drouth have left their impact on the land. As essential environmental factors both soil and water must be managed wisely to fulfill the continuing needs of people and plants.

SOIL, THE STOREHOUSE OF PLANT NUTRIENTS

Kinds of Soils　Soils are classified on the basis of particle size, relative amounts of the various sizes, and the content of organic and inorganic materials.

Inorganic soils (also called mineral soils) generally have less than 10 percent of organic matter in the surface soil. The inorganic portion consists of varying amounts of sand, silt and clay.

Sandy soils are coarse textured and have large pore spaces. They have poor water holding capacity and low nutrient retaining ability. Thus they leach and dry out readily. They are early soils and warm soils due to the large pore spaces and good aeration which favor evaporation and heat convection. Therefore they are especially well adapted for certain vegetables in which earliness and rapid root growth are important factors. It is the most common medium for propagating cuttings. Irrigation is usually needed on sandy soils. Horticultural crops on sand need to be supplied with considerable quantities of nutrients for best production.

Loam soils form the largest soil group in agricultural use. They include a wide range of textures in variable amounts. All loams have significant amounts of sand, silt, and clay. When one texture predominates or is present in sufficient quantity to markedly alter physical properties of the soil, a further descriptive term is given the loam soil. Thus a sandy loam contains considerable sand, whereas a clay loam has a high content of clay. Characteristics of loam soils vary with their relative amounts of each texture group. The sandy loams are earlier soils than silt or clay loams. Their water and nutrient holding capacities are greater than with sandy soils but less than that of the finer textured classes. Loams of all types are used considerably in horticultural production.

Clay soils are fine textured and have very small pore spaces. They are not well adapted to production of most horticultural crops unless they

FIGURE 4-1 This is evidence of mismanagement of Nature's heritage, the soil. (Courtesy Iowa State University, Ames, Iowa.)

are "opened up" with large quantities of organic matter. Clay is slow to dry out in the spring because of its poor aeration and high water holding affinity. It is a late soil and a cold soil. Because of its tiny particle size, it puddles when wet and bakes to a hard crust as it dries. It becomes seriously compacted if worked when wet. In spite of these many undesirable attributes of clay soils, the component clay performs extremely important functions in most soils. Its adhesive nature is helpful in forming soil aggregates which contribute to good soil structure. As a colloidal material it attracts, holds, and releases nutrients for plant growth.

Organic soils are high in organic matter, usually 20 percent or more. They are found in swamps, bogs, shallow lake bottoms, and river beds. Drainage is usually essential to their development for crop use. The organic soils are classed as *peats* and *mucks*. Although the state of decomposition of organic matter is sometimes used to distinguish between them, the principal means of distinction is by percent of organic matter.

Peat is upwards of 50 percent organic matter and is sometimes as high as 95 percent. It has a reddish brown color as "raw" peat and a dark brown to black color in its more advanced stages of decomposition. It has an extremely high water folding capacity because of its organic matter content, which in effect acts like a sponge. Peat is also well aerated. It is a soil rich in nitrogen and well adapted to the production of vegetables which efficiently utilize large quantities of nitrogen. Onions, potatoes, carrots, cabbage, and lettuce are some of the principal peat-land crops. Blueberries and cranberries are the principal fruits grown in peat bogs. Peat is widely used in greenhouse operations.

Muck ranges from 20 to 50 percent organic matter. As peat "wears out" its organic matter becomes depleted, more mineral soil becomes incorporated, and it becomes muck. This is not the only origin of muck. A shallow layer of organic matter in a drained swamp may be incorporated into the upper mineral soil to produce an organic soil of less than 50 percent organic matter. Its characteristics are similar to those of peat, but its period of productivity is usually shorter, its water holding capacity not as high, and its aeration more limited. It does combine advantages of both organic and inorganic soils.

Soil Fertility

The inherent or native fertility of soils is the basis for high production and large returns in the great agricultural lands of the world. But the inherent fertility must be bolstered by a sound fertility program if productivity is to be maintained. Ability of a soil to support plant life and produce abundant harvests is dependent on available nutrients in the soil and the rate of release of additional nutrients that are present but not available to plants. The rate of release of "tied-up" or combined nutrients is dependent on such factors as bacterial action, soil temperature, soil moisture, and aeration. The inherent fertility nutrients are thus being released gradually from the vast storehouse of the soil by processes of Nature. Good soil management aids these natural processes and promotes conservation practices in addition to supplying extra materials for the current crop.

Depletion of soil fertility occurs as a result of four specific factors. These are: (1) crop removal, (2) erosion, (3) leaching, and (4) volatilization. When crops are removed from the land, not only the foods manufactured from carbon dioxide and water (carbohydrates) are removed, but mineral nutrients as well. Thus continuous cropping for generations has a very detrimental effect on availability of essential nutrients. Unfortunately the losses of nutrients are frequently greater from mismanagement and negligence than by crop removal. Erosion, the washing or blowing away of soil, results in losses of tremendous quantities of nutrients as well as good topsoil which has taken centuries to build. Leaching of nutrients down through the soil profile, by water on its way to the water table, is an important loss of fertility. Rendering nutrients available, or supplying them, in greater quantity than can be utilized by plants invites loss by leaching resulting from heavy rainfall. Volatilization of soil nutrients is most apt to occur with nitrogen in the ammonia form. Lack of incorporation of organic matter into the soil

FIGURE 4-2   Erosion has an important role in soil formation. *Left*, tree roots have been exposed by washing water. *Right*, rocks are a primary source of soil particles. (Photographs by K. A. Denisen.)

prior to decomposition may result in release of valuable nitrogen to the atmosphere.

Replacement of soil fertility is a joint endeavor of Nature and the good soil manager. Upon it depends the maintenance of soil productivity and increasing the supplies of food for the world. The storehouse is replenished by the addition of chemical commercial fertilizers and plant and animal wastes and remains. It is further maintained by soil building and soil conserving crops which take available nutrients from the soil and hold them in organic form until released gradually by decomposition. The action of bacteria and other soil microorganisms is instrumental in rendering both organic and inorganic materials available to plants. One family of plants, the legume or bean family, which includes peas, beans, alfalfa, and the clovers, is capable of adding nitrogen to the soil from the atmosphere. This is accomplished only when specific bacteria are present on the roots of the legumes and the two live in symbiotic relationship, i.e., reciprocal cooperation, with each other. Another group of microorganisms, the micorrhiza, have similar capabilities with other families of plants, especially the trees.

FIGURE 4-3   **Soil erosion, by wind or water, results in loss of good topsoil and tremendous quantities of nutrients. (Courtesy Iowa State University, Ames, Iowa.)**

Good conservation practices, such as plowing under trash or debris instead of burning it, help in returning many nutrients to the soil. Vines, stalks, leaves, roots, and other plant organs are often left as discards of production. These can often be returned to the soil. This practice reduces the drain of nutrients, for the harvested portion of the crop contains the only nutrients not returned to the land.

Organic Matter

The organic fraction of a soil is that which previously had life. It consists of dead and decaying vegetation, animal wastes and remains, and dead and decaying soil microorganisms. As organic matter, it is a reservoir of nutrients not yet released but potentially capable of supplying plant food gradually as it decomposes. It is this gradual release of nutrients which makes organic matter especially valuable. Available nutrients supplied as chemical fertilizers in excess of plant needs can be readily lost by leaching; whereas, the nutrients in organic matter are held against leaching.

Decomposition of organic matter is dependent on the presence and activity of soil microorganisms, principally bacteria. Proper conditions of moisture, temperature, soil pH, and oxygen supply are essential for normal decomposition to proceed. As organic matter breaks down and loses its identity as plant or animal remains, it is called *humus*. Humus gives the dark color and mellow texture to inorganic soils in which it is present. As humus continues in the decomposition process nutrients are released to the soil solution where they are available for absorption into plant roots. Nitrogen is the soil nutrient provided in greatest abundance by decomposing humus. Considerable carbon dioxide is also released both to the atmosphere and to the soil in the oxidation of humus. This product forms carbonic acid and carbonates in the soil which may easily leach. However, most of the carbon dioxide produced by oxidation of organic matter escapes to the atmosphere where it is available to the plant for photosynthesis. The nitrogen fixation process is accomplished by soil bacteria which break down the proteins of organic matter to ammonia, to nitrites, and finally to nitrates. In nitrate form the nitrogen is highly soluble and readily available to the plant. It can also be leached readily from the soil. Since the decomposition of organic matter is gradual, large quantities are not made available at one time and nitrogen is conserved. This is one of the principal advantages of high humus content in a soil.

Improved physical condition of the soil is a principal role of organic matter in soil improvement. Light soils are improved by increasing their water holding capacity. Pore spaces are filled with decaying organic matter which acts as a sponge. Nutrient holding capacity is increased in these open, coarse-textured, and light soils. The humus adsorbs nutrients and holds them against leaching. In heavy soils, organic matter can play an equally important role. In the clays, silts, and their loams, water holding capacity need not be increased as with sand. It is often imperative that they be "opened up" or better aerated. Organic matter can do this, again by its sponge-like characteristics. It occupies space which otherwise would contain fine soil particles. Because of its organic structure it provides drainage and reduces the puddling effect of compaction.

Composting is a method utilized to produce a highly organic soil mixture for specific uses. Compost is produced by alternating layers of

FIGURE 4-4   **Sweet clover being plowed under as a green manure crop to increase organic matter content of the soil. (Courtesy Iowa State University, Ames, Iowa.)**

manure, leaves, plant refuse, or other organic materials with layers of soil. The layers are kept moist to favor decomposition. Either pits or frames are used for composting. After several months to a year in the compost pile, most organic matter has lost its identity due to decomposition. The humus this produces is thoroughly mixed with the soil layers to complete the composting process. It is used as a source of nutrients and an aid to soil structure when used as top dressing or worked into the soil. Compost finds great and varied uses in greenhouses, gardens, lawns, and homes.

In spite of the many virtues and important role of organic matter in horticultural soils, it is not the sole means of maintaining fertility and soil structure. The supply of organic matter that can be returned to the land is not adequate to meet the nutrient needs of all crops. Other sources of nutrients are needed. Careful tillage to avoid compaction and to reduce cultivation to its essential needs aid in the maintenance of good soil structure. Organic matter is extremely helpful with its cementing action to form soil aggregates. But this improved structure can be rapidly destroyed by cultivating when wet, cultivating when not needed, and compacting with extremely heavy equipment.

Soil Reaction   The acid, alkaline, or neutral reaction of a soil is expressed on the pH scale. A pH of 7 is neutral (as for distilled water). Points below 7 are acid and those above 7 are alkaline. As the pH of soils are closer to the neutral point, the soils are less acid or less alkaline. As the pH is further from the neutral point, the soils are more acid or more alkaline. The pH of a soil has an important effect on plants growing in that soil. Some plants grow best on an acid soil, others are favored by alkaline conditions, and many prefer the near neutral range. Just as there is an optimum range of other environmental conditions, there is also an optimum pH range for plant growth. This range may vary from one crop to another and is one of the bases for determining crop adaptability to an area. Most horticultural crops are adapted to a pH range (Fig. 4.5) just below neutral, that is, of a slightly acidic nature.

The pH of soils can be adjusted. It is one of the techniques the horticulturist uses to adapt his soil to the specific needs of a crop or a group of crops. Soils can be made more acid (or less alkaline) by applying acid producing chemicals such as aluminum sulfate, sulfur, and gypsum (calcium sulfate). Soils can be made more alkaline (or less acid) by applying hydrated lime or limestone. In areas of heavy rainfall leaching will remove many soluble carbonates and the result is a more acid soil. Often this land is underlaid with limestone which has been deposited through many years of leaching.

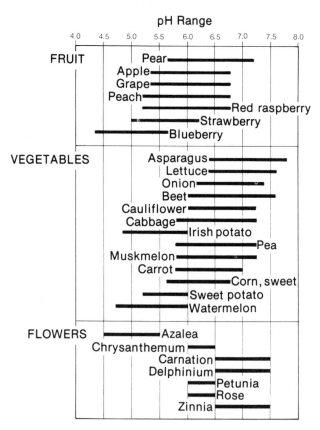

FIGURE 4-5   Optimum pH range for several fruits, vegetables, and ornamentals.

SOIL MANAGEMENT PRACTICES

Fertilizing   The obvious and almost immediate results frequently obtained by fertilizing soils has directed much attention to this practice. Tillers of the soil have long known the beneficial effects of animal manures on the productivity of most mineral soils. More recently gardeners and other agriculturists have accepted and adopted the use of chemical fertilizers.

Commercial fertilizers are now the principal source of supplemental soil nutrients for home gardeners, vegetable farmers, fruit growers, and nurserymen. Because of the great outlet for their products, the miners, manufacturers, formulators, and distributors of chemical fertilizers have developed a tremendous industry. This has contributed to the relatively low cost of supplying nutrients to the land. Most horticulturists now obtain commercial fertilizers at a lower cost per unit weight of nutrients than for the equivalent in animal manures.

The *analyses* and *ratios* of commercial fertilizers are expressed as N-P-K (nitrogen, phosphorus, potassium) always in that order. Analysis refers to the percent of available nutrients. A 6–10–4 fertilizer is 6 percent available nitrogen, 10 percent available phosphorus (actually P_2O_5) and 4 percent potassium (actually K_2O). When these figures are reduced to their lowest common denominator, the resulting figures are the ratio of the fertilizer. The 6–10–4 analysis has a ratio of 3–5–2. A 5–10–5 fertilizer has the same ratio as a 10–20–10, that is, 1–2–1. The ratio is of special importance in calculating the respective amounts of two different analyses of fertilizer to apply for obtaining the desired rate of nutrients per unit of area. High analysis fertilizers are those whose total percent of available nutrients is 30 or above. A 12–24–12 fertilizer is high analysis. A 6–12–6 is low analysis. The trend is toward greater use of the high analysis type. The cost per nutrient is usually less because of reduced handling and shipping costs. However, greater accuracy is needed in application.

The *components* of commercial fertilizers may be inorganic or organic. The inorganic forms are generally more immediately available to the plant. However, since they are so readily available, they are also subject to loss by leaching. Consequently organic forms are coming into greater prominence especially as controlled release fertilizers. The organic form undergoes changes in the soil which renders it available to plant roots. These changes or reactions are gradual so the nutrients become available at a rate approaching the current needs of growing plants.

Application of commercial fertilizers varies by method, rate, time, and placement. The methods are principally by broadcast and working-in prior to planting, broadcast surface application without working-in, row application at planting time, and row application during growth which is also called *side-dressing*. Fertilizer is sometimes drilled-in prior to planting instead of being broadcast and worked-in. The type of crop usually determines the method of application. With an established perennial crop such as an orchard, vineyard, berry patch, or lawn, surface applications without working-in are common. With vegetables and annual flowers, broadcast applications worked-in and a later side-dressing are standard procedures.

Rates of application vary with inherent fertility of the soil and needs of the crop. Soil tests by a soil testing laboratory are usually a reliable guide for determining nutrient needs. Quick soil tests have been devised. Tissue tests are being employed to an increasing degree for certain perennial crops such as apples. Analysis of nutrients in leaves or twigs is associated with appearance, vigor, and production to determine plant food needs.

Time of application is determined by type of crop, time of planting, land preparation, and influence of fertilizer on the productive and vegetative phases of growth. Most annual and biennial crops are fertilized prior to seeding or transplanting. They may also be side-dressed during the growing season. Fertilizers are frequently applied during land preparation, at which time they can be readily worked into the soil. In the orchard and vineyard, fertilizers are generally applied in the spring so the spring rains will help move the soluble nutrients into the root zone for early availability. Most fruit trees differentiate their fruit buds for the following year during late spring or early summer. At that time most shoot growth has occurred and the potential photosynthetic activity of the plant has been determined. Abundant nitrogen for the bearing tree will contribute to large leaf area, high rate of photosynthesis, balanced carbohydrate utilization for shoot growth, and carbohydrate accumulation for fruit development.

Fertilizer placement has received considerable attention in some crops. Band and hill placement locates the fertilizer closer to the plants. Since plant roots grow into the surrounding area, this close placement may be only a temporary advantage. In general, broadcast applications at planting time give as good results as band or hill placement. Side dressing in a trench beside the row during the growing season is generally more quickly available than a broadcast application at that time.

Carry-over of fertilizer nutrients from one season to the next is sometimes the goal of a horticulturist. On peat and muck soils, the principal fertilizer nutrients are phosphorus and potassium. They do not leach as readily as the nitrates and will usually provide continued supply the next season if applied in excess of current season needs. In mineral soils, phosphorus does not move readily. If applied in large amounts and worked in prior to establishing a perennial enterprise, the carry-over can last for several years unless the soil reaction is quite acid. An acid soil tends to "tie-up" phosphates.

Special techniques of fertilizing employed by the horticulturist include starter solutions, foliar feeding, and liquid and liquefied gas applications. A *starter solution* is merely a fertilizer in solution. About 1 cup of a readily soluble fertilizer in 5 gallons of water, or 4 pounds in 50 gallons, makes a good starter solution for transplanting vegetables, strawberries, and herbaceous ornamentals. The solution is applied following transplanting to give the plants water and quickly available nutrients for a good start. It is also done on a commercial and mechanized scale. Many transplanting machines have special tanks and equipment for applying starter solutions. *Foliar feeding* is fertilizing through the leaves. The fertilizers applied in this manner must be highly soluble and non-caustic to leaf surface. Organic forms of plant nutrients such as urea are usually better adapted to foliar feeding than inorganic materials; since, they are more readily absorbed and less likely to cause

FIGURE 4-6   **Application of sewage sludge by chisel injection.** *Bottom,* **shows close-up of chisels. (Courtesy Metropolitan Waste Control Commission of the Twin Cities area.)**

Soil and Water Management

burning. The principal argument for foliar feeding is more rapid absorption of nutrients to the needed areas, the leaves. Some of the micro elements can be supplied in "chelated" form in foliar sprays. Iron chelates applied to foliage deficient in iron are readily absorbed. In a relatively brief period the foliage develops normal color, correcting the deficiency symptoms. Since many horticultural crops are sprayed several times during the growing season for pest control, foliar fertilizer components may be added to the sprays. This provides foliar feeding without adding an extra spray to the schedule. It is important that the foliar fertilizer and the spray chemicals be compatible, i.e., do not interfere with normal activity of each other. *Liquid fertilizer* applications are sometimes used to give greater depth penetration. Also it makes fertilizer application possible through an irrigation system. Since these fertilizers are highly soluble and soon penetrate to absorbing areas of the roots, they give rapid response to the growing plants. *Liquefied* gas fertilizing refers to the application of ammonia gas to the soil. The gas is injected under pressure into the soil by long tubes terminated by blades. The soil must be in good tilth so the furrow or groove left by the blade fills in immediately to avoid escape of ammonia gas. This method of fertilizing is not generally practical for orchards and other perennial crops. The blade and injector are restricted by large roots and will prune small feeder roots in their path.

ANIMAL WASTES. Before the advent of commercial fertilizers, the principal method of supplying essential nutrients to the soil was by incorporating animal manures. These manures vary widely in nutrient content depending on the kind of animal and the kind and amount of bedding material included. The significant feature is that they are all relatively good sources of nitrogen, phosphorus, potassium, and the micro elements. Regardless of amount of bedding, they are an excellent source of organic matter. Because of their organic nature and also because of the digestive processes they have undergone, there are both immediately available and unavailable nutrients present. Response on crop plants can be both rapid and lasting.

In some parts of the world, human excreta is an important fertilizer. The principal hazard of this material is the danger of spreading communicable diseases. It is only recently that sewage commissions of some of the larger cities have processed, dried, and pulverized sewage sludge for fertilizer as a by-product of sewage disposal. Disease organisms that may have been present are killed, objectionable odors are greatly reduced, and the granular material thus produced is of use to both amateur and professional horticulturists.

Liming

The principal reason for liming soils is to raise the soil pH. An acid soil can be made less acid, neutral, or alkaline depending on the quantities of lime added. Prior to liming, one should first determine the need. Several reliable pH tests are available. Most pH field kits and some laboratory techniques utilize indicator solutions. Soils testing laboratories usually include a pH reading along with the nutrient analyses.

These readings may be made on an electric pH meter, which is a rapid and accurate method of determining pH.

The source or form of lime is usually either calcium oxide (quicklime), calcium hydroxide (hydrated lime) or calcium carbonate (limestone). Relatively large quantities of lime are needed to amend the soil pH for some crops. To raise the pH of the upper 6 in. (15 cm) of a loam soil from 5.5 to 6.5 may require about 3 tons of pulverized limestone per acre, or 2 tons of hydrated lime, or $1\frac{1}{2}$ tons of quicklime. If leaching is considerable and the soil relatively sandy, it may be necessary to repeat the liming operation every three to five years. To delay the need for liming one may use fertilizers that give an alkaline reaction such as ammonium nitrate and sodium nitrate. Ammonium sulfate supplies nitrogen but tends to produce acid conditions.

Lime is usually broadcast on the surface and worked into the soil. It is often applied in the fall to raise the pH for the following growing season.

Cultivation   There are many reasons why cultivation is used for horticultural crops. Not all reasons given for cultivation are sound soil management. A frequently heard avowed purpose is "to stir up the soil." Actually stirring the soil may do more harm than good. It tends to pulverize soil aggregates and be detrimental to good structure. It increases evaporation from the soil surface. It increases compaction below the surface soil which restricts air movement.

Sound purposes of cultivation include weed control, aeration, water penetration, and incorporation of organic matter. Even these factors can be overdone. It has been frequently demonstrated that the most important reason for cultivation is to control weeds which otherwise would compete with crops for moisture and nutrients. The use of herbicides for weed control reduces this need for cultivation. Aeration of the soil is important to good root growth. It aids soil microorganisms in rendering nutrients available. Cultivation may actually reduce aeration of the soil by increasing compaction and pulverizing soil aggregates. Air is incorporated into the soil, however, by plowing, disking, and harrowing prior to planting. These operations turn surface soil and crop residues under and leave an uneven surface and loose structure. Improved water penetration of a soil occurs with plowing, disking, and harrowing if the surface tends to be uneven, the structure loose, and soil aggregates of varying sizes. On the other hand an overly-cultivated, compacted, powdery soil is more apt to puddle when wet and actually restrict water movement and penetration. Incorporating organic matter is not only a sound reason for cultivation, but also aids both aeration and water penetration. Plowing and disking are the most effective methods of incorporating organic matter.

Depth of cultivation has a very influential effect on crop response. Most vegetable, flower, and small fruits plants are shallow-rooted and have fine feeder roots near the surface. Deep cultivation between rows does considerable root pruning by cultivator shovels or knives. Shallow cultivation can be fully as effective in weed removal and does less root pruning on the crop plants. There is also less evaporation from the surface soil, less pulverizing of the soil, and reduced power requirements with shallow cultivation. Total growth and yield can be drastically reduced by root pruning from deep cultivation.

Plowing as a tillage method deserves special consideration. Its advantages and disadvantages and the arguments *pro* and *con* are much under discussion. As an agricultural practice, it is still widely used. Properly used, it improves aeration and water penetration and incorporates organic matter. Improperly used, it can create compaction, cause puddling, and ruin soil structure.

Fall plowing is generally better than spring plowing because the furrow ridges and depressions catch and hold rain or snow. Alternate freezing and thawing tends to cause aggregation of soil particles. The soil which has been fall plowed will need less spring tillage, and the good structure gained during winter months will last longer. Spring plowing immediately causes some loss of aggregation due to pulverizing and compacting forces. Where annual plowing is practiced, it is well to vary the depth of plowing. This avoids formation of a "plow sole" which is an artificially created zone of compaction caused by the plow bottom taking the same path year after year.

Rotation

Crop rotation on our land is practiced to equalize depletion of nutrients, to avoid toxic effects on the succeeding crop, and to control horticultural pests. Since crops vary in their content of minerals, degree of succulence, and amount of dry matter produced, it follows that they should require nutrients in varying quantities and in different proportions. A crop like tomatoes is a heavy user of phosphorus. When followed by cabbage, which uses only moderate amounts, the soil has an opportunity to adjust itself. Unavailable phosphorus from the soil's inherent fertility reserve can undergo transformation to the available form during the soil's "rest" from a high phosphate user. Some crops have a toxic effect upon the same crop if followed on the same land. This may be due to substances in the crop residues or exudations of the roots during the previous season. Insect and disease pests are most serious in areas of intense production of the host crop. Many pests tend to overwinter in adjacent fence-rows or vegetation, on the refuse of the crop, or in the soil. In each instance relocating the next season's crop should help in eliminating or reducing the seriousness of the pest. Crown gall, cucumber beetle, and stalk borers are only a few examples when crop rotation aids in pest control.

Annual and biennial plants do not have or require as long a rotation period as perennials. Most vegetables grown as annual crops fit nicely into a four or five year rotation plan. For example, on either a peat or upland soil, the following is a satisfactory rotation: potatoes, cabbage, onions, legume. On upland (mineral) soil the legume is usually plowed under for green manure. On peat the legume is usually harvested. Strawberries, vegetables, and legumes can be used for a rotation series over a four to eight year period. The bramble fruits require a longer time to complete the rotation since a bramble planting in itself is usually of six to ten years duration. The fruit or nut orchard is a long term crop.

The orchard itself may last 30 to 50 years or more. If the land is level enough for cultivation, the old orchard site can be renovated by turning under green manure crops, growing fibrous rooted crops which open up the soil, and devoting the land to row crops where good weed control methods can be practiced.

Intercropping   Many perennial horticultural crops have wide spacings between rows and between plants. When these plants are small and before they fully utilize the land, other short term crops may be interplanted. Vegetables and some of the small fruits are often planted between the rows of trees in a young orchard which is on fairly level land. The principal objective is for monetary returns before the trees begin fruiting. Produce from the intercrop helps to defray some of the costs of bringing a new orchard into bearing. Fuller utilization of the land and available nutrients are other reasons for intercropping. It tends to avoid loss of nutrients by leaching in soil areas not yet reached by the young tree roots.

Several precautions should be observed when intercropping. (1) The main consideration should be for the primary or basic crop. Monetary returns at the expense of the long term enterprise is poor economy. (2) Avoid competition between the primary crop and the intercrop. The primary crop should be allowed to grow and develop as though the intercrop was not present. (3) Leave adequate space for cultural operations. Trailing vegetables such as melons and pumpkins make it difficult for spray machinery to move between tree rows without injury to the intercrop. (4) Use short term crops. It is then easy to quit the intercrop when the time comes and concentrate fully on the primary crop.

Mulching   The purposes of mulches are many and varied. They are regularly utilized to conserve moisture, regulate temperatures, control weeds, and improve the quality of produce. These uses will be discussed in greater detail in succeeding chapters. In soil management relationships, mulching influences organic matter content, activity of microorganisms, availability of soil nutrients, erosion and soil compaction.

Most mulches in common usage are some type of organic material. They contribute to the humus content of the soil while they undergo decomposition. Leaves, straw, and hay decompose readily. Large quantities are used on the surface of the soil to prolong the effectiveness of these rapidly decomposing materials. Pulverized corn cobs, peanut hulls, and cottonseed hulls are more resistant to decomposition and last longer as mulches. Wood sawdust and wood shavings last even longer, sometimes remaining effective as long as two seasons. When incorporated into the soil, their decomposition is hastened and occurs at a rate dependent on the source of material. Plastics and aluminum foil used as mulches do not contribute to the organic content of the soil. Soil microorganisms are active in the plant residue types of mulches especially at the soil surface where both the soil and the mulch are moist. The nitrogen fixing bacteria require nitrates in their metabolism while breaking down organic matter. This leads to a "tie-up" of nitrogen and

FIGURE 4-7   **Straw mulch in the vineyard.**

temporary nitrogen starvation. Application of a nitrogen-containing fertilizer prior to mulching tends to prevent a nitrogen shortage.

After a mulch of organic materials settles and weathers it becomes more fixed in location. The fusion of soil and mulch at the soil surface serves as a deterrent to erosion. The soil holds more water due to absorptiveness of the mulch while there is less free flowing of water to start erosion. In areas of considerable foot traffic such as between rows in strawberry or raspberry plantings, compaction of soil is reduced by a mulch of plant remains. The mulch absorbs the weight and distributes it over a larger area. Earthworm activity, which aids in opening the soil, is increased under mulches.

SOIL CONSERVATION METHODS

Soil conservation implies the maintaining of soil nutrients, soil structure, soil microorganisms, soil organic matter, and the physical existence of the soil. Losses from erosion, leaching, compaction, run-off, and over-cropping are held to a minimum. Soil conservation techniques are also good soil management practices for those areas and situations where conservation programs are essential.

Grass Sod   In orchards and waterways a good turf greatly reduces soil erosion and water run-off. Thus, both the soil and moisture are conserved. The grass sod provides a firm base for machinery. In the orchard it cushions falling fruit and reduces bruising.

FIGURE 4-8   **Grass sod in the orchard.**

Because of its competition for nutrients, the sod is not usually started in a young orchard until trees are well established. As tree roots go deeper into the soil, there is less competition between sod and trees; because grass roots are mostly in the upper 6 to 12 in. (15–30 cm) of surface soil. The fertilizer program of the orchard is beneficial to both the grass and trees. The grass is usually mowed with the clippings allowed to remain on the orchard floor. Here they serve the function of a temporary mulch and a raw material for organic matter.

Contour Planting   Row crops planted and cultivated on the contours of sloping, hilly land greatly reduce the amount of soil erosion and water run-off. Most vegetables, nursery crops, and small fruits can be readily adapted to contour planting. Millions of tons of good fertile topsoil are lost annually from fields of row crops which are cultivated up and down the slope instead of along the contour. Gardeners of Europe, Asia, and Africa have long recognized the importance of safeguarding their soil against the destructive forces of erosion. Americans have only recently begun to practice effective soil conservation.

FIGURE 4-9   Alternate contoured strips of row crops and grass with grassed waterways help prevent erosion by water. (Courtesy Iowa State University, Ames, Iowa.)

FIGURE 4-10   Newly constructed earth terraces on the contour prior to planting grapevines on the terrace ridges. (Courtesy H. L. Lantz, Iowa State University, Ames, Iowa.)

Strip Cropping   Alternate strips of row crops and soil retaining crops on the contour are effective for holding soil on hilly or rolling land. The retaining crop usually is a non-cultivated biennial or perennial legume or grass. These strips are also a source of mulching material and organic matter for other soils and crops.

Terracing   One of the most effective methods of retaining soil and moisture on steep and rolling land is with earth terraces constructed on the contour. Water is held by the terraces until it penetrates the soil. It thus not only prevents loss of good topsoil but builds up a reservoir of soil water as well. Excess water is channeled to grassed waterways which allow the water to flow away without appreciable soil loss. Orchards and vineyards are both easily adapted to terraces. Trees may be planted on the parapet of the terrace or between terraces. Vines are usually planted on the terrace ridge. Grass sod, rye, and small seeded biennial legumes are frequently used for ground cover and organic matter between rows of fruit trees and grapes.

Cover Crops   The small grains, small seeded legumes, and annual grasses make very desirable cover crops in small fruit plantings, vineyards, and nurseries. In the dormant season they are effective in reducing both wind and water erosion. In cold climates they hold snow which adds moisture. Cover crops open up the soil with their root growth, add organic matter, and, if legumes, build up the nitrogen content of the soil. They are also used in hardening woody plants.

WATER, TOO MUCH

Flooding   Prolonged rains, cloudbursts, and spring thaws may cause flooding in local areas. Rivers and streams serving as outlets for several local areas may overflow their banks and cause floods far from the sources of excess water. Both types represent ravaging effects of excess water. Flooding is destructive to most plants. It may completely kill all vegetation in an area, depending on how long the plants are completely submerged or how long the soil has been completely saturated. Flooding kills by lack of aeration. Reduced photosynthesis, restricted root growth, general stunting, and reduced yields are the effects of prolonged submersion due to flooding. If the outlets are adequate for rapid water removal, there may be little damage from flooding. Many drainage outlets are adequate for most situations but unfortunately fall short of the needed capacity in unusual cases. This is sometimes found in areas reclaimed by drainage, e.g., peat beds.

Rivers or streams which overflow their banks at times can be kept under control by levees. These are constructed by mounding soil along the banks or a short distance inland from the stream. Much crop land with many acres of growing plants can escape the devastation of floods by such precautions.

FIGURE 4-11   *Upper*, erosion can be serious in a young cultivated orchard. *Lower*, a cover crop of oats holds the soil in place and aids in hardening the trees for winter. (Courtesy Iowa State University, Ames, Iowa.)

Skills and Practices in Horticulture   **92**

Water-logging   Soil underlaid with an impervious layer of clay or rock has poor subsoil drainage. When water collects above this impervious layer, or hardpan, the soil becomes saturated with water. This condition is known as "water-logging." Plant roots unable to get sufficient oxygen become stunted, the plant becomes stunted, and yield is markedly reduced. A plow sole in clay soils is also conducive to water-logging; since the plow sole is actually a man-made hardpan. Deep tillage with a sub-soiler or lister can break up a plow sole. A natural hardpan is usually too deep to break by tillage operations so the water-logged area must be drained either by vertical tile outlets into the soil sub-strata or horizontally to lakes or streams.

Sometimes the water table is so high that the soil is permanently water-logged or swampy. It may be best to grow crops adapted to aquatic or semi-aquatic conditions on land of this type.

WATER REMOVAL AND STORAGE

Natural Removal of Water   Most drainage is accomplished by Nature without the aid or ingenuity of man. Rivers and streams remove tremendous quantities of excess water in a gradual and orderly fashion. The force of gravity is responsible for movement of water from a watershed to the streams and rivers. It also pulls considerable water downward through the soil. This water and the run-off from the watershed are *gravitational* water. The water

FIGURE 4-12   **Flooding due to upstream rains and overflowing river banks caused severe damage to this potato field.**

DRY SOIL   SATURATED SOIL   DRAINED SOIL

FIGURE 4-13   **How water is held in the soil. A dry soil has only a thin film of water around soil particles which is unavailable to plants. A saturated soil contains gravitational water and has little or no aeration, typical of water-logging. A drained soil contains capillary water up to field capacity, the soil's capacity to supply water to growing plants.**

left in the soil, after gravitational forces have acted, is *capillary* water. It is held physically by capillary forces and is available to plants. A soil which is full of capillary water and has lost all gravitational water is at its *field capacity*. This is its capacity to supply water to growing plants. When plants have removed all the moisture they can from the soil resulting in permanent wilting of the plants, the percent of water left is called the *wilting percentage* of that soil.

Plant life is thus another means of water removal by Nature. The plant processes of absorption, translocation, and transpiration remove tremendous quantities of water from the soil while utilizing it for a solvent, a carrier, and a raw material.

Storage for Future Use   Dams and reservoirs are commonly used in flood control. Since they also play an important role in many irrigation projects, they combine usefulness both for controlling and utilizing water. These projects may be on the small scale of a single home or farm enterprise or of national scope, including parts of several states. Community cooperative effort is frequently the initial nucleus for developing the water potential of an area. Government aid, on both a national and state level, and privately owned public utility organizations have served the public interest in many areas. They have reduced the danger of floods, harnessed water for power, made it available for plant growth, and stored the excess for future use. The healthy economy in many areas where water is at a premium is due to sound water management.

In some areas both excess rainfall and drouth conditions occur during the same season. *Ponds* play an important role in conserving water during these periods of heavy rainfall. Then, as a moisture deficit occurs, the stored water is available for irrigation. The pond should be so located as to receive the run-off from a large area. It is important to have an impervious layer of earth on the bottom. This, in effect, is like a hardpan which prevents water loss from the pond by seepage. The impervious layer can be constructed by compacting several inches of clay with heavy equipment. Excavation to an impervious layer is sometimes feasible. It is also important to make mounded sides of the pond impervious to seepage.

The *soil* plays an important role in water storage. Good soil management also implies good water management. Terraces to prevent erosion also serve as traps for water and hold it in place so more can penetrate the soil. Organic matter with its great absorptive capacity contributes greatly to water conservation. *Weed control* is important to water conservation. However, care must be exercised or large quantities of water can be lost from a soil if cultivation is deep or carried to excess. The use of shallow cultivation or herbicides where feasible reduces water loss from weed transpiration and excessive surface evaporation.

Legal Aspects
of Drainage

Regulations, ordinances, laws, and codes have been imposed or enacted to protect property rights and individual rights. Many legal problems arising over drainage can be avoided by good human relations prior to installation. Consultations with neighbors concerning ditches, feeder ditches, tile lines, feeder lines, streams, outlets, and even probable joint endeavours will usually improve community feeling on a drainage problem. A knowledge of local and state regulations is vital to a drainage program. Often it is advisable to get legal advice for working out the intricacies of lawful drainage procedures.

NOT ENOUGH WATER

Arid and
Semi-arid
Regions

Many areas do not receive adequate moisture during the course of a year to grow most horticultural plants. These areas must make special provisions to have satisfactory home gardens or commercial production. Although dry land farming is employed in the growing of small grains and other farm crops, it is not generally adaptable for most horticultural crops. Irrigation is thus a necessity for the success of most home and commercial horticultural enterprises. The prerequisite for irrigation is an abundant supply of water. Some arid areas are not likely horticultural potentials for this reason. Yet some other arid locations have been successfully developed. The presence of underground water in abundant quantities or more frequently the flow of large rivers and streams has led to the reclaiming of sub-marginal land for fruit and vegetable production. Because of the high cost of getting water to the land and of land preparation, horticultural crops are especially well adapted to irrigated areas. These crops provide a high cash return per acre. This hastens the amortization of the irrigation development. Horticultural crops are characteristically crops of intensive culture. Irrigation, with its high cost of installation and operation, places the land in the category of intensive agriculture.

Areas Subject
to Drouth
Conditions

Most non-arid regions of the temperate zone are subject to occasional periods of drouth. These areas vary in the frequency, duration, and intensity of drouth. In most cases, damage due to drouth is reflected in low yields, poor quality, or stunted growth rather than complete crop failure. Drouth may occur during a critical period of the season and drastically affect horticultural plants. Fruit, tuber, and root enlargement requires considerable water for expanding cells; they would suffer drastically from a water deficit during this period. Adequate shoot growth for sufficient leaf area is essential to normal production and efficiency of fruit and ornamental plants. Drouth during spring and early summer results in reduced shoot growth and reduced area for photosynthesis. Reduced photosynthesis during any part of the growing season means decreased carbohydrate production and lower yields for annual, biennial, and perennial plants. There is less reserve food for perennial plants, reduced growth and production the following season, increased danger of winter injury, and reduced flower and fruit bud formation. For many of the salad vegetables, drouth results in poorer quality due to toughened tissues, bitterness, or early seedstalk development.

IRRIGATION

Supplemental irrigation is a means for counteracting the detrimental effects of drouth periods. Except under extreme drouth conditions, the quantity of water needed for irrigation is not as great as is required in the arid and semi-arid regions. However, it must be available in sufficient quantity to provide an abundance of water at each application. It is extremely important to apply supplemental moisture before drouth inflicts injury. Irrigation should be planned to avoid adverse conditions or any interruption in normal growth and development. A good philosophy of irrigation can be summed up as follows: let rainfall supplement irrigation, do not use irrigation to supplement rainfall. When irrigation is available it should be applied on a regular schedule and only omitted if precipitation is adequate. When applied as a supplement to rainfall, it often is applied after plants have already suffered from a deficit. There is also a tendency under this plan to delay irrigation in anticipation of expected rain, which rain may not come.

Water Rights   The legal aspects of irrigation are extremely important to both the commercial horticulturist and the home owner in the arid and semi-arid regions. They are becoming increasingly important in other areas as supplemental irrigation becomes a more widely adopted practice for commercial enterprises, lawns, and home gardens. As growers produce more crops and larger yields, water requirements are increased. An expanding population has increased water requirements. The enlargement of manufacturing industries increases the utilization of water. In modern living the general trend is for greatly increased water usage. These various facets of our civilization are reaching a stage of competition for water. Thus the need for legislation with an established code of water rights.

Areas highly developed for irrigation found one of the first problems of water allocation was in the individual's legal right to request or

FIGURE 4-14   Evidences of drouth conditions, cracks in soil, and curling of sweet corn leaves.

demand his quota. Consequently these regions have led in water rights legislation. Many codes on water rights include *prior usage* clauses. This establishes the right of the original developer to continue to utilize the facilities, source, and amount of water which he originally tapped or developed. Principalities which are enacting legislation on water rights for the first time are faced with considering prior usage claims. A grower who contemplates irrigation in such an area can often eliminate future problems of water supply by establishing an irrigation system and thus acquire prior usage rights.

In most irrigation areas, water rights are registered with the land and remain with the land as part of the real estate. The privilege of irrigating an orchard, a field of vegetables, or even a home lawn is granted or controlled by the water rights provisions of the deed to the property. Water rights codes not only establish the property owners right to use water, but they may specify the amount that can be supplied at one time or during the entire season. They may establish a priority of usage and a sequence of application. Also, they may decree the manner in which water is used during periods of extreme shortage.

Water rights involve not only the privilege of using water, but also imply conservation. Water is essential for all manner and forms of life. As a limiting factor to plant growth and production, water must be regulated by both voluntary action and legislative processes to insure its efficient and judicious use.

Determining the Need

Several factors influential in determining the need for irrigation are: (1) annual precipitation, (2) periods of moisture shortage, (3) types of crops and cropping, (4) expected increase in production with irrigation, and (5) balancing of costs and returns. Since an irrigation installation is generally a big investment, it justifies thorough planning.

Annual precipitation often determines the type of production. If rainfall is high, intensive production of horticultural crops is usually feasible, provided the rainfall is distributed regularly through the season. If rainfall is low, a grower may have the option of extensive production of drouth tolerant crops or, when irrigation is available, intensive production of high value crops. The relative humidity of the region and the frequency of fog and dew can markedly alter the growth of plants when moisture is at a premium. The ability of the soil to absorb winter precipitation is governed by depth of freezing, amount of organic matter present, particle size, and extent of saturation of the surface soil prior to freezing.

Periods of moisture shortage, which may be regular or intermittent, frequent or occasional, are the principal reason why supplemental irrigation is desired in areas of usually adequate total rainfall. A period of drouth results in reduced yields, poorer quality vegetables, smaller fruits, and reduced growth of all plants. Supplemental irrigation in anticipation of such critical periods is becoming more common. During most years in nearly all horticultural areas, there are occurrences of moisture shortage. They may not last for long periods or create serious water deficits, but they usually cause some injury to the crop. If accompanied by extremely high temperatures, strong winds, or both, the drouth period will be even more injurious.

Types of crops have an important bearing on the need for irrigation. Most horticultural crops have high moisture demands. This fact alone does not indicate that irrigation is a necessity. Some plants, such as fruit

FIGURE 4-15   Cabbage heads often split open following an abundance of moisture after a prolonged dry period. Tomatoes and the tree fruits will crack under these conditions. (Upper photograph courtesy W. L. Summers, Iowa State University, Ames, Iowa.)

trees, have deep root systems. During brief periods of drouth, they may suffer very little or not at all if the subsoil moisture is at a high level. However, even with deep rooted crops, many feeder roots which are located near the surface may be injured unless good soil management is practiced. In some cases, where there are no precautions against run-off and evaporation, a good mulch for reducing surface evaporation can be more effective than supplemental irrigation.

Many horticultural crops are shallow rooted and cannot tolerate even short periods of drouth without damage to production, or quality, or both. Most fruits and many vegetables produce food products which are succulent; consequently, for maximum size and quality, abundant supplies of water are essential. Moisture shortages during critical periods of cell enlargement will greatly reduce fruit size and edible parts of vegetable plants. It will also reduce the growth potential of perennial plants. Onions and other bulb crops are relatively shallow rooted. The bulbs are also high in water content. Moisture shortage during the "bulbing" stage results in much smaller bulbs. The tissue has become hardened through differentiation. If moisture suddenly becomes abundant, reactivated meristematic tissue produces double-bulbs or "over-growths." Similar results occur with other crops. Potato tubers may crack or produce knobs which are secondary growths. Apples, peaches, plums, and strawberries may split open because their protective tissues are unable to expand sufficiently for the increased water intake. Supplemental irrigation before a serious deficit occurs can avoid these conditions.

Some horticultural crops have an extremely high cash or aesthetic value. Berry crops, nursery stock, flowers, and many market garden items have a high cash value per acre. If water is available, it is generally advantageous to irrigate during drouths to avoid a nearly complete crop failure. In the home, the lawn, large trees, shrub plantings, and other ornamentals are of monetary and aesthetic value. Supplemental irrigation to avoid loss or severe injury of these plants is commonly practiced.

Expected increase in production is an important economic consideration for a grower contemplating irrigation. It is reasonable to expect that supplying additional quantities of water will result in increased production. Irrigation experiments have shown repeatedly that additional water, up to optimal amounts, will give increased returns during seasons of moisture shortage. Many variable factors affect the results of irrigation. Each season has its own peculiarities in respect to time, duration, and severity of moisture shortage. These are further affected by humidity, temperature, and light intensity. Nevertheless, increased production usually results from irrigation. Some growers favor use of irrigation during seasons when moisture is not considered in short supply. Even though water is not considered critical, additional quantities often result in increased production. This is almost always found in highly succulent crops which thrive on luxury consumption of water. Irrigation between harvests of strawberries will increase production merely by increasing berry size.

Balancing costs and returns is the ultimate economic basis for the practicability of irrigation. Costs of applying each acre-inch of water can be calculated. These costs must consider amortization of equipment and initial installation costs based on the probable period of operation prior to replacement or repair. Increased production can also be based on the returns from each acre-inch of water. A valid estimate of the feasibility of irrigation can thus be obtained.

Determining the Frequency of Irrigation

The nature of soil itself is an important criterion in determining the frequency of irrigation. Soils of fine texture hold moisture longer than soils of coarse texture. Deep soils hold larger quantities of water than shallow soils. Organic matter increases the water holding ability of light soils and provides better aeration in heavy soils. Soils that puddle and

FIGURE 4-16   Excavated root system of a Concord grapevine in a loess soil. Roots were traced to a depth of 14 feet ($4\frac{1}{2}$ meters) but most of the feeder roots were in the upper 4 feet ($1\frac{1}{4}$ meters). (Courtesy C. C. Doll, Iowa State University, Ames, Iowa.)

crack on the surface lose more moisture by evaporation than do soils with an unbroken surface. When the water holding ability of soils is increased, the interval between irrigations can be extended, because more water can be applied at each application, since the storage facilities are greater.

Rate of absorption by plants shows very little variation within the level of readily available moisture. The plant utilizes about the same amount of moisture when the soil is just above the permanent wilting level as it does at field capacity. On this basis water can be applied at any time between these extremes. However, fewer irrigations are needed when the soil moisture is permitted to approach the lower level. There is less likelihood of soil saturation or water-logging with less frequent irrigation. Excessive leaching of soil nutrients also accompanies too frequent application of water.

Transpiration of the crop plants affects the rate of absorption of water and consequently influences frequency of irrigation. Those plants with large leaf surfaces require more water than those with reduced leaf surfaces. Environmental factors that increase transpiration, high temperatures, wind, low humidity, also increase the water needs of plants with a resulting increase in the need for frequent irrigation.

The root system of the crop has considerable influence on the frequency of irrigation. The amount of water in the soil also influences the extent to which a plant's root system develops. A saturated or water-logged soil does not favor root development because of restricted

| TABLE 1 | **Approximate Depth of the Principal Feeder Root Zone in Several Horticultural Crops Growing in a Medium Textured Soil** |

Crop	Feeder root depth	
	ft	*cm*
Beans	2	60
Beets	3	90
Berries (brambles)	3	90
Cabbage	2	60
Carrots	2	60
Cucumbers	2	60
Grapes	4	120
Lettuce	1	30
Melons	3	90
Nuts	4	120
Onions	$1\frac{1}{2}$	45
Peas	$2\frac{1}{2}$	75
Potatoes	2	60
Strawberries	$1\frac{1}{2}$	45
Sweet corn	$2\frac{1}{2}$	75
Sweet potatoes	3	90
Tomatoes	3	90
Tree fruits	$2\frac{1}{2}$	75

aeration. A very dry soil is unfavorable to root development because of limited water for absorption. The most extensive root systems on crops are obtained when the level of soil moisture favors aeration, absorption, and root growth through the soil for additional moisture and nutrients. Plants absorb moisture both through their deep tap roots and their more fibrous feeder roots. Feeder root depths for several horticultural crops are given in Table 1. Most of the water which enters a normal growing plant is supplied by feeder roots and a smaller amount by tap roots. Thus, it is apparent that the water intake of a plant is seriously reduced as the water level of the surface soil becomes depleted. Accompanying the reduction in absorption, there is a loss of some feeder roots because of desiccation. These factors emphasize the need for a level of water in the upper regions of the soil ranging between its field capacity and the wilting point. Irrigation after wilting occurs is too late for maximum benefit to the plant, since some permanent damage will have already occurred. Therefore, it is important to avoid a water deficit.

There is no absolute method for determining when it is time to irrigate. However, there are well-established guides which growers can effectively adapt to their own conditions. An experienced person can tell by the "feel" of the soil at various depths whether the level of soil moisture is near the critical stage and needs replenishing. Soil augers are helpful in the "feel" test not only before irrigation but during irrigation as well. The depth and rate of penetration can be observed at intervals while the water moves down through the soil. Since the "feel" test is difficult to describe and requires considerable skill, instruments have been developed for rapid measurement of soil moisture. These instruments are designed to operate on the principle of conductivity of an electric current through soil moisture. A needle-gauge galvonometer directly indicates the moisture content or irrigation requirements. The

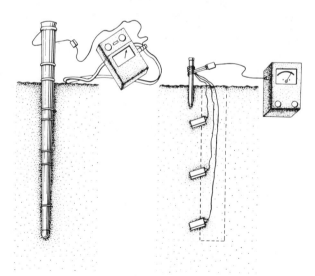

FIGURE 4-17   **Electrical soil-moisture meters properly installed can give reliable indications of available soil moisture. (Adapted from USDA.)**

Soil and Water Management   **103**

soil-moisture meter has plaster blocks which are buried in the soil. The *tensiometer* has a porous clay cup that is placed in the soil. Both types give readings on soil moisture conditions at the depth of placement.

Overhead Irrigation

Portable pipe sprinkler irrigation is the most widely used of the overhead systems. It is the principal type used on non-leveled arid and non-arid land or in areas where irrigation is supplemental to natural rainfall. Advantages are portability, light weight, adjustable-height nozzles, and large output. Its effect is a simulated rain with sufficiently small droplets, consequently soil compaction is not a factor. Unless sprinkling exceeds the rate of penetration, puddling of the soil and erosion are not of serious concern. Important precautions with sprinklers include proper spacing distances and adequate overlap of the fringe areas. Approximately 40 percent overlap of sprinkler patterns is required for equal moisture penetration near the periphery of each circle. The moving of pipes may entail considerable labor. However, various labor saving devices such as wheels and flexible couplings have been improvised to reduce handling time. Portable pipes are usually made of aluminum or other strong, lightweight materials. Various types of plastics and rubber hoses are used on lawns and other turf areas.

Wind interference is one of the problems of sprinkler irrigation. Wind not only changes the shape of the sprinkler pattern but also reduces sprinkler efficiency. Evaporation is increased with moving air. With a six

FIGURE 4-18   **Portable pipe sprinkle irrigation in a young pear orchard. (Courtesy Rainy Sprinkler Sales, Peoria, Illinois.)**

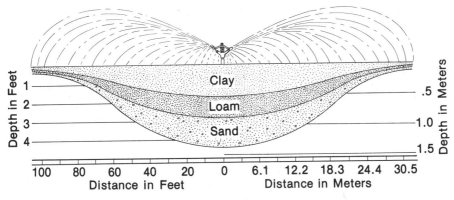

FIGURE 4-19 Depth of penetration of water decreases with increased distance from the sprinkler head. A 40 percent overlap of sprinkler patterns gives nearly uniform application of water.

mile-per-hour wind, as much as 40 to 50 percent of the water may be lost to the atmosphere. Under these conditions, sprinkler arrangement needs to be modified for uniform coverage. A hot dry climate has greater evaporation from overhead irrigation systems than does a more humid or a cooler climate.

Some form of power is required to provide water pressure sufficient to turn the sprinklers by centrifugal force. Also pressure is needed to force the water in height and breadth for good distribution. Rarely is the force of gravity adequate to operate the system unless the water is stored in a sufficiently high reservoir, tank, or dam. Pumps are usually used in large

FIGURE 4-20   Portable pipe sprinkler irrigation in a field of cabbage transplants. (Courtesy National Rain Bird Sales and Engineering Corp., Azusa, California.)

sprinkler systems. Municipal water pressure is usually adequate in urban areas where irrigation is on a smaller scale.

Oscillating pipe irrigation has been used considerably by market gardeners, florists, and nurseries. Cost of installation and operation is usually greater than with most irrigation systems and necessitates intensive use of the land and facilities. They are often operated off the municipal water supply. Water pressure operates a piston at the inlet end of a long pipe. This converts the pressure to an oscillating, back and forth rolling motion. The pipe has jet openings at 24- to 30-in. (60–76 cm) intervals which emit streams of water during oscillation of the pipe. The pipe lines are usually located about 50 feet (15 meters) apart for adequate overlap. They are permanently located on posts. Wind can cause even greater problems than with the rotary sprinklers because of this permanent placement. Portable types have been devised to use with wheels, skids, and cross-arms. However, the oscillating feature then has been lost, since it depends on straight pipes rigidly placed.

Perforated hose and pipe irrigation is used on the home grounds and for small-scale operations. Plastic hoses with a flat underside and a

FIGURE 4-21   **Perforated hose sprinkler on a home lawn.**

pattern of perforations on top are effective for irrigating in strips rather than in circles as with rotary sprinklers. Perforated pipe differs from oscillating pipe in that the holes or jets are located within a ninety degree area on the upper side to permit distribution without oscillation.

Surface Irrigation

Applying water to the soil surface without aerial application characterizes surface irrigation. It is an important method of irrigation in arid and semi-arid areas and where the land is nearly level or can be leveled without exhorbitant expenditure or considerable loss of fertile topsoil. Surface irrigation depends on gravity for spread of the water over the area. It is not as efficient in water distribution since more water than is needed is supplied at the inlet of a field or plot and much less reaches the lower end. However, that lost by evaporation is considerably less than with sprinklers especially in hot, dry climates. The power requirements for surface irrigation are much less than for sprinklers since water is not forced into the air. Some surface systems rely entirely on gravity distribution from the head ditch to all parts of the area under irrigation.

Furrow or corrugation irrigation is the most common of the surface irrigation methods. The crops are planted on ridges between the furrows. Water seeps into the soil to plant roots and does not wet the foliage or other aerial parts of the plant. While water is running in the furrows, it requires constant supervision to avoid formation of a channel from several furrow streams uniting. Obstructions in a furrow have to be removed, and the depth of penetration in the soil must be checked frequently for adequate watering and to avoid excess application. Some erosion occurs with the moving water. Where the slope is greater, rows should be shorter to slow down water movement and reduce erosion. More puddling of the soil will occur than with sprinkler irrigation. Water

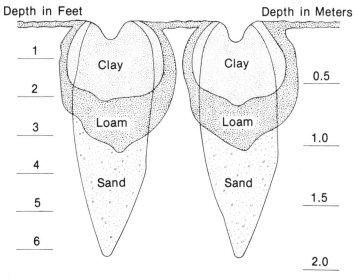

FIGURE 4-22   Relative depth of penetration of water in different soils from furrow irrigation.

that stands tends to separate the soil particles into strata with the fine particles at the surface. These puddles will first restrict aeration, then bake or harden, and finally crack and cause drying of the soil. Soils that readily absorb water are less apt to puddle. Application of more water at longer intervals reduces the amount of puddling.

Flooding irrigation is limited in its usefulness. It involves the use of small fields or plots surrounded by dikes. It is best adapted to crops with a high tolerance of water for brief periods. Cranberries and lowbush blueberries can be flooded. Onions can be grown successfully with flooding irrigation, but the foliage must not be completely submerged. Constant supervision is a requisite for irrigation by flooding.

Trickle irrigation and *soakers* are used for gradual applications of water in berry plantings, nursery operations and vegetable enterprises. They are very efficient in water use and distribution. The water leaks from a punctured canvas hose and seeps into the soil at a slow and uniform rate. The perforated hose used for overhead irrigation can be inverted, the water pressure reduced, and it becomes essentially a soaker for surface irrigation.

Sub-irrigation

Irrigation from below is limited in its application, but is an important method for specific situations and with certain crops. Plants in greenhouses require frequent watering. They are growing in a relatively shallow layer of soil, and their rate of transpiration is usually quite high. Sub-irrigation has been adopted by some greenhouse operators as a means of keeping adequate water available to the plants at all times and to reduce the labor costs of frequent watering. Water reaches the plant roots by capillarity from the bottom of the tank-type bench. Immediately above the water level is a layer of gravel, then sand, and then surface soil, which is high in organic matter. A float at the desired water level operates an automatic shut-off valve.

Sub-irrigation is also used in drained areas, such as swamps and lake bottoms, during drouth periods. The outlet of the drainage tile or ditch is closed and water allowed to remain or back-up in the system. This is especially effective in peat and muck soils where capillary forces move water faster in the highly organic soils. The level of sub-irrigation water must be watched closely or water-logging may result.

STUDY QUESTIONS

1. Why does the public get very concerned when topics such as soil and water conservation come under discussion? How does it relate to them?
2. How do the characteristics of leguminous bacteria and micorrhiza differ? What is their potential for future supplies of soil nutrients?
3. What is the source of peat and muck soils? How can a soil be built up with these materials? Are the sources inexhaustible?

FIGURE 4-23 Trickle irrigation of raspberries (*left*) and strawberries (*right*). (Courtesy H. G. Taber, Department of Horticulture, Iowa State University, Ames, Iowa.)

4. Make a sketch of a composting technique using either a pit or an above-ground container.
5. Devise a satisfactory crop-sequence rotation system for vegetables. For fruit trees. For a nursery. For a flower garden. For small fruits. Suggest length of time for each crop.
6. What is a good plant material for building the organic matter composition of a soil? What are the advantages and disadvantages of that material? Suggest alternatives.
7. From an economic standpoint when should a gardener irrigate his lawn? His vegetable garden? His ornamental plantings?
8. How important is it to lime the average lawn or garden area? Explain.
9. How would you evaluate the reliability of a totally new soil material for improving the water-holding capacity of a soil, increasing the amount of available water, improving the drainage of a soil, and increasing soil volume?

Soil and Water Management   **109**

SUGGESTED REFERENCES

1. Black, C. A. 1968. *Soil-Plant Relationships*, 2nd ed. Wiley, New York.
2. Buckman, H. O., and N. C. Brady. 1964. *Nature and Properties of Soils*. Macmillan, New York.
3. Dowdy, R. H., R. E. Larson, and E. Epstein. 1976. Sewage sludge and effluent use in agriculture, in *Land Application of Waste Materials*. Soil Conservation Society of America, Ankeny, Iowa.
4. Dowdy, R. H., and W. E. Larson. 1975. The availability of sludge-borne metals to various vegetable crops. *Journal of Environmental Quality*, *4*(2). Soil Science Society of America, Madison, Wis.
5. *Drip Irrigation News*. 1975. Vol. 3, No. 2. *Controlled Water Emission Systems*, El Cajon, Calif.
6. Robinson, D. W., and J. G. D. Lamb. 1975. *Peat in Horticulture*. Academic Press, London.
7. Thompson, L.M. 1975. *Soils and Soil Fertility*, 3rd ed. McGraw-Hill, New York.
8. U.S. Department of Agriculture Yearbooks of Agriculture:
   1957. *Soils*.
   1955. *Water*.

5 Light and Temperature Management

We have noted that light strongly influences photoperiod, photosynthesis, and in some cases even the process of germination. Insofar as temperature is concerned it is also strongly influenced by light, since the energy of light is rather readily converted to heat energy. Because of this relationship we are discussing them as a very closely related pair of essential environmental factors.

LIGHT ENERGY

Light energy, also called radiant energy, includes those parts of the spectrum that are not visible to the human eye. The spectrum includes all colors of light as seen in a rainbow or as separated out by a glass prism. A humid atmosphere can screen out some of the wavelengths of light. Thus a cloudy sky does considerable screening of light before it strikes the earth or plant leaves and results in partial shade. If further screening is desired to reduce light intensity in a greenhouse, certain pigments may be incorporated into the glass or plastic or painted on the surface, thus reducing intensity or quantity of light and, depending on the color of the pigment, changing the quality or color of light. Figure 5-1 shows the *spectral intensity* of sunlight at 40°49′ N Latitude and 96°42′ W Longitude (lower left) compared to a 100-watt tungsten incandescent lamp (upper left), 40-watt Sylvania Gro-Lux wide spectrum lamp (left center), daylight fluorescent lamp (upper left), 400-watt high-pressure sodium vapor lamp (left center), and 400-watt white mercury lamp (upper right). The spectral intensity can be measured and plotted for the various sources of light in units of lumens which by further calculation can be expressed as *foot-candles* or *meter-candles* of light. The wave length unit of measure can be defined as the distance between two similar points on adjacent waves.[1] Light in itself is a fascinating study in physical phenomena and visual sensation. Horticulturists are very dependent on light for plant growth and production. Its relationship to heat is extremely important in greenhouse production.

The discovery of the pigment, *phytochrome*, by a team of research workers of the U.S. Department of Agriculture at the Beltsville, Maryland Station, has revealed an intensely productive avenue of study in plant response to light. There are two forms of phytochrome which

---

[1] *Lighting for Plant Growth*, 1972. Bickford, E. D. and S. Dunn, Kent State University Press.

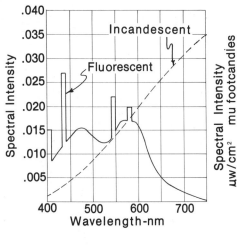

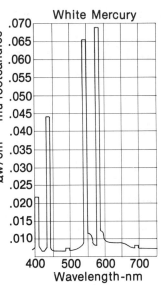

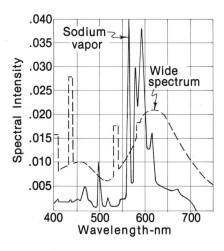

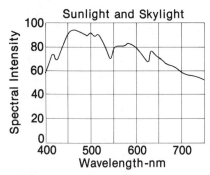

FIGURE 5-1
Spectra intensity curves at 40°49′ N. Latitude and 96°42′ W. Longitude of sunlight (*lower left*) compared to incandescent and fluorescent lamps (*upper left*), sodium vapor and wide spectrum lamps (*center left*), and white mercury (*upper right*). Note the much higher spectral intensity of sunlight compared to light from the various lamps. This is one reason for the tremendous need for energy in operating growth chambers. (Acknowledgment is due ISCO, General Electric, and Sylvania for spectral curves of the various lamps.)

they called P_{660} and P_{730} on the basis of activity at the wavelengths 660 nanometers (red light) and 730 nanometers (far-red light). The presence and activity of phytochrome explained much of the phenomenon of photoperiodism. The two forms of phytochrome are converted one to the other in light or in darkness. For example, P_{660} absorbs red light at a wavelength of 660 nanometers (nm) and becomes P_{730} which is the

FIGURE 5-2   Kalanchloe plants subjected to varying photoperiods show flowering response at short day lengths and vegetative response at long days. (Courtesy M. P. Garber.)

active form of phytochrome that stimulates flowering in long-day plants but prevents it in short-day plants. The P_{730} form is converted back to P_{660} after being exposed to several hours of darkness. Phytochrome has been called the "chemical basis of photoperiodism."

Plant growth structures such as greenhouses, sash houses and phytotrons serve as sites for plant growth under modified conditions. These modifications show us that light and temperature control can regulate the artificial environment in which we have placed plants. An important use of these modifications is for conducting experiments under specifically controlled environmental conditions. Such examples are the "phytotron" at the California Institute of Technology at Pasadena, California and at the St. Louis Botanic Garden, St. Louis, Missouri. Other very complete conservatories such as at Denver, Milwaukee, Winnipeg, and Boston provide the public with views of tropical settings during all seasons of the year. Because of the interest of government and the public in environmental controls many more of these will be built in other cities in the future. This will increase the knowledge and interest of the general public in plant growth and production as well as giving public insight to laboratory development. An interesting development of such an occurence is the installation of "Gro-Lites" in a kitchen with no windows because the occupants wanted to grow house plants. The growing of flowering plants in intensely lighted shopping malls and highway cloverleafs present problems unless careful selection of plant materials is made with regard to photoperiod.

LOCATION

Man's efforts to modify the severity of climates and protect plants against unfavorable light and temperature conditions has greatly increased the bounds within which many horicultural plants can be grown. Many warm-season vegetables can be grown in temperate areas by starting the plants indoors to lengthen the warm part of the growing season. Cool-season vegetables can be grown in sub-tropical areas by utilizing the cooler, winter part of the year. Special equipment, special practices, and special cultivars have been developed for growing plants in environments that differ greatly from their place of origin.

Latitude   Broadly speaking, in the northern hemisphere temperatures are colder in the north and warmer in the south. This relationship of temperature to latitude is reversed in the southern hemisphere. When we approach the poles of the earth, it becomes colder, and when we get closer to the equator, it is warmer. There are many factors of the environment, however, that tend to modify these relationships. The modified climates are utilized by horticulturists to grow and produce certain crops that, a look at the globe might indicate, were not adapted. We can take advantage of the presence of lakes, seas, mountains, and prevailing winds in locating our horticultural crops and enterprises.

Large Bodies   The presence of large inland lakes, such as the Great Lakes of North
of Water   America, have a profound influence on the temperatures of adjoining land. This phenomenon of nature occurs because the temperature of water changes much more slowly than temperature of air. Latent heat is stored in water. Large bodies of water or deep bodies of water have considerable latent heat which is released slowly. This extends the mild temperatures of fall and reduces the hazard of fall frosts. If the water freezes during the winter, there is still a stabilizing effect on surrounding areas. With the coming of spring, the slow temperature changes of the ice and water tend to depress the atmospheric temperatures. Plants remain dormant and growth is delayed. By the time temperatures are favorable for growth, the season has progressed to the extent that the danger of late spring frosts is averted. Summer temperatures are also stabilized by the presence of lakes.

Prevailing winds have an interacting influence with the presence of inland lakes. The windward side of a lake may have very little temperature modification. The leeward side may have a considerably modified climate because the invading air mass brings warmth in the fall and coolness in the spring. The frequency and velocity of the prevailing winds and the size of the lakes determine the extent of the area which receives a modified climate.

Crop adaptations to areas of modified climates have resulted in "islands of production" of cool season vegetables, and "fruit belts" of warm season fruits which do not generally thrive at those latitudes.

Coastal areas often show an extremely marked response to the modifying effects of water. The Gulf of Mexico produces a climate of extremely mild winters along southern United States from Texas to Florida. It requires more than proximity to water, however, to significantly modify the climate. Topography of the coast and prevailing winds have an important role. A nearly level and gradually rising coastline with prevailing winds from the sea will show the greatest influence

FIGURE 5-3   **Large bodies of water, either in breadth or depth are very influential in stabilizing temperatures of an area both in spring and fall. They also tend to moderate extremes during both summer and winter. (Photograph by K. A. Denisen.)**

because of location. If the incoming air is warm, very mild climate results.

Many areas of the world have become important garden areas because of the influence of the sea. Lands on all sides of the Mediterranean Sea have mild winters. Many of these areas grow citrus fruits,

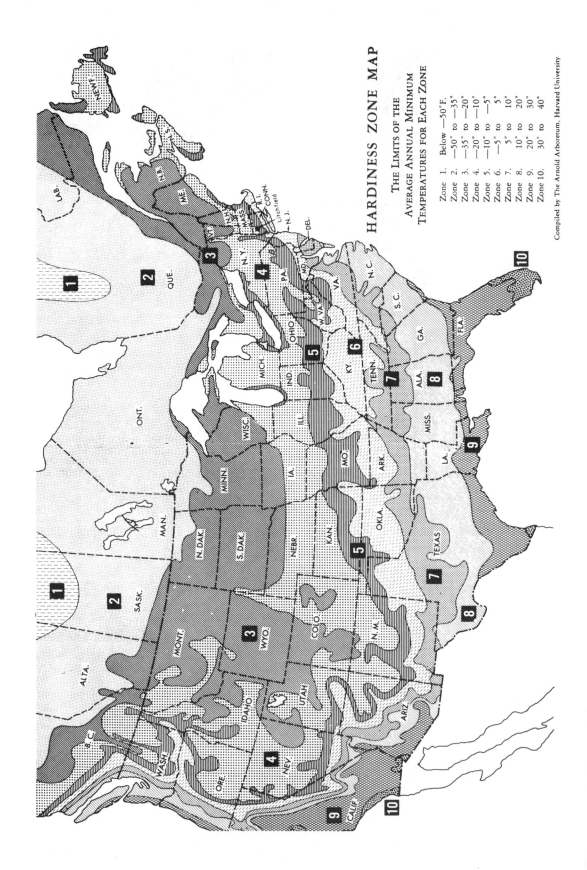

HARDINESS ZONE MAP

THE LIMITS OF THE
AVERAGE ANNUAL MINIMUM
TEMPERATURES FOR EACH ZONE

Zone 1. Below —50° F.
Zone 2. —50° to —35°
Zone 3. —35° to —20°
Zone 4. —20° to —10°
Zone 5. —10° to —5°
Zone 6. —5° to   5°
Zone 7.   5° to 10°
Zone 8. 10° to 20°
Zone 9. 20° to 30°
Zone 10. 30° to 40°

Compiled by The Arnold Arboretum, Harvard University

FIGURE 5-4

olives, figs, and other sub-tropical crops that are not adapted further inland. The west coast of North America has a warmer climate than lands inland from the coast. The same is true for the west coast of Europe.

Elevation   Increasing height above sea level means lower temperatures, all other factors being equal. High mountains or abruptly rising heights along lakes or coastal areas tend to oppose the warming effects of the large bodies of water. The pronounced influence of elevation on temperatures can be seen in the snow-capped peaks of mountains in areas close to the equator in contrast with the green, productive river valleys near the Arctic Circle in Alaska. High elevations also influence the temperature of surrounding areas to the leeward of prevailing winds.

FIGURE 5-5   **Winterkill in "Latham" raspberries. This cultivar is extremely hardy to low temperatures, but does not stand the severe shock of widely fluctuating winter temperatures.**

Light and Temperature Management   **117**

Growing Season The length of the growing season is determined by the number of days between beginning growth in the spring and cessation of growth in the fall. Its duration will vary for different crops in the same locality because of the relative hardiness of plants to low temperatures. The growing season for an apple tree in a temperate climate is usually longer than for a tomato plant. Apple shoots will start growth early in the spring and will survive late frosts. The outdoor growing season of such tender crops as the tomato is limited to frost-free days.

Dormant Season For many woody plants the growing season may be completely adequate both from the length and temperature standpoint. The critical limitation may be on the severity of the dormant season. The peach cannot be grown easily in northern parts of the temperate zone. The length of growing season and average summer temperature may both favor good peach production. However, the minimum winter temperatures will usually kill the fruit buds and may even kill the trees.

Much of the cold injury which occurs during the dormant season is due to extreme fluctuations in temperature. This is especially true of fluctuations which occur during the latter part of the dormant period when most plants have completed their rest period. Mild weather at this

FIGURE 5-6 A tamarack swamp provides the acid soil condition favorable for blueberries (*Vaccinium* spp.). The snow cover is heavy and continuous during the extremely cold winter and provides the blueberry plants with a natural mulch.

Skills and Practices in Horticulture   **118**

time may result in physiological activity within the cells in preparation for growth. Cells lose some of their hardiness as a result of this activity. Since winter is still in progress, lower temperatures again return, often quite suddenly. Winter injury is the result. Some varieties of apricots can withstand winter temperatures as low as −40°F (−40°C) in extreme northern United States. Yet these same varieties may lose their fruit buds, and sometimes the entire tree will winterkill in milder climates. Apricots have been known to bloom in February or early March in north central United States during a prolonged mild period. The severe temperature drop which follows results in fruit or tree losses.

For many low-growing plants, winter protection is often obtained through the insulating effects of snow. Many brambles, blueberries, and native shrubs will withstand severe winters of extreme northern United States and Canada, and yet they will winterkill in the central Great Plains. In the colder areas, the insulating effect of snow stabilizes the temperature and actually keeps the plants at nearly constant temperatures throughout the winter. Such plants are not subjected to the low or high winter temperatures as they are in the milder areas with less constant snow cover.

SITE

Location refers to a geographical area. *Site* implies a selected area within a locality. It is the local place of operation in a climatic area. Site can have considerable influence on the success of a horticultural enterprise.

Exposure The *topography*, or the "lay of the land," may indicate exposure to prevailing summer and winter winds. The amount of slope and direction of slope have considerable influence on the success of an orchard crop. If the *prevailing winter winds* are from the northwest, a north and west slope will give greater exposure to the coldest winds. Hardy cultivars are needed, spring blooming dates will be retarded; however, there will be less likelihood of injury from late spring frosts. With the orchard site on a slope to the north and east, the wind will be somewhat less severe. An east and south slope will afford protection of the hill or mountain from the direct and maximum blast of cold air. This site is, in effect, to the leeward of an obstruction to the wind. Temperatures are slightly milder and spring bloom is usually earlier. The effectiveness of air drainage may determine its likelihood of frost and frost injury. The south and west slope would receive some of the fury of cold winter winds, and may subject the trees to a condition known as sunscald or "southwest disease." The warmest part of the day occurs in mid-afternoon when the sun is in the southwest. Its rays strike the trunks of trees and heat is absorbed. The cambium may start growth activity. When the sun sets, the cold air reduces the temperature of the tree trunks. Death of the cells may follow and result in large areas of dead bark on the trunk or

FIGURE 5-7 **Sunscald on a peach tree. Although the tree is low-headed, there are no branches to shade the trunk from afternoon sun during winter months.**

lower branches. The resulting sunscald can greatly reduce the production of an orchard.

The *prevailing summer wind* greatly influences the shapes of trees. Constant winds tend to push the foliage and force the direction of growth in the opposite direction.

Air drainage is an important consideration for those areas and crops subject to unseasonal frosts. Convection currents of the lower atmosphere are the instruments of air drainage. When air is moving the likelihood of frost is greatly reduced. Still air can be called "dead" air. During a cold night, air gives up its heat, so that if there is no air drainage the cold air collects in valleys and near barriers of various types, and frost will occur when the temperature falls low enough. Vegetables growing on peat soils are sometimes subjected to frost during the growing season. Because of their location in swamps and low places, the still, cold air collects and frosts occur. Air drainage is difficult to have in a lake or swamp which also has natural topographical barriers formed by surrounding hills and high ground.

Wooded areas and windbreaks are factors for consideration in selecting sites for horticultural plants. In early spring, a grove, shelter belt, or

Skills and Practices in Horticulture

windbreak may retain cold air longer. The proximity of horticultural plants to a wooded area may delay their growth in the spring and decrease the danger of frost. The presence of woods and shelters will decrease wind velocity and reduce exposure. In selecting sites near windbreaks it is important that the crop and sheltering plants be sufficiently far apart to avoid competition.

Cities and towns in areas of temperate climate afford moderation of temperatures during the winter months. Because of proximity of heated buildings, of a smoke blanket over the area, and of reduced wind velocities, temperatures do not drop as low as in the surrounding countryside. There are numerous examples of less hardy plants growing and producing satisfactorily under these circumstances in climates considered extremely unfavorable. The temperature moderating effect of the homes and industries within the city limits is apparently adequate for these "borderline" plants to survive. Plant sites on the leeward side of buildings also increase winter survival of the less hardy plants.

HARDINESS

Maturation of Tissues
While the season progresses and food manufacture continues, the food is either used for further growth or accumulated in living cells. Maturation of tissues is characterized by accumulation of solids. The cell wall is toughened. Starches, sugars, and other foods are stored in the cells of storage organs, stems, and roots. The water content of the cells is decreased, and there is an increase in the relative amount of *bound water*, which is that form of water held by the cell against freezing. Water which leaves the cell to form ice crystals in the intercellular spaces is known as *free water*. Tissues high in free water content are susceptible to frost and freezing temperatures.

The maturing of tissues is accompanied by arrested cell division and cessation of cell enlargement. Unless this occurs, and is followed by accumulation of foods in the tissues of perennial plants, cold injury is likely to result. Prolonging active growth by such practices as late summer fertilization of woody plants and excessive fall irrigation delays maturation of tissues. A historical incident of inadequate maturation of tissues occurred in midwestern United States in the fall of 1940. There was a mild fall with no killing frost through early November. Temperatures, moisture, and light favored vegetative growth. On November 11, winter descended with all its pent-up fury. Thousands of fruit trees were killed and thousands more were severely injured which seriously reduced future production. Lack of sufficient maturation and inability of the cultivars to take the sudden shock of extreme cold, had a disastrous effect on many orchards. In a survey of damage, it was observed that those apple trees on hardy stocks survived the onslaught to a large extent.

Die-back of the tips of roses, raspberries, and many shrubs commonly occurs each winter in severe climates because growth continues at the apical meristem until low temperatures stop all growth. Maturation occurs in the older wood, however, and die-back is limited to the more tender extremities. Late summer and fall pruning often result in winter injury.

State of Rest Period Woody plants in temperate climates which are deep in their rest period are less subject to injury from extremely low temperatures than are those plants whose rest period has been broken. Low temperatures in early winter are thus less injurious to woody plants and herbaceous perennials than similar temperatures during late winter.

LOW TEMPERATURE PRECAUTIONS

Horticulturists through the years have initiated, invented, and adopted numerous practices to guard against extremely hot and extremely cold temperatures. These techniques employ various structures, methods of culture, materials, and ingenuity. They can mean the difference between success or failure of a crop or enterprise.

FIGURE 5-8   A technique for winter protection of tree roses in climates of severe winters. On the *left*, soil is being removed from the aerial portion of the plant which was tipped over and buried in the fall. *Right*, tree rose in erect position.

Skills and Practices in Horticulture   **122**

Hardy and Adapted Cultivars   Horticulturists and plant breeders are constantly exerting their efforts toward better cultivars and hybrids. In temperate climate areas, there is continuous search for greater hardiness of woody plants, cold-tolerant herbaceous perennials, and frost-resistant annual cool-season plants. One of the common sources of parental material used by plant breeders is the wild or native form. If these native plants have been adapted to

FIGURE 5-9   **Painting or white-washing tree trunks for reflecting light rather than absorbing it reduces the amount of sunscald.**

rigorous and severe winters, they may be a good source of hardiness as parental types. However, this ability to survive severe climates should be studied and observed in its native environment. Plants may survive because of the insulating effects of snow cover. Their environment in nature may be in wooded areas where other vegetation has helped them survive. Or the natural supplies of water, nutrients, and light may restrict excessive vegetative growth and cause early and complete maturation of cells. In many cases the native plants, when placed under cultivation, are less hardy than in their natural enivornment.

Native plants are by no means the only source of parental material for new and hardier types. Numerous cultivars, species, and strains of plants from other regions of the world contribute to the complex pedigree of many cultivars. Reports of cultivar trials and hardiness tests, conducted by state and federal experiment stations, are good sources for such information. The adaptability of cultivars to climate, soil, and other factors is often suggested or indicated by these reports.

Topworking on Hardy Stocks

Crown and crotch injury are common forms of winter injury that woody plants sustain when grown in areas where they may not be completely hardy. The apple is a good example of a fruit where cultivars show a broad range of hardiness. Many of the less hardy types, which are subject to crown and crotch injury, if grafted on hardy stocks can be grown successfully in quite severe climates. The Virginia crab and Hibernal cultivars have exceedingly hardy trunks, crotches, and scaffold branches.

Low-headed Fruit Trees

Lowering the head of fruit trees results in a shorter trunk and a reduction in the exposed area subject to sunscald. Trees that are low-headed not only have the lowest branch closer to the ground, but also have closer scaffolds which further shorten the trunk. Although the subsidiary branches are naked of leaves during the dormant season, they are nevertheless quite effective in reducing the amount of light that reaches the surface of the trunk. Early training of the tree should include the objective of low-headedness and a network of twigs and branches which will cast broken shadows on the trunk and scaffold branches. Peaches are quite susceptible to sunscald because of their spreading scaffold branches. These trees should be encouraged to produce lateral branches to keep their silhouette low-headed.

Proper Season for Transplanting Woody Plants

Transplanting operations weaken plants to varying degrees. Some plants lose many roots and root hairs in the digging, handling, storage, and shipping operations. Plants that are weakened are less resistant to further adverse conditions such as severely cold temperatures. This is the principal reason for the low percent of survival often encountered with fall transplanting in northern United States.

In southern and eastern United States and in other areas of relatively mild winters, fall transplanting of woody plants is common. Deciduous dormant plants are transplanted during fall months so as to allow sufficient time for establishment of roots before winter arrives. If planted too early in the fall, bud-break may occur and tender shoots thus produced are subject to winter injury. Well established plants set out in the fall have a time advantage over spring set plants. They can start growth as early as environmental conditions permit in the spring. No time is lost in establishing a root system, since it became established the previous fall.

In areas of severe winters, early spring is generally recommended for transplanting. The hazards of fall transplanting are so great that it is not worth the risk of losing the plants. Spring planting avoids frost heaving

FIGURE 5-10 **The blooms of this magnolia tree were protected from frost injury by irrigation. As icicles were formed, the water released its latent heat. (Courtesy *Des Moines* (Iowa) *Tribune*, photograph by Jervas Baldwin.)**

of the soil and consequent severing of newly established roots. It avoids death by desiccation so common with cold and windy weather, frozen soil, and a limited root system. If fall transplanting must be done under these adverse conditions, it is well to plant sufficiently early in the fall, firm the soil around the roots, water well before the ground freezes, and mulch to reduce soil heaving.

Irrigation   The latent heat of water suggests another purpose for irrigation, to ward off frosts. Both flooding and sprinkling are used effectively at various times during the season in areas subject to frosty conditions. Sprinkling water into the cool air on a frosty night releases the latent heat to the atmosphere. This warms the air sufficiently that frost does not occur in the immediate area. It is not unusual for the water itself to freeze as it gives up its latent heat. The freezing of the sprinkled water is often an indication of what may have happened to the crop had irrigation not been used. Small fruits and vegetables are occasionally protected in this manner.

FIGURE 5-11   **Strawberry blossoms injured by frost. The dark centers are dead pistils. Berries formed before the frost and blossoms which opened after the frost were not injured. Irrigation, mulch, moving air, or foam could have prevented this injury.**

Flooding or furrow irrigation releases latent heat from its surface but more slowly than sprinkling. Complete submersion of the crop is used to ward off frosts in cranberry bogs. These plants will tolerate complete submersion if drained a few hours later. In flooding and submersion the latent heat mostly remains with the water. The water thus serves as both source of heat and insulator for the submerged plants and fruits.

Irrigation during the dormant season is sometimes essential to winter survival. Winter injury is frequently due to desiccation of cells. Soils that are extremely low in moisture content, as encountered with drouth conditions, provide very poor conditions for root survival. Lack of snow cover or mulch increases the likelihood of root injury under these conditions. Roots do not have a rest period like stems and buds. Neither do they undergo a winter maturation or hardening process to the same extent stems do. Shrubs and trees should be watered well before the ground freezes if drouth conditions prevail. Evergreen trees and shrubs may also need irrigation during the winter months under extremely dry and windy conditions. This is especially true of types and cultivars not native to the area.

Artificial Heat, Smudge, Fog, and Foam

During periods of threatened late spring frosts, orchardists often employ techniques which supply heat or trap heat for the fruit trees. In supplying heat, circulation is of prime importance. It is usually only a matter of a few degrees difference in temperature between loss of a crop and crop safety. Gasoline, oil, and wood-burning space heaters or open bonfires are used. To be effective they must provide for heat distribution before it is lost to greater heights. Smudge pots, smoke generators, fog, and foam distributors produce particles of soot or water which remain suspended for brief periods. On a still night the suspended particles dissipate slowly. Smoke or fog blankets the area and traps the daytime heat in the orchard. In areas with good air drainage, the smoke or fog moves through the trees to the outlet and gives good distribution for the "anti-frost blanket." Citrus groves and deciduous orchards both benefit greatly from the use of smoke and heat to avoid frosts and consequent blossom injury.

Plant Growing Structures

Greenhouses, sashhouses, plastic houses, hotbeds, and cold frames are used for propagation. The glass and plastic house structures are also used for out-of-season production of flowers and vegetables. Cold frames can be used during the dormant season for mulching tulips, lilies, and other potted bulb crops that are to be forced for late winter or early spring bloom.

Greenhouse structures, whether constructed of glass or flexible plastics, present many problems not encountered in other types of plant production, especially during the winter months. Maintenance of a uniform and adequate temperature is basic to good greenhouse management. Thermostatic control and circulation of air to all parts of the house are especially helpful in establishing optimum temperature conditions. Air conditioning in greenhouses is receiving increased attention as its beneficial effects on plant growth becomes more

FIGURE 5-12   **Polyethylene plastic greenhouse helps to establish an outdoor bed of flowers. (Photograph by K. A. Denisen.)**

FIGURE 5-13   **A low silhouetted lath-house for decreasing the intensity of sunlight during bright sunny days of spring. (Photograph by K. A. Denisen.)**

apparent. Warm season and even strictly tropical plants flourish during the winter months in well-managed greenhouses.

Hardening-off Transplants   The maturation process has been mentioned frequently for increasing the hardiness of woody plants. The principles of maturation or hardening can be applied to vegetable plants that are commonly transplanted to the field or garden. Adverse conditions result in a slowing down of growth. This slowdown is accompanied by less succulence of the tissues, hardening of the cell walls, more concentrated cell sap and protoplasm, more bound water, and a general increase in hardiness to cold or other factors.

Cabbage plants can be "hardened-off," by artificially induced unfavorable conditions. Withholding moisture from succulent growing plants in a greenhouse until near permanent wilting occurs is effective in hardening. Reducing the temperature below the optimum range for growth is another effective procedure. Plants grown in a cold frame may be easily hardened-off by reducing the amount of solar heat, by admitting cool air, and by reducing the amount and frequency of watering. Cabbage can be hardened sufficiently to withstand temperatures 6 to 8°F (3–5°C) below freezing. However, tomato plants cannot be altered appreciably in their ability to survive freezing temperatures. Hardening will increase their ability to resist the wilting effects of constant winds, intense sunlight, and the ravages of sucking insects.

Hardening causes a reduction in the rate of growth. When subjected to favorable conditions after transplanting, it may take longer for the hardened plants to resume growth. However, there may be considerably higher mortality of nonhardened plants, especially when they are under adverse growing conditions out-of-doors.

An alternative or supplement to hardening-off is the use of *plant protectors* or "hot caps" placed over the tender and succulent plants in the event of cold or freezing weather. These are made of translucent rigid paper or plastic and must be weighted down with soil around the edges. Placing the protectors over the plants and removing them after the anticipated frost can become tedious and time consuming.

Competing Cover Crops

In addition to their usefulness as good soil management practices, cover crops can be effectively used to hasten maturation of woody tissues. Oats and other annual small grains or grasses are often planted between the rows of bramble fruits and grapes. Competition for moisture and nutrients between the crop plants and cover crop slows down fall growth and tends to harden the tissues of the woody plants. This is especially helpful for increasing the hardiness of these plants when there is abundant moisture. Weeds can be used in lieu of the cover crop. Care must be taken to guard against the hazard of re-seeding of the weeds or cover crop. Too early planting of the cover crop may result in severe competition, reduced growth, decreased fruiting area per cane, and seed maturity of the cover crop. Late summer or early fall seeding of the competing crop will usually slow down woody plant growth adequately, yet prevent maturing of cover crop seeds.

Winter and Spring Mulches

Mulches are good insulators. Straw, hay, cornstalks, and other rather coarse or stemmy materials are good winter and spring mulches. Leaves are satisfactory for mulching unless there is danger of smothering the crop. Use of twigs or cornstalks under the leaves will usually give

Light and Temperature Management

129

adequate aeration under leaves and other compacting types of mulch. A winter mulch is applied to stabilize temperatures, not to prevent freezing. Strawberries, roses, herbaceous perennials, brambles, shrubs, and vines may be given winter protection by mulching. With strawberries a common practice is to allow the ground to freeze before application of the mulch. However, exposure to temperatures below 20°F (−7°C) before mulching may cause injury to the fruit buds already present in the dormant crowns. The mulch is generally 2 to 3 in. (5 cm) deep. Wide fluctuations in temperature of outside air will not appreciably change the temperature under the mulch because of its insulating effect. Prolonged periods of mild temperatures in midwinter may possibly thaw the frozen soil. However, the mulch keeps the temperature sufficiently low that a resumption of growth is avoided. Sharp dips in temperature are likewise appreciably reduced in intensity by the winter covering.

When roses, brambles, and vines are mulched, soil may be utilized for temperature stabilization. The soil is mounded around the base of the rose plant to a height of 8 to 10 in. (20–25 cm) before they lose their leaves and become dormant in the fall. The mulch is not applied until the mounded soil freezes. The brambles (raspberries, blackberries, etc.) and vines are mulched by laying their canes on the ground and covering with soil. After the soil freezes the organic mulch is placed over the canes and soil. Tender shrubs may also be mounded, or mulched without mounding.

Spring mulches are used on strawberries and herbaceous perennials to delay warming of the soil and prevent early spring growth. By this

FIGURE 5-14   **Dormant roses mounded with soil.**

FIGURE 5-15   An improvised use of polyethylene and dry autumn leaves for winter mulch of roses. Leaves will stay dry unless the film is punctured.

practice the growth or blooming date may be delayed, avoiding injury from late spring frosts. Later the mulch on strawberries may be pulled between the rows to form an aisle for the pickers and to keep the berries clean. It is thus available, if needed, to put back over the row if frost is anticipated. Some berry growers do not mulch during the winter but rely on snow cover for insulation. Then as spring approaches, they mulch for delayed growth, frost protection, and clean berries. Mulches in herbaceous perennial and shrub borders can be used in the same way.

Woody plants that are fall transplanted, usually benefit from a mulch around the base of the plant to prevent heaving of the soil. Heaving occurs in soils that have areas of different degrees of compaction which is frequently found with newly set plants. When the soil heaves, roots may be broken or snapped and the plants may actually be moved upward in the soil. Death of the plant usually results. The mulch to avoid heaving is actually used on the soil rather than on the plant.

HIGH TEMPERATURE PRECAUTIONS

Shading and Screening   When transplants are set out or cuttings are placed in a propagating frame, they may require partial shade. The root system is often reduced in size with transplants and it is completely eliminated with stem and leaf cuttings. The absorptive capacity of the plant or cutting is thus reduced. Transpiration will continue as before unless an attempt is made to reduce it. Reducing the light intensity, wind velocity, and temperature will reduce transpiration. This may be accomplished by the judicious use of shades and screens.

Shading can also be used to avoid bleaching of pigments in flowers and fruits. Cheesecloth placed over a rose bed during bloom in climates of intense summer sunlight will prolong the beauty of the flowers. Sunscald or sunburn on tomato fruits can be reduced or avoided by shading. Cultivars of tomatoes with dense foliage cover over the fruits have fewer sunburned fruits and redder color.

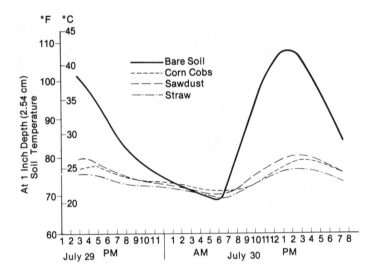

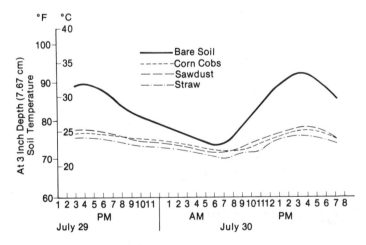

FIGURE 5-16   **The temperature-stabilizing effect of summer mulches. Note there is less fluctuation in bare soil at the three-inch depth than at one inch. (Courtesy R. H. Shaw and E. L. Denisen, Iowa State University, Ames, Iowa.)**

Summer Mulches   The insulating properties of mulches can also be used to advantage when they are a precaution against high temperatures. Sawdust, chopped corn cobs, and straw as mulches have a very stabilizing effect on the soil temperature during a 24-hour period. In periods of intense light and heat, mulches can often hold soil temperature within optimum conditions for root growth, absorption, and microorganism activity. In warm climates strawberries, tomatoes, and roses grown in soil covered with a temperature stabilizing mulch of plant organic matter are invariably more thrifty. The combined effects of temperature stabilization, moisture holding ability, weed control, and clean fruits make mulches popular for home gardens and for many commercial enterprises.

Rapid Cooling for Quality Control   In the presence of enzymes, sugars are converted to starches at high temperatures. With certain vegetables such as peas and sweet corn, high sugar content is a component of high quality. The harvest period for both peas and sweet corn is often during the warmest days of the year. Unless the harvested products are cooled immediately, they begin to deteriorate in quality. For the canner of corn and peas, immediate processing is advocated. This stops the conversion of sugars to starch by inactivation of the enzymes. However, for fresh market or home use, rapid cooling in a cold storage or refrigerator will greatly retard conversion. Harvested ears of sweet corn are often held for several days at temperatures just above freezing with little loss of quality.

Irrigation   Evaporation has a cooling effect. Sprinkling plants with water can be helpful in reducing leaf temperatures and increasing the humidity. Both these factors reduce transpiration and wilting. This method of cooling and transpiration control must be closely watched and carefully controlled. Sprinkling on warm leaf surfaces during bright sunlight can also result in scorching or burning of the leaves.

Adjusting Planting Dates   Cool-season crops with a short growing period do not thrive in hot weather. Winter or early spring planting of lettuce, radishes, peas, and spinach in warm climates will usually yield high-quality produce. Early fall planting of these same vegetables will subject them to cool temperatures and again give high-quality produce. However, when planted during the warmest part of the season, at above the optimum range in temperature, poor quality results, flowering of lettuce and radishes occurs, and tissues become tough and fibrous.

Ventilation in Plant Growing Structures   During winter months, artificial and solar heat combined may raise temperatures in greenhouses above the optimum range. Ventilation is used to reduce these high temperatures but must be used carefully to avoid chill to the plants especially on very cold days. Circulation of the cool air over the heating unit and elimination of the hot air at an upper vent will lower the temperature gradually and without shock or chill to the plants. Cool-season crops are more readily injured from high

FIGURE 5-17   **Irrigation is essential for container-grown stock in a nursery. (Photograph by K. A. Denisen.)**

temperatures in artificially and solar heated greenhouses than are warm-season crops.

STUDY QUESTIONS

1. If we follow the concept that neither matter or energy can be created or destroyed, are we assured of adequate food and power for the world of the future? Will we ever exhaust our potential for food production? Discuss thoroughly.
2. Assume you are a producer of chyrsanthemums in North America. How can your operation be adapted to photoperiod requirements during long days? During short days? During hot weather? Indoors? Out-of-doors?

Skills and Practices in Horticulture   **134**

3. How does extremely hot weather during the harvest of peas affect quality of the peas? What can be done to avoid unfavorable effects of temperature on quality of peas?
4. Irrigation was discussed in the previous chapter (Soil and Water Management). How does it apply to the current chapter (Light and Temperature Management)?
5. Mulches have been mentioned and described frequently. Make an outline to show the functions, materials used, and effects of mulches on horticultural plants.
6. Why is it important to prevent a cover crop from producing seeds? Give examples.
7. Develop a planting calendar for a home horiculturist including shrubs, trees, vegetables, annual flowers, lawn, and perennial herbaceous plants. Include approximate dates for each.

SELECTED REFERENCES

1. Bickford, Elwood D., and Stuart Dunn, 1972. *Lighting for Plant Growth*. The Kent State University Press, Kent, Ohio.
2. Borthwick, H. A., S. B. Hendricks, and M. W. Parker, 1952. The reaction controlling floral initiation. *Proc. Nat. Acad. Sci.*, **38**:929–934.
3. Cathey, H. M., and H. A. Borthwick, 1961. Cyclic lighting for controlling flowering of chrysanthemums. *Amer. Soc. Hort. Sci. Proc.*, **78**:545–552.
4. Curtis, Helena, 1972. *Invitation to Biology*, Chapter 5–4, Biological clocks: Circadian rhythms and photoperiodism. Worth Publishing, New York.
5. Denisen, E. L., and H. E. Nichols, 1962. *Laboratory Manual in Horticulture*, Exercise 2, Environmental Factors Affecting Horticultural Plants, Iowa State University Press, Ames, Iowa.
6. Downs, R. J., H. A. Borthwick, and A. A. Piringer, 1958. Comparison of incandescent and fluorescent lamps for lengthening photoperiods. *Amer. Soc. Hort. Sci. Proc.*, **71**:568–578.
7. Edmond, J. B., T. L. Senn, F. S. Andrews, R. G. Halfacre, 1975. *Fundamentals of Horticulture*, 4th ed., Chapter 1, Fundamental crop-plant reactions and processes; Chapter 5, Heat and temperature; Chapter 6, The light supply. McGraw-Hill, New York.
8. Janick, Jules, 1974. *Horticultural Science*, 2nd ed., Chapter 6, Controlling the plant environment. W. H. Freeman and Co., San Francisco, Calif.
9. Laurie, A., D. C. Kiplinger, and K. S. Nelson, 1968. *Commercial Flower Forcing*, 7th ed. McGraw-Hill, New York.
10. Newburn, Lorance H. 1975. *Iscotables, Handbook of Data for Biological and Physical Scientists*, 6th ed. Instrument Specialties Company, Lincoln, Nebraska.
11. U.S. Department of Agriculture Yearbook of Agriculture, 1941. *Climate and Man*.

6 Growing Plants from Seeds

Plant propagation is the perpetuation or increase in numbers of plants. When the flower is involved as an essential phase of increase in numbers, propagation is *sexual*. That is, sex organs of the flower have functioned to produce seeds. Sexual propagation of horticultural plants is also called seed propagation or seedage. If the flower is not involved and other plant parts are used for increasing numbers, propagation is *asexual*. Principles and methods of asexual propagation are discussed in future chapters.

SEEDS AND SEED PRODUCTION

What Is a Seed?
A seed is a living entity which serves as a bridge between generations of a plant. It is formed in the pistil of the flower and develops from the ovule following fertilization. As the fertilized egg grows and develops, it becomes the *embryo* of the seed. The embryo is a complete plant in miniature, for it consists of leaf, stem, and root primordia. Concurrent with the development and growth of the embryo, provision is made for storing food for future use. This food, which is high in carbohydrates, is stored in the *endosperm* of many seeds as with sweet corn, coconut, and onion. In other cases the food is channeled from the endosperm to the seed leaves or *cotyledons* for storage as with the bean, peanut, almond, and watermelon. In either case, the embryo has a source of food on which to draw when it resumes growth at the time of seed germination. Encasing the embryo and endosperm is the *seed coat* which serves as a protection against mechanical injury, excess water loss, and other unfavorable conditions.

Flowering and seed formation are differentiation processes during which the ovary and its developing parts have high priority in their food needs. The plant is in its productive phase of growth and foods accumulate in such storage organs as seeds and fruits. Good growth and a high rate of photosynthesis favor production both in quantity of seeds and seed size.

Cultivars, Strains, and Hybrids
In glancing through a seed catalog, one notes frequent usage of the terms "cultivar" and "hybrid" and to a lesser extent the term "strain." A *cultivar* is a group of closely related plants of common origin which have similar characteristics. Special types selected from the cultivar which are considered to be superior in some way are called strains of the

136

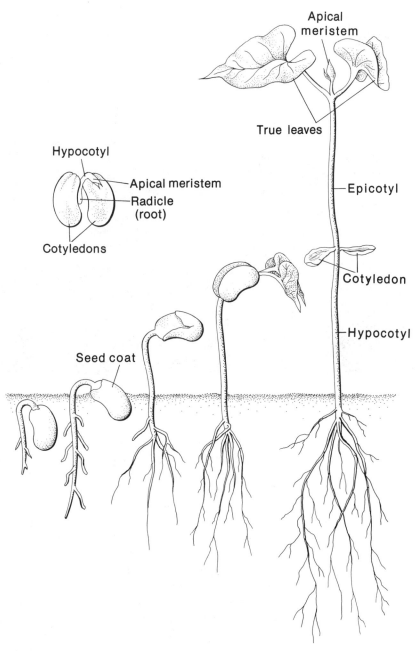

FIGURE 6-1 **Bean seed and germination. (Adapted from Helena Curtis, *Invitation to Biology*, Worth Publishers, N.Y.)**

cultivar. The first generation progeny from a cross of different cultivars, strains, or inbred lines is a *hybrid*.

To clarify the relationship of these terms it is helpful to have an understanding of the principles of pollination, selection, and hybridization. Many plants are *self-pollinated*, that is, the pollen from the anther is transferred to the stigma of the same flower. The tomato is an

FIGURE 6-2　　　　Onion seed production occurs during the second year from this biennial crop.

example of such a plant. Since pollination occurs within the flower, contamination by foreign pollen is very unlikely. Cultivars and strains of self-pollinated plants are thus relatively easy to maintain in a pure state and true to type. *Cross-pollinated* plants, conversely, are more difficult to maintain in a pure state. Their more usual source of pollen is from other flowers and from other plants. Sweet corn is normally cross-pollinated. Pollen is produced abundantly by the tassels (anthers); and wind, as the agency of pollination, scatters pollen throughout the field. A very heterogeneous lot of pollen may settle on the silk (stigma) of each ear and result in a very non-uniform group of progeny. Squash and other members of the cucumber family are also normally cross-pollinated. By the nature of their imperfect flowers, monoecious and dioecious plants are largely cross-pollinated. Perfect flowers may be commonly self-pollinated or cross-pollinated depending primarily on flower structure and agency of pollination. The principal agency of pollination among perfect flowered types is the honey bee, although other insects will transfer pollen during their search for nectar. Since cross-pollination occurs readily among certain plants, and since the agencies of pollination may be difficult to control, the most practical method of producing seed of certain cultivars and strains is by *isolation* of the crop.

Careful management of parental lines, crossing techniques, isolation, and labeling are essential to the successful production of hybrid vegetable and flower seeds. Because of the outstanding performance of

FIGURE 6-3   A cabbage plant producing seeds. The plant produces a "head" the first year, and the second year produces seed.

hybrid corn, hybrid cucumbers, and hybrid tomatoes, the term hybrid has assumed magic significance to many people. Actually not all hybrids are superior. A hybrid is merely the progeny of a cross, and not all progenies of crosses are superior to their parents or to other hybrids. Since diverse cultivars, strains, or inbred lines go into production of a hybrid, seed saved from hybrid plants will produce plants which segregate into types having characteristics of either parent. There is normally a loss of vigor when seed from hybrids is used for the next generation; consequently, it is not practical to save seed from hybrids. A knowledge of heredity is essential to successful seed production and to maintain purity of cultivars, strains, and hybrids.

USING GOOD SEED

Cost account records of flower growers, nurserymen, vegetable growers, turf specialists, and other plant growers show repeatedly that the best seed is the most economical seed. When considering the costs of machinery, labor, land, irrigation, pest control, and numerous other

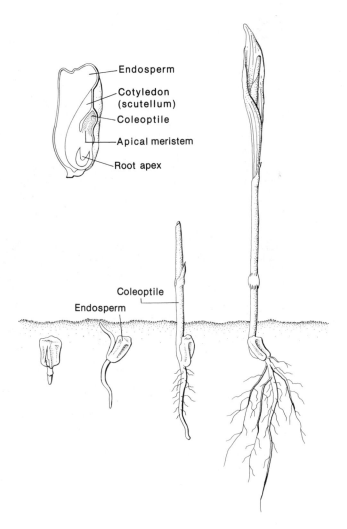

Endosperm

Cotyledon
(scutellum)

Coleoptile

Apical meristem

Root apex

Coleoptile

Endosperm

FIGURE 6-4   **Sweet corn seed germination. (Adapted from Helena Curtis, *Invitation to Biology*, Worth Publishers, N.Y.)**

items, the cost of seed is a very minor item of the total. Yet the quality of seed can mean the success or failure of the enterprise. This is also true for the home or part-time gardener. Poor seed can mean wasted effort and ultimate discouragement.

Characteristics   Good seed should be (1) true to name, (2) free of diseases and insects, (3) clean, (4) of good viability, and (5) graded to size. The reliability of the seed-house is an important factor in obtaining good quality seed.

Trueness to name is an important attribute of good seed if for no other reason than to give the customer what he wanted. It goes further than that, however. Mixtures of cultivars or a substituted or incorrectly labeled cultivar may mean lack of adaptability to an area. It can mean late maturity when earliness was desired. It can mean a disease-susceptible cultivar when a disease-resistant cultivar or strain was ordered. It can mean a loss to the grower, and if justice prevails, it will ultimately mean a loss to the seedsman. Paralleling trueness to name are claims made in advertising. Some advertisements do not actually state untruths but by implication make their products sound better than they really are. This has been noted among a few less reliable merchandisers of lawn seed. A great flourish was given the fact that a seed mixture contained a highly advertised and very expensive cultivar of lawn grass seed. Upon examination of the contents listed on the label, it was observed that less than 2 percent of the mixture was of the highly prized and highly advertized grass. If the buyer did not read the label, he was under a misapprehension.

Freedom from infestation of diseases and insects has received considerable attention from seedsmen. Seed fields should be inspected and plants eliminated which harbor diseases that are carried in the seed. Disease organisms carried on the surface of the seed can be controlled by seed treatment. Some seed dealers prefer to treat their seed before they sell it. In such cases, seed should be so labeled in order that the purchaser may use due caution in handling and not use the excess for livestock or poultry feed. The principal insect problem is apt to be the weevil, a storage pest. Since weevils burrow into seeds, their damage is

FIGURE 6-5   **Tomato plants for transplanting. Since the tomato is self-pollinated, it is relatively easy to maintain as true-to-type cultivars.**

Growing Plants from Seeds   **141**

expressed in reduced germination percentage. Most seedsmen fumigate at regular intervals as a precaution against weevils and other insects.

Clean seed refers to its freedom from dirt, chaff, and weed seeds. Dirt and chaff are objectionable mainly because they are inert materials which add to the weight and reduce the number of seeds per pound. Weed seeds also displace the good seeds, but even more important, they introduce additional pests to the planting. Weed seeds planted with vegetables, flowers, and especially lawn grasses create a serious problem while the desired seedlings are still extremely small. Some seed processors have elaborate and efficient equipment for removal of foreign seeds and inert matter. Prior to harvest and processing of seed, the crop should be grown under as complete weed-free conditions as possible.

Good viability or germination is essential for a complete stand of plants. A missing plant means no return for that area. The handling of the seed through harvest, threshing, and storage affects viability. Injury due to rough mechanical treatment, heating as a result of high moisture content, and excessively high temperatures in drying equipment exact their toll in reduced seed viability.

Seed *graded* according to size is extremely important when machine planting is used. Large seeds should be planted at a higher rate (in weight or volume) per area than small seeds of the same cultivar for an equivalent stand. In general, large seeds get seedlings off to a faster start than small seeds because of the greater amount of stored food. This statement does not apply when comparing seed size of hybrids to cultivars.

Testing Seed

When testing seeds, the seeds are placed in an environment favorable to growth and within a few days the percent germination can be calculated. The seeds to be tested are usually placed in a chamber on moist blotting paper, the temperature is kept at near optimum for the species, and the chamber is ventilated to insure a good supply of oxygen. The seeds so tested are a representative sample of the lot for which germination percentage is desired. Known quantities of seeds may also be planted in containers of sand or soil and the germination determined from stand counts of growing seedlings. Still another method consists of placing seeds on a flannel cloth, moistening the cloth, and rolling from one end to form a "rag doll." The dolls are placed in shallow pans and kept moist until germination, when counts are taken and percentages determined.

Many of the state Agriculture Experiment Stations have seed testing laboratories. They offer a public service to both growers and seedsmen. For a nominal fee to cover expenses, they determine such information as percentage of germination, percent of weed seeds and other foreign matter, and the general vigor of seedlings.

Longevity of Seeds

Many seeds, particularly weed seeds, have been known to remain alive for many years. Actually, the longevity of most flower and vegetable seeds is relatively short, usually about two to five years. Since a seed is a living organism, it is continually carrying on respiration. Under ideal storage conditions, the rate of respiration is very low and less of the food reserve of the seed is consumed. Optimum storage conditions require adequately dried seeds; that is, with moisture content of about 10 to 12 percent. For most species a cool temperature is desirable for storage. A rodent- and insect-free storage is another requisite for maximum seed longevity. It is well to keep in mind that it is not only the number of live seeds that is important. The vitality and vigor of seedlings are extremely important adjuncts to a good stand. Seeds stored for several years may still be mostly viable, but because of depleted food reserves will produce weaker seedlings and slower early growth. It is well to remember that the cost of seed is a small part of the total cost of producing a crop.

Conditions for Germination   When seeds germinate, they are actually resuming growth to develop into mature plants. Germination of viable seed is dependent upon favorable environmental factors of temperature, moisture, and oxygen. Cool-season and warm-season types have their own optimal range of temperature for germination. Adverse temperatures will slow down germination and may result in irregular emergence of seedlings. Moisture is essential for seeds to swell, for cells to enlarge, and for food to be translocated within the seeds. The rate of respiration is high in actively growing regions of plants. Consequently, an abundant supply of oxygen is needed for germinating seeds. With a few species, including celery, light also appears to be essential for germination.

GROWING PLANTS FOR TRANSPLANTING

Objectives of transplanting are for earliness, for economy of space, to grow long-season crops in short-season climates, to encourage cool-season crops to mature in the spring before hot weather begins, to produce larger plants for summer production, and, in general, to increase the length of the growing season and ultimately to increase production. Many garden vegetables such as tomatoes, cabbage, peppers, and broccoli are started indoors. Annual flowering plants like petunia, marigold, zinnia, and aster will bloom earlier and longer if started indoors and transplanted to outside locations when danger of frost is past.

Plant Growing Structures   Structures for starting plants from seed for later transplanting may be of a permanent or temporary type. In some cases windows of homes, garages, and basements are converted into temporary hotbeds for starting seedlings.

Cold frames are widely used for growing plants from seed. They are simple and easy to construct, and most types now used are portable or easily dismantled. A cold frame consists of a bed on the ground enclosed by side walls which are usually from 8 to 12 in. (20–30 cm) high. The

FIGURE 6-6   **Annual flowers started in pots for transplanting.**

frame is covered with window sashes of glass or other transparent or translucent material which admits sunlight. The source of heat in a cold frame is solar energy. Temperatures inside the cold frame are regulated by raising or removing the sashes for cooling, and placing straw, mats, or other insulating material over the top for retaining warmth. Since the cold frame is entirely dependent on radiant energy as the source of heat, its season of use is limited primarily to early spring. Two- to four-weeks time can usually be gained by starting vegetable and flower seedlings in this type of structure.

Hotbeds are similar in design to cold frames but differ in that they are provided with a source of heat to supplement solar energy. Earliness of starting plants can be extended even further than with a cold frame. Methods and sources of supplemental heat include fermenting organic matter, hot water, steam, hot air, and electricity. Of the fermenting organic matter type, the manure-heated hotbed was formerly very common. Horse manure is particularly useful because of considerable heat generated during decomposition. Other materials, which will produce enough heat to markedly warm a hotbed, include corn stalks, silage, and wet hay. However, this method of heating has largely been replaced by other types of hotbeds.

Hot water, steam, and hot air are similar methods of heating in that a furnace or heating plant is needed. Pipes or flues are placed beneath the

FIGURE 6-7   A cold frame is useful for hardening, protection from wind, and display, in addition to plant production.

soil of the hotbed and the hot water, steam, or hot gases flow through them. The soil is heated by conduction. A fairly even temperature can be maintained with these types if placed on thermostatic control. One of the problems which may arise is the loss of heat from the intake to the outlet of the pipes or flues. This is usually compensated for by raising their height at the outlet so there is a shallower layer of soil to heat at the far end. These types of hotbeds and the manure-heated type all require a pit or excavation below the soil level for the heating element.

The many advantages of electrically heated hotbeds have made them very popular. A specially coated electric cable is the source of heat. It can be placed on the bottom of an ordinary cold frame, thus eliminating the need for a pit. The layer of soil is placed over the cable and a thermostat located in the soil. Very uniform temperatures can be maintained and after the need for supplemental heat has passed, the cable can be disconnected and the bed handled like a cold frame. It is easy to install and operate and requires less constant attention than the other types of hotbeds. However, if electric rates are extremely high, its use may not be feasible.

FIGURE 6-8   Fiber-glass greenhouse under construction. (Photograph by K. A. Denisen.)

Greenhouses are important permanent-type structures used in propagating plants from seed. They are more costly to construct, maintain, and operate than hotbeds and cold frames. Unless they are used during most of each year for continuous cropping, their use for seedling production in the spring is not justified. High overhead demands that they be in use most of the time. Greenhouses provide an excellent environment for producing vegetable and flower transplants. They are usually equipped with good ventilation, lighting, heating, and watering facilities. When earliness is an important factor, the greenhouse has distinct advantages over the other types of plant-growing structures.

Overhead has been reduced, while many advantages of greenhouses are retained by using modified types. These include the sash greenhouse and the plastic covered greenhouse. The sashhouses use the type of sash found on cold frames and hotbeds. The pitch of the roof is formed by the sashes laid against a center ridge or support. The sashes can be removed and stored during the summer months to prevent breakage of glass while not in use. Most suitable plastics are inexpensive compared to glass and can be installed by tacking onto the wood frame rafters at very low cost. The plastic will weather and ultimately need replacing. Both the sash greenhouse and the plastic type may be heated by space heaters or central heating as supplements to solar energy.

Planting the Seeds   Seeds for transplanting may be sown in flats or pots or directly in the soil of a cold frame, hotbed, or greenhouse bench. If the seedlings are to be potted and grown for increased size before setting out-of-doors, they are usually started in flats. For flat planting, the rows are about 2 in. (5 cm) apart and seeds are planted closely, so that each seed touches adjacent seeds. Since they will be transferred to pots soon after emergence, overcrowding is not a factor. For those sown directly into benches and cold frames, the rows are usually about 6 in. (15 cm) apart. These plants have more space for growth before overcrowding and are usually not potted before transplanting to the field or garden.

Flats will vary in size but are commonly about 15 by 22 in. (37 × 55 cm), outside dimensions. A soil mixture containing about $\frac{1}{3}$ sand and $\frac{2}{3}$ loam soil provides ample aeration and water retention for most vegetable and flower seeds. The soil should be moist but not wet when placed in the flat. It is leveled and compressed slightly so that small marks or furrows can be imprinted on the surface. A marker with 8 to 12 beveled laths can be used to designate the rows, and when pressed down will make furrows for the seeds (Fig. 6-9). For accurate and uniform distribution of seeds into the furrow, a test tube or an open-flap seed packet is very useful. Seeds can be planted at an even rate of flow by tapping the tube or packet with the forefinger and moving the hand horizontally down each row. After sowing, they are covered with soil either by closing over each row or sifting additional soil over the entire flat. The flat is labeled, placed in its proper location, watered, and covered with a paper or cloth to prevent drying out.

When sowing directly into the soil of the greenhouse bench or cold frame, the seeds are planted slightly deeper. This prevents washing out during watering as a coarser spray will usually be used for the larger area. The soil in the bench or frame is usually a loam with no sand added. Since the rows are spaced wider and the plants remain longer before transplanting, their demands for nutrients are greater. Sand improves the aeration of a seed bed but is low in nutrient content. The

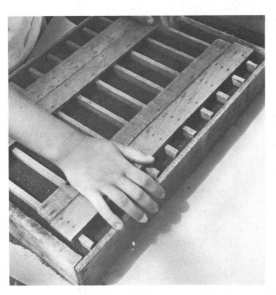

FIGURE 6-9   **Marking the soil and planting seeds in a flat.**

Growing Plants from Seeds

FIGURE 6-10   **Peas on left treated with Captan. Those on the right untreated.**

soil may also need to be weeded between rows unless sterilized prior to planting.

Whether started in flats, benches, or frames, an extremely important factor is controlling seed decay and damping-off organisms. Certain bacteria and fungi in the soil may cause seed decay, root rots, or girdling of the stem at the soil line. The latter condition is known as damping-off because of its occurrence with wet soil, wet plants, and high humidity. Seed treatment with a chemical such as Captan will markedly reduce these diseases. The seed is coated with the dust or slurry of this material. The chemical then acts as a shield or protectant against the invasion of organisms from the soil. Sterilization of the soil with steam or such chemicals as Vapam and methyl bromide will kill weed seeds and damping-off organisms. Most greenhouse operators have pipes and covers for steam sterilizing the soil directly in the benches. For smaller amounts live steam can be directed into a barrel or other tight container in which the soil has been placed. In the home, small quantities of soil can be placed in a water bath on the kitchen range for 15 to 20 minutes with good results. A small potato placed in the soil may be used as an indicator; when it is cooked the soil is sterilized.

Seedling Care   Immediately following germination and emergence of seedings, partial shade is desirable, especially during days of intense sunlight. Watering should be done either with a sprinkling can or a fine spray head on a hose at low pressure. This gives uniform watering without splashing or washing. The forenoon is preferred for time of watering, since it prevents moisture shortage during the warmest part of the day. Also, there is ample time for the foliage and soil surface to evaporate or absorb excess moisture and thus reduce likelihood of damping-off. If damping-off should occur, its spread can often be prevented by applying a mixture of dry sand and Captan (about 1 percent) sprinkled along the rows at the base of the stems. It is a good practice to allow the surface of the soil to become dry before doing additional watering. Sphagnum moss spread thinly over the soil surface is also effective in reducing damping-off. Good ventilation of the greenhouse or cold frame is extremely important to seedlings.

Transplanting When flowers and vegetables are started several weeks before setting in the field or garden, it is usually necessary to transplant them from the seedling row to containers. This permits greater top and root growth; there is less competition between plants. The seedlings may be transplanted to pots, bands, or flats. Types of pots include those made of clay, fiber, paper, peat, and plastic. All are satisfactory for plant growth; however, some of the organic matter types may disintegrate before their desired span of usefulness is over. Roots may penetrate the fiber, paper, and compressed peat types. As they become water-soaked, they will tolerate less handling. When transplanting to flats, plants are usually spaced in a square pattern about 2 in. (5 cm) apart. A spotting board is very useful for marking the location for each plant. There is no division or barrier between plants, and the root systems will become interlaced after a few weeks in the flat. Cutting through the soil with a knife between each row in both directions will confine the roots of each plant to its own area. This "blocking" prunes the roots and branch roots

FIGURE 6-11   Four kinds of pots: (*left to right*) clay, plastic, paper, and compressed peat.

Growing Plants from Seeds   **149**

develop. If done about a week before setting out, the plants will recover and show very little set-back from final transplanting.

Transplanting is a shock to the seedling. Therefore it should be accomplished as soon as possible after the first true leaves appear. The ability of the plant to recover and replace its root system is greater and more rapid at this stage of growth. The soil for transplanting should be of good fertility and high in organic matter. It must provide nutrients for the growing plant and be well aerated and water retentive for root development. Care should be used in handling the seedlings, both in lifting from the flat and transplanting to the containers. A hole is made in the moist soil with a dibble or the forefinger, the seedling is placed in the hole slightly below its former depth, and the soil is compressed lightly around the root and stem. After transplanting to containers the plants should be watered, placed in the bench or frame, and shaded to prevent or reduce wilting. After recovery from wilting, they can tolerate and utilize full light. They are watered more heavily but less frequently to promote good root growth.

When transplanting to the field or garden, it may be desirable to harden the plants as a precaution against low temperature, wind, or drouth. Hardening of the tissues is usually accomplished by reducing the amount of water supplied or lowering the temperature. A cold frame is often used to harden plants that have been growing in a warm greenhouse. Plants in containers should be given a thorough watering before setting out as a precaution against dry weather.

Pot-grown plants are removed from the pots by placing the hand over the soil, with fingers on each side of the stem, and inverting the pot.

FIGURE 6-12 **Broccoli plants spaced in flats. They are blocked into individual squares several days prior to the next transplanting.**

FIGURE 6-13   **Method of removing plant from pot for transplanting.**

With a flick of the wrist, the ball of soil with roots intact will come free. Since the principal advantage of pot-grown plants is in avoiding root injury and disturbance, the ball of soil is placed intact in a trench or hole and soil firmed around it. Pots made of compressed peat can be planted with the ball of soil and will decompose and furnish nutrients to the plant.

Plants which have grown in the seedling row without transplanting to other containers may be transplanted directly to the field. However, hardening is desired with this method since these plants are otherwise apt to be leggy and weak stemmed. The root systems of plants grown in the seedling row will be intertwined and much loss of absorbing area will occur in separating them. For this reason, pruning of some top growth to equalize tops and roots may prove beneficial. With exceptionally leggy plants, it may be desirable to lay the root and lower half of the stem horizontally in a trench, curve the growing point upward, and cover the root and lower stem with soil. With this method wilting and drooping are less likely to occur, and as the plant becomes established roots will arise from the buried portion of the stem.

Some plants are not adapted to the usual methods of transplanting because of slow root replacement or regeneration. Examples include cucumber, melons, squash, pumpkins, and sweet peas. When it is deemed essential that these plants get an earlier start than can be

Growing Plants from Seeds   **151**

FIGURE 6-14   **Cucumbers seeded directly in pots can be transplanted without disturbing the roots.**

obtained by direct sowing out-of-doors, seeds can be planted in pots. After emergence they are thinned to one or two seedlings per pot. Plants so started are then set out when proper environmental conditions prevail. By growing the roots in the same undisturbed volume of soil these hard-to-move plants can be started indoors for early production. Another method of early production for such plants as cucumbers and melons is to sow the seed directly into the soil under portable cold frames or hot caps. As growth progresses and temperatures rise, the cold frames or hot caps are lifted and a considerable advantage in earliness can be realized.

PLANTING SEEDS OUT-OF-DOORS

Many flower and vegetable seeds are planted directly into their permanent location. Important factors in outdoor planting are (1) date of planting, (2) seedbed preparation, (3) seed treatment, (4) rate of planting, (5) depth of planting, (6) spacing for efficiency, and (7) care of seedlings.

Date of Planting   If the grower is interested primarily in earliness of production, then he wants the earliest practical planting date. This date is determined by a crop's thermal requirements, warm-season or cool-season, and average

TABLE 2   **Effect of Date of Planting on Yields of Two Cultivars of Onions at Clear Lake, Iowa**

Date planted	Yield in bushels per acre	
	Early Yellow Globe	Brigham Yellow Globe
May 11	1248	1255
May 17	1084	1055
May 23	922	897
May 31	610	581

date of the last killing frost for a specified area. As the date of planting approaches it may be further hastened or delayed by current weather conditions. Date of planting is also influenced by date or time of maturity, harvest, or bloom and can be governed accordingly. A market gardener wants to supply fresh sweet corn during early summer, midseason, and late summer, so he spreads out planting dates of the same variety accordingly. A homemaker may be planning a garden reception during the summer and wants lots of bloom from annual flowers. One can be reasonably certain of bounteous bloom if date of planting and days to flowering are taken into consideration. Most seed catalogs will state average number of days to bloom, edible stage, maturity, etc. Also, information on cool-season or warm-season characteristics is usually mentioned in seed catalogs and grower's guides. Timeliness of planting may be extremely important for some crops. The onion is a cool-season crop, and unless planted early, yields are reduced sharply as shown by the data in Table 2. Yields were reduced over 50 percent by a 20-day delay in planting. Early growth under cool conditions gave the earlier planted onions a great advantage. Warm-season crops, conversely, may react favorably to delayed planting especially under conditions of a late spring or unusually cold weather.

Seedbed Preparation

A good seedbed will provide optimum conditions of moisture and aeration for seed germination and seedling growth. There must be intimate contact of soil and seed for moisture absorption by the embryo and endosperm. At the same time oxygen is also required in the soil for the release of energy during germination and root growth. Soil texture is a governing factor of moisture and aeration in the seedbed. Many flower and vegetable seeds are extremely small. These small-seeded types require a finer textured seedbed in order to obtain the intimate contact of soil and seed. Larger-seeded types will actually respond favorably to a coarser textured seedbed. A soil which is ideally prepared for planting is said to be in good *tilth*. In preparing the soil for seeds and striving for good tilth it is important to avoid compaction. Using heavy equipment frequently on the same area and working the soil when wet are apt to result in compacted soil which limits air and water movement.

Seed Treatment

The same materials are used for treatment of seeds planted out-of-doors as in plant-growing structures. Seed decay and damping-off organisms are present in most soils and may cause serious reductions in stand. Seed treatment is considered low-cost insurance against losses resulting from poor stands and unoccupied crop land. Small quantities of seed can be treated with the fungicide dusts in the packet, a paper bag, or a fruit jar. Larger quantities may be treated in barrels, drums, or mills designed specifically for dust or slurry applications.

The pelleting of seeds is sometimes done by seed processors. It involves the incorporation of a seed treatment fungicide and a small quantity of fertilizer, around each seed. Pelleted seeds are larger which makes sowing and handling of small seeds easier and more accurate. However, it increases the cost of seeds, may reduce the rate of water absorption, and may delay germination. It has not become a widely used practice for most vegetable and flower seeds.

Rate of Planting

If seeds are planted too sparsely, the land is not fully utilized. If planted too heavily, crowding results, and the plants will need to be thinned or there will be reduced size, quality, or yield. Numerous types of mechanical planters have been developed which are generally quite accurate. Other than mechanical failure, principal errors in rate of planting are due to variation in seed size (ungraded seed), poor germinating seeds, and lack of proper adjustment of openings or plates. A trial run of a power-driven or hand planter will usually be worth the effort. The number of seeds per foot or meter should be determined and proper adjustment made before proceeding with the planting. Thinning is costly of labor, thus every effort is generally made to plant at the rate which will give the desired stand. In some crops, such as melons and squash, where each plant occupies a large area, it is usually advisable to plant more seeds than needed. Thinning is less costly in these crops because of wider spacing, and the use of more seed is a precaution against large vacant areas.

Depth of Planting

In outdoor planting, seeds are sown deeper than in the greenhouse or cold frame. This tends to avoid difficulties arising from environmental adversities. Washing or floating of seeds to the surface following rain or sprinkler irrigation and blowing out of seeds during wind erosion may occur with shallow planting. Under conditions of good soil moisture most seeds are planted at depths equal to about three or four times their greatest thickness. This is only a general rule, however, and those seeds which remain in the soil and send up shoots, like sweet corn and peas, may be planted slightly deeper than the kinds that push up the seed leaves or cotyledons like beans, marigolds, and watermelons. If dry weather prevails at planting time and the soil is dry near the surface, seedling emergence can sometimes be hastened by planting at a slightly deeper level where moisture is more abundant. Seeds may be planted deeper in light, sandy soils than in heavy, clay loam soils.

Spacing for Efficiency   In a garden where hand labor is used in planting and cultivation, rows of the various kinds of plants may be spaced according to their needs. However, if machine cultivation is used in the garden, it may be well to adjust most of the vegetables and flowers to the same width between rows. Many farm gardens are planned in this manner. In large commercial planting, distance between rows is quite specific for each crop. Sweet corn is planted in rows 30 to 40 in. (75–100 cm) apart; onions and carrots, 12 to 18 in. (30–40 cm); snap beans, 18 to 24 in. (45–60 cm); etc. Multi-row cultivators require seeding by multi-row planters of the same size. Width of tractor wheels, length of sprayer booms, and manner and type of harvesting are other considerations when spacing for efficiency.

Care of Seedlings   After seedlings have emerged in outdoor plantings, their primary needs are water, nutrients, and light. The extent to which these environmental factors are available to the young plants is greatly influenced by competition from weeds. Thus, weed control is of considerable importance in seedling care. Cultivation and chemical weed killers play an important role in this regard. Drainage, irrigation, fertilization, adverse temperatures, and pests present problems for young seedlings. These shall be discussed later in connection with the similar problems of more mature plants and the asexually propagated plants.

SPECIAL PRACTICES IN SEEDAGE

Stratification   The seeds of most woody plants have a rest or after-ripening period. For hastening germination and increasing uniformity of emergence, a cold treatment process known as stratification is employed. It consists of placing the seeds in a moist medium of sand, peat, or loam soil and holding at temperatures slightly above freezing. Time required will vary from one to six months depending on species.

Stratification is actually a simulation of Nature's method of breaking the rest period in seeds. Alternate freezing and thawing of the soil with continual moist conditions found beneath fallen leaves and other organic matter are favorable conditions for producing seedlings of native plants. Plant breeders of woody plants, nurserymen growing trees from seed, and plant propagators growing seedlings for root stocks use stratification as a tool for producing large numbers of seedlings.

Scarifying Seeds   Scarification is breaking, scarring, or injuring the seed coat to promote the absorption of water and aid germination. Seeds of the peach, lupine, and almond will germinate sooner and more uniformly if scarified. Methods of scarifying include the use of abrasives and acids for small seeds and hammers or vises for large hard-shelled types.

Embryo Culture   Plant breeders have utilized the technique of excising the embryo from the seed and placing it on nutrient media to obtain seedlings from otherwise non-germinable seeds. When the developing embryo is taken from immature seeds and placed on suitable culture media, it by-passes the rest period, drying, and after-ripening. It should be transplanted to soil as soon as possible so the roots will become adapted for water and mineral absorption.

STUDY QUESTIONS

1. Why is propagation by seeds called sexual propagation?
2. What is a cultivar? Why has the use of this term replaced the term "variety"?
3. What are the characteristics of good seed? Why is each important to the gardener?
4. Distinguish between a hot-bed and a cold frame. When is each type used in seed propagation?
5. Polyhouses are widely used in many parts of the world. What are specific advantages over glass houses? Disadvantages?
6. Peat pots are used very widely for bedding plants and vegetables. Why have they replaced clay pots or spaced planting in flats?
7. Why are seeds of some warm-season plants seeded directly into fields or gardens before danger of frost is over? Is it worth the gamble?
8. Why are seeds planted out-of-doors usually sown deeper than the same kind planted in a greenhouse?
9. What is "after-ripening" of seeds?
10. Explain the technique of embryo culture.
11. Why are lath-houses used for transplants?
12. How can one reduce the amount of wilting of transplants?
13. What is wilting?
14. Does photosynthesis proceed in a wilted plant?

SELECTED REFERENCES

1. Copeland, L. O. 1976. *Principles of Seed Science and Technology.* Burgess Publishing Co., Minneapolis, Min.
2. Denisen, E. L., and H. E. Nichols, 1962. *Laboratory Manual in Horticulture*, 4th ed. Ex. 3, "Growing Plants from Seeds." Iowa State University Press, Ames, Iowa.
3. Harrington, J. F. 1961. The value of moisture-resistant containers in vegetable seed packaging. *Calif. Agr. Exp. Sta. Bul. 792.*
4. Hartman, H. T., and D. E. Kester, 1975. *Plant Propagation*, 3rd ed. Prentice-Hall, Englewood Cliffs, N.J. A very complete reference source.
5. Mahlstede. J. P., and E. S. Haber. 1957. *Plant Propagation.* Wiley, New York.
6. U.S. Department of Agriculture Yearbook, 1961. *Seeds.*

7 Growing Plants by Layerage, Cuttage, and Specialized Structures

VEGETATIVE REPRODUCTION

The regeneration of plants merely from structural plant parts has been practiced for centuries. Yet we still marvel at the ability of many plants to be propagated from root, stem, or leaf cuttings or by means of specialized plant structures and techniques. There are several reasons why asexual propagation of plants is used regularly for some species instead of seedage. (1) Some plants do not produce seeds. Notable among this group are the banana, seedless grape, seedless orange, and seedless grapefruit. It would be impossible to perpetuate these plants were it not for vegetative propagation methods. (2) Some plants do not come "true" from seeds. This is characteristic of most varieties of tree fruits, small fruits, and many herbaceous and woody ornamental plants. These plants are heterozygous, that is, they have such a mixture of heritable factors in their chromosomes that they show great variation of type when grown from seed. Consequently, by maintaining their original type through asexual propagation, their cultivar characteristics remain the same. It is a method of maintaining cultivars. (3) Asexual propagation may be faster than seed propagation. This is especially true for many extremely small-seeded types of perennial plants. In addition to the fact that they do not breed true, the strawberry, geranium, and potato produce very small seeds which on germination produce small seedlings. Growth and production of these seedlings is slow compared to plants produced by runners of the strawberry, stem cuttings of the geranium, and cut tubers of the potato. (4) It may also be more economical to produce plants asexually, or (5) in some cases easier than by seed propagation.

Environmental Conditions   As with seedage, vegetatively propagated plants require optimum conditions of temperature, oxygen, and moisture. In addition, they require a type of medium favorable to good root development and water absorption. Unlike seeds which house miniature plants, vegetative plant parts used in propagation represent more mature growth. They must either develop roots or shoots or both to form a complete plant progeny. Adapting the environment to favor development of these essential organs is of primary concern to the plant propagator. Temperature is important for the occurrence of normal plant processes. The rate of respiration is high in regions of high meristematic activity. Since root and shoot development are vital to propagation and require cell

FIGURE 7-1

A propagation pot contains a rooting medium and a moisture supply. Water seeping through the porous clay pot in the center provides the necessary moisture for rooting.

division, oxygen demands are great. Likewise, moisture is vital as needed in transpiration, cell enlargement, and photosynthesis. The propagating media, whether it be sand, soil, peat, or some other material, has an important influence on temperature, moisture, and oxygen. Compaction of the media, its porosity, and retentiveness of nutrients and moisture all influence the environmental conditions for vegetative propagation.

Internal Conditions
Food supply, rest period, and hormone balance are important internal criteria for propagating plants asexually. In addition to plant parts having favorable environment to produce new plants, they must be in the proper stage of maturity or development to become reactivated for the meristematic processes of root and shoot production. These factors are much more critical for some species than for others.

LAYERAGE

When plant parts to be rooted remain attached to the parent while rooting, the process is called layerage. It is in effect like a young man's getting financial help from home until he becomes established in his own business or profession. The plant progeny is dependent on the parent plant until it has its own root system and can carry on its own life process.

Tip Layerage   Certain plants, such as the black raspberry, have the ability to produce both roots and shoots from the stem tip when placed in the proper environment at a specific stage of maturity. The current season's growth of these plants reaches a stage in late summer, which is characterized by long internodes and small leaves at the tip. When this tip is placed in the soil at a depth of 1 to 3 in. (2–5 cm), it is highly meristematic. Heavy fibrous root development occurs and a vigorous shoot is formed which usually emerges from the soil before dormancy.

The following spring, the "tip" is dug and the new plant is severed from its parent by cutting the cane several inches from the rooted tip. It is transplanted to its new location, set at about the same depth as previously, and the remaining stub of old cane, which has been used as a handle, is removed.

Simple
Layerage   Many woody shrubs and herbaceous plants can be propagated by placing their branches in continuous contact with moist soil. This process known as simple layerage can be more effectively accomplished if the underside of a lower branch is scarred, scratched, or otherwise

FIGURE 7-2   **Tip layering the black raspberry. The shoot tip is placed in opening made by spade. Soil is compressed against tip.**

Growing Plants by Layerage, Cuttage, and Specialized Structures   **159**

wounded to encourage the development of wound tissue. The formation of wound tissue for healing is a meristematic process. As cells become activated for meristematic activity, the moist medium of soil encourages root development. After several weeks or months of growth the branch can be severed. It is then a plant on its own roots.

Simple layerage occurs naturally with many spreading, prostrate forms of plants. Tomatoes often form adventitious roots along the undersides of branches when they have prolonged contact of moist soil. Dogwood (*Cornus* sp.) and certain other shrubs will also react in a similar manner.

The principal difference between simple layerage and tip layerage is that the tip must not be covered in simple layerage. Roots are formed along the stem rather than from meristematic areas of the tip as in a tip-layered plant. Shoot growth proceeds from the exposed tip and forms the stem of the new plant.

Trench Layerage A length of cane or branch is placed in a shallow trench, the underside is scarred, and it is covered lightly and gradually with soil. In principle, each bud can give rise to a new plant. As the buds break and shoots emerge through the soil, more soil can be placed over the trench-layered stem. Roots may arise either from the scarred stem or from the basal area of the new shoots which arise from the buds. The tip of the branch should be left exposed so the flow of water, nutrients, and manufactured food will continue through the stem.

The main advantage of trench layerage is the potentially large number of new plants that can be propagated from a single branch or cane. Not all plants will respond favorably to trench layerage, however. The buds of many plants will not break if buried. It can be used successfully with the grape, blackberry, willow, viburnum, and dogwood.

A modification of trench layerage known as *serpentine layerage* employs the technique of placing alternate buds above and below the soil. The underside of the stem opposite the buried bud is cut slightly to promote rooting. Shoots develop from the alternate buds above ground. The method is especially well adapted for propagating most vines.

Mound Layerage Ornamental shrubs and perennial-canned bush fruits such as goose-berries and currants are frequently propagated by mound layerage. A dense compact shrub can produce as many new plants as it has branches or stems. Wounding or scarring the basal portion of each stem, as with simple and trench layerage, will hasten rooting. The shrub is mounded with soil in the spring and is left undisturbed until fall or the following spring. Roots develop at the base of each stem, these stems are severed below the new roots, and new plants are then ready for transplanting.

Young stems root more readily than old ones. By severe pruning of shrubs or other plants the season previous to mounding, shoot growth is increased and more young wood is available at mound level for rooting. This technique called "stooling," is used for producing root stocks for fruit trees. Young trees of the desired root stock are pruned back to

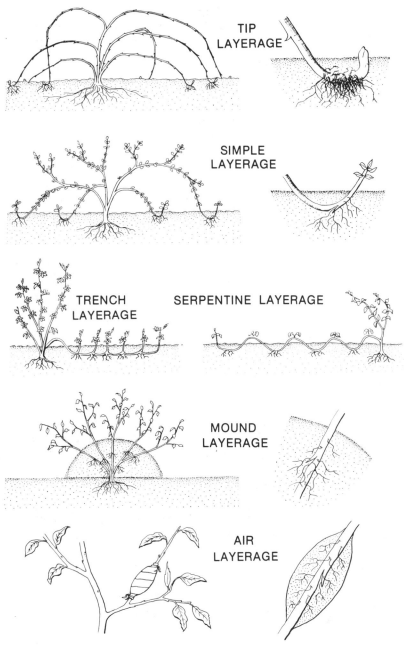

FIGURE 7-3   **Methods of layerage.**

within a few inches of the ground. Rapid growth of lateral branches near soil level follows, and the shoots are mounded. At the end of the season, these shoots are well rooted and they may be dug at that time for grafting during the winter months. Severe pruning of shrubs to be propagated by mound layerage will also result in more vigorous shoot growth and more rapid rooting.

Air Layerage Some plants are very difficult to propagate by the usual sexual or asexual methods. Air layerage has proved to be effective in a wide range of species as a means of increasing their kind. It is a method which was originated by the Chinese several centuries ago. Basically it consists of producing a plant from aerial branches which remain in position while rooting. The moist medium for rooting is usually sphagnum moss which is placed around the wounded area of the branch. It is held in position by placing a wrapping around the periphery of the ball of sphagnum. Polyethylene film is one of the most useful types of material for this, since it eliminates the need for watering.

Young branches $\frac{1}{4}$ to $\frac{3}{4}$ in. ($\frac{1}{2}$–2 cm) in diameter are most desirable for air layerage. They should be growing vigorously and have large leaf areas so they can supply carbohydrates to the newly developing roots. The scar or wound is produced by making a slanted cut into the branch or by removing a section of bark. It is essential that the cut remain open and not be permitted to heal. This procedure severs the phloem and tends to trap the manufactured food in the branch near the wound resulting in more rapid root development. After roots have grown to the outside of the ball of sphagnum moss, the branch may be removed and planted in a pot on its own roots.

Greenhouse, conservatory, and house plants such as rubber plant, fig, azalea, and croton are often propagated by air layerage.

CUTTAGE

The term cuttage implies the severing or cutting of plant parts prior to rooting. Just as the comparison was made between layerage in plants and help from home for progeny getting established in business, a similar analogy applies to cuttage. The plant parts—stems, roots, and leaves—are severed from the parent plant and forced to adapt themselves to the change much like progeny with no inheritance and no possibility of getting outside help. Both are left to their own resources and must establish their "roots" while carrying on their life processes.

Types of Media Cuttings have varying amounts of stored food on which to rely while developing roots. The nutrients provided by the media or material used for anchorage are not of great consequence unless the plant remains in the medium for a long period beyond root establishment. However, the types of media do have an effect on aeration, water holding capacity, temperature, and disease organisms and thus influence the rate, amount, and type of rooting.

Many a plant in the home has gotten its start from a "slip" or cutting placed in a cup of *water* until roots formed. In this instance, water is the medium for rooting. Although widely used in amateur horticulture, it is not generally practiced for commercial use. Aeration is very limited and a very succulent type of root growth results. If not transplanted soon after roots are formed, the plants have difficulty in adjusting their root system to soil conditions.

FIGURE 7-4   Propagation of jade plant in water.

Sand is the most widely used media for propagating plants from cuttings. It is well aerated, thus providing for extensive root growth. Its cost is generally low and it is usually available in adequate quantities. Sand is low in water-holding ability, however, and must be watered frequently. Since it contains little or no organic matter it does not harbor most disease organisms. It is low in nutrients, so plants should be transplanted soon after rooting.

Peat is sometimes used for propagating cuttings. It is very high in organic matter, and consequently has a high water-holding capacity and good aeration. It is high in nitrogen. Peat is usually more expensive than sand. Acid peat is especially useful in propagating cuttings of blueberry, azalea, and rhododendron.

Vermiculite or expanded mica is used considerably as a medium for rooting cuttings. Since it is inorganic, it does not harbor disease organisms. It has good water retaining properties as well as good aeration. It supplies little if any nutrients to growing plants. Although it is generally more expensive than other types of media, it is very popular among many plant propagators because of its clean handling properties and its effectiveness in rooting cuttings.

Volcanic ash (e.g., Perlite) is also an inorganic material and is light in weight like vermiculite. It is white in color, clean, and has good aeration and water-holding properties. In most areas it is slightly more expensive than vermiculite.

Loam soil is not to be forgotten as a propagating medium for cuttings. It can be used successfully, especially if the cuttings are to remain in the propagating media for some time after rooting. Mixtures of loam and sand are frequently used to combine some of the advantages of each. Loam is higher in nutrients and more retentive of water. It is not as well aerated and requires a longer period for rooting than sand. Sterilizing is desirable for avoiding disease.

Herbaceous Cuttings

Softwood cuttings and greenwood cuttings are other terms also given to this group. These cuttings are sections of either young or mature stems of herbaceous plants or the soft current season's growth of woody plants. Many commonly grown potted plants such as geranium, coleus, artillery plant, and irisene are propagated by herbaceous cuttings.

Although age of stem influences the rate of rooting to some degree, most plants propagated by herbaceous cuttings will root regardless of age. The very tip stem of growing plants may be so succulent and have such a minimum of stored food that rooting may be delayed with a resulting small root system because of water or food deficit. One must keep in mind that a herbaceous cutting is a severed portion of the stem with leaves. While rooting it must absorb moisture to meet the needs of transpiration, cell turgidity, and to a limited degree, photosynthesis. Older and harder stems will generally root also, but not as readily as stem tissue intermediate between the tip and basal portions of the plant. This is an area of carbohydrate reserve which is without old age, without overmatured tissues, and without reduced meristematic activity.

Size of cutting will vary with species. Geranium cuttings range from 3 to 6 in. (7–15 cm) in length for commercial growing (Fig. 7–5, page 165). When the tip section of the stem is used, the cutting should be nearer 6 in. (15 cm) for greater carbohydrate concentration at the basal part of the cutting. Midsections can be shorter since there is less succulent tissue and more food reserve. Cuttings should be at least two nodes in length so one is in the rooting medium and the other can give rise to shoot growth.

The basal cut should be at an angle, rather than perpendicular, to the central axis of the stem. This makes it easier to cut, easier to "stick" into the media, and decreases the likelihood of damage to the xylem vessels in the stem. Making the cut immediately below a node will generally result in a greater concentration of roots at the more basal part of the cutting. This is especially true of coleus and irisene, which have long internodes.

The leaf area of a herbaceous cutting is generally reduced by about one third to one half. This permits closer planting in the bench or propagating structure, reduces water loss from transpiration, and makes handling of the cuttings easier. In trimming cuttings, the lower leaves are removed entirely and outer portions of upper leaves may be snipped.

Cuttings are stuck fairly deep, 1 to 2 in. (2–5 cm), to reduce the danger of drying and to give support. They are placed as close together in the row as leaves will permit. Rows are usually about 4 to 6 in.

FIGURE 7-5   **Geranium. *Left*, herbaceous stem cutting. *Right*, leaf-bud cutting.**

(10–15 cm) apart to accommodate the developing root systems without serious intertwining.

Subsequent care of cuttings after sticking includes shading to reduce light intensity, heat, and transpiration. Frequency of watering is dependent on type of medium and other environmental factors. It is essential that the medium be kept moist as the root primordia are extremely succulent and should be favored with an adequate supply of water in their early development.

Transplanting of rooted cuttings can be done as soon as a ball of medium clings to the root system when dug. Rooted cuttings should be planted in a good fertile soil and kept well watered. When the root system has become firmly established in good soil, thrifty top growth will develop.

Evergreen Cuttings   Both narrowleafed and broadleafed evergreen plants are often grown from cuttings. The primary reason for this is the much more uniform progeny than is obtained from seeds. Cuttings from the conifers consist of matured current season's growth with the cut made in one-year-old wood to form a "heel" for the cutting. The piece of one-year-old wood or heel has an active cambium which produces more callus or wound tissue than the new wood. Although callusing is not considered essential for rooting, it usually results in plants with sturdier root systems.

Broadleafed evergreen cuttings from plants such as holly, azalea, and gardenia are taken from the tip sections of current season's growth with a heel from older wood as with conifers. They are then trimmed of lower leaves in much the same manner as herbaceous cuttings.

Evergreen cuttings are stuck in sand or other media as are herbaceous cuttings. They are stuck at a slanting position approximately 30° from vertical. This permits easier observation, reduces drying, and increases the area for rooting. Hormone dips or powders are sometimes used to increase the rate or amount of rooting. However, these preparations do not always give positive results and may be quite exacting in their use.

The roots that develop on evergreen cuttings are usually few in number compared to herbaceous cuttings. They consist of strong laterals that penetrate to considerable depth or have a broad expanse. Extreme care should be taken in transplanting to extract all the roots and avoid drying of root tips and root hairs. Transplanting should be done soon after roots have formed and before they become intertwined with those of adjacent plants. It is a good practice to transplant to pots before setting out in the open. This makes it easier to water, shade, and protect them until they reach larger size. They may then be set in their permanent location by resetting the ball of soil and roots with very little disturbance to the plant. Root replacement by evergreens is slow, and transpiration is continuous, so loss of roots and reduced absorbing area can be serious. Evergreens, whether transplanted from pots or from a nursery row, should have their roots encased in a ball of soil.

FIGURE 7-6   **Rooted pine cutting. Note broad expanse of roots. (Courtesy Iowa State University, Ames, Iowa.)**

Woody Cuttings   Woody or hardwood cuttings are taken from the mature tissues of woody plants, generally one-year-old wood. The twigs or canes used for cuttings are dormant when collected. If they have not completed their rest period they are stored in a moist medium, such as sand, at temperatures just above freezing for a few weeks to two or three months. If the cuttings are prepared for sticking before being stored, they will form callus during storage and may root more rapidly when placed in the propagating medium.

Cuttings are usually 5 to 7 in. (12–19 cm) in length and $\frac{3}{16}$ to $\frac{3}{8}$ in. ($\frac{1}{2}$–1 cm) in diameter. Thicker diameter cuttings of the same age wood result in larger roots and more extensive shoot growth. The tips of branches or canes are younger, have less stored food, and shorter internodes. They root less readily, have smaller root systems, and may have more buds attempting to produce shoots. Plants propagated from woody cuttings should be transplanted before foliage growth becomes heavy to avoid severe wilting.

Propagating woody plants from hardwood cuttings is not always feasible. Many trees such as walnut, apple, peach, elm, oak, and ginkgo are difficult to propagate in this manner. Other trees such as willow and poplar can be propagated quite readily from woody cuttings. Many commonly grown shrubs like honeysuckle, barberry, cotoneaster, spirea, and mock orange are propagated by this method. Grape, ivy, and wisteria vines also may be started from cuttings.

Leaf-bud Cuttings   With the stem cutting, shoots arise from aerial buds, and roots are formed from the lower regions of the stem. With the leaf-bud cutting, the bud itself is the source of shoot growth and roots arise from the small segment of stem tissue that accompanies the bud. More plants can be propagated from a herbaceous stem by this method than by stem cuttings. The plants produced are smaller for a given period of time, however, since less stored food is available for root and shoot production.

The leaf-bud cutting is prepared by making a cut into the stem below the base of the petiole, cutting under the bud, and finishing the cut above the bud. The leaf and petiole are left intact to continue transpiration, which promotes absorption through the cut surface. Photosynthesis also can be continued to provide food, since very little stored food accompanies this type of cutting.

Geranium, philodendron, German ivy, peperomia, coleus, and many other house plants can be propagated from leaf-bud cuttings. The depth of placement in the media which ranges from $\frac{1}{2}$ to $1\frac{1}{2}$ in. (1–3 cm) is determined to considerable degree by length of petiole. If planted shallow, watering should be frequent. Following rooting and emergence of the shoot, the new plant should be transplanted to a pot or its permanent location. The old leaf and petiole can be left until they disintegrate or may be removed since the plant is now on its own root and shoot. A very uniform and well-proportioned plant can be obtained from this type of cutting since no old, cut pieces of stem are visible. Some woody plants can be propagated in this manner from current season's growth. In this instance the bud actually by-passes its rest period.

Leaf Cuttings

In the leaf-bud cutting, a living bud is essential for regenerating the plant. With leaf cuttings, only the leaf blade or a portion thereof is needed to produce a new plant. Propagation by leaf cuttings is thus limited to those species which have regenerative tissue in their leaves. These species vary widely in their manner of shoot and root development from leaf cuttings. The bryophyllum leaf has tiny vegetative embryos at the indentation of each serration. Severing the leaf from its stem is the only stimulation required for the embryos to start growth and produce roots and shoots. When placed on sand or other propagating media, the tiny plants emerging from the edge of the leaf will take root and grow rapidly. The thick leaf of the bryophyllum has an abundant reserve of both food and water. Even if suspended in the air by a string, the bryophyllum leaf will produce young plants with both leaves and roots. They will continue to grow until the food and water of the parent leaf is exhausted. Some bryophyllum (*Kalanchoe*) leaves produce new plants in the serrations of the leaf margins while still attached. New plantlets fall to the ground and root.

Sansevieria, or snake plant, may be propagated by cutting its long leaves horizontally into sections and placing each section base downward in the propagating medium. Roots will form at the base of each

FIGURE 7-7   **Bryophyllum producing plantlets at serrations of leaf.**

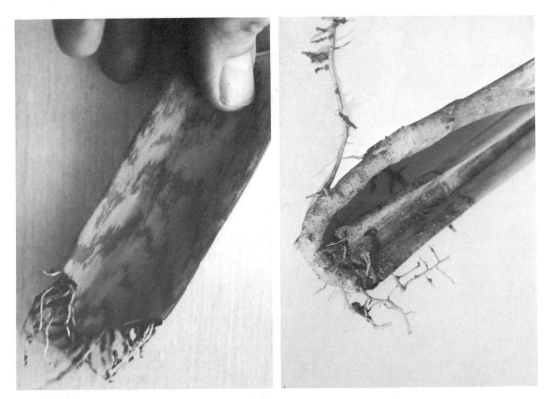

FIGURE 7-8   **Sansevieria leaf cuttings. *Left*, roots at basal end. *Right*, shoot arising at juncture of leaf and roots.**

section. A shoot is formed later near the mid-rib of the leaf, arising at the juncture of leaf and roots.

The jade plant, echeveria (hen and chickens), and African violet are propagated from leaf blades placed upright in the propagating medium. The Rex begonia leaf placed flat on moist sand, with the leaf veins severed near the petiole, will produce new plants at the severed veins.

These examples of leaf cuttings are cited to show the specificity of plants for this type of propagation and to point out the need of special techniques for obtaining progeny of hard-to-propagate plants.

Root Cuttings   Some species of plants can be propagated from roots or sections of roots. The phlox and red raspberry are typical of the type propagated from root sections. The larger, more vigorous portions of the root are used and are cut into 3 to 6 in. (7–15 cm) sections. These are placed horizontally in a trench and covered with soil to a depth of 1 to 2 in. ($2\frac{1}{2}$–5 cm). Shoots arise from reactivated latent buds formed prior to sectioning. In some cases these buds grow without sectioning of the roots and appear as "suckers" near the parent plant. Rather than waiting for sucker development, the number of plants produced from a root of either the raspberry or phlox can be greatly increased by digging the roots and making root cuttings.

FIGURE 7-9   **Rooted African violet leaf cutting with shoots arising at base of petiole.**

The sweet potato root is a storage organ and it is also capable of producing new plants. The whole root is placed in moist sand at a warm temperature, and shots or "slips" are produced at the proximal end of the root. The shoots grow up through the sand and develop roots at the basal end. The entire slip with its basal roots is snapped or cut from the parent root in preparation for transplanting. Following removal of the slips, another series of slips is produced from the parent root. This procedure can continue until the food reserve of the parent root is exhausted. However, this is not usually done because of the wide range in maturity of the crop from early- and late-formed slips.

Special Cuttage Practices   Bottom heat in a greenhouse bench or propagating frame will often stimulate more rapid and extensive root growth. The heat should be confined to the bottom side of the bench, however, as increased air temperature will tend to produce more top growth. In rooting cuttings, good root growth is of foremost importance and top growth should be secondary. After rooting and transplanting, top growth should proceed at a rate equal to root growth. Supplying bottom heat to cuttings involves special equipment; consequently it is not universally used for propagation.

FIGURE 7-10   **Propagation of sweet potato. "Slip" at right has been snapped from potato for transplanting.**

The use of fluorescent lights is another technique to get more rapid and extensive rooting of cuttings. Some species, notably evergreens, respond to this treatment, presumably because of added light for food manufacture. In addition, the light is detrimental to certain fungus diseases which may be present.

Mist propagation is a recent development for increasing the rooting percent of cuttings. Some of the hazards in propagating from cuttings are drying out of the medium and low humidity. These are eliminated by using either a constant or an intermittent mist over the propagating bed or bench. Full light is allowed to reach the beds and since light is detrimental to bacteria and fungi, disease is not generally a problem as might be expected under the extremely high humidity. The mist is produced by fine nozzles located either among or above the cuttings. Neither the constant or intermittent mist requires excessive use of water. The water droplets are very fine and remain suspended in air briefly before settling on the cuttings and propagating media. The intermittent mist may be electrically controlled to operate at regular intervals, or it may be controlled by a solenoid diaphragm which responds to humidity changes.

Growing Plants by Layerage, Cuttage, and Specialized Structures   **171**

SPECIALIZED PLANT STRUCTURES AND ORGANS

Layerage and cuttage methods utilize plant parts whose primary function is not propagation. The stem, leaves, and roots are actually encouraged or forced to produce new plants when used for cuttage and layerage. Many species of plants are provided with a natural means of vegetative propagation through adapted or specialized structures and organs. In addition to propagation, some specialized parts may also serve as storage organs. They possess the ability to produce complete plants and many make provision for carryover from one generation to another much as seeds do.

Stolons or Runners A stolon is a horizontal stem, located above ground, which produces plants with roots and shoots at the nodes. A well-known example is the strawberry which is propagated by runners. The strawberry runner is a stolon which has very long internodes. The strawberry may produce plants as much as 12 to 18 in. (30–45 cm) apart in a runner series. Any time after a runner plant becomes rooted, it can be severed from the mother plant. However, this interrupts the development of a long runner series, since the mother plant may be furnishing food for several developing plants through the channel of one runner. After plants are

FIGURE 7-11 **A runner series of the strawberry.**

established, runners eventually disintegrate. Strawberries which tend to be runnerless are more inclined to develop branched or multiple crowns. These crowns can be separated or divided to form additional plants.

Some grasses, notably bent and Bermuda, produce stolons which are responsible for rather rapid spread and a tight turf. Strains of bent grasses are propagated by plugs consisting of soil and grass from turf nurseries. These plugs, usually about 2 in. (5 cm) in diameter, are inserted at various points in a lawn or golf green, and spread is accomplished by the stolons.

Some house plants, *Episcia* (flame violet) and airplane plant, produce stolons which not only are useful in propagation, but also add interest and beauty.

Rhizomes

These horizontal underground stems differ from stolons primarily in being located below the soil surface. Although considered serious weeds, quackgrass (*Agropyron repens*), Johnson grass, and field bindweed (*Convolvus arvensis*), are some of the best examples of rhizome-propagated plants. Horticultural examples include the iris, sansevieria, and lily-of-the-valley. These and other rhizome types are encouraged to produce an abundance of rhizomes which are dug,

FIGURE 7-12   **A Sansevieria plant propagating itself by rhizomes.**

Growing Plants by Layerage, Cuttage, and Specialized Structures   **173**

FIGURE 7-13   **Iris is propagated by rhizomes. (Photograph courtesy of** *Better Homes &* *Gardens* **magazine.)**

separated, and replanted in their new locations. Rhizome production can be increased by flower stalk removal, cultivation, and fertilization.

The nodes of rhizomes are capable of producing both roots and shoots. The internodal areas are smooth and unbranched which distinguishes them from roots. A reserve of stored food is located in the rhizome to help the new plant get a good start.

Tubers   As modified under ground stems, tubers are similar to rhizomes. They differ in that shoots, but not roots, arise from the nodes. Roots in turn arise from the basal part of each shoot. The white potato is a typical example of a tuber. It is an enlarged and foreshortened stem. The nodes are identified by the buds, or "eyes," which may each have several growing points. The potato tuber is planted either whole or divided into two or more parts, depending on size. The usual size "seed piece" weighs 1 to 2 ounces (28–57 grams). It must contain at least one eye. The food for early growth of the shoot comes from the tuber which

FIGURE 7-14   Potato tuber cut into "seed pieces."

continues to provide food even after the potato plant has its own root system and is carrying on photosynthesis. Excised eyes from potatoes are sometimes planted instead of sections of tubers. They are cheaper to ship but do not get the resulting plant off to as early or vigorous a start as larger seed pieces.

Tubers have a rest period and are adapted to storage from one growing season to the next, like true seeds.

Bulbs   Most members of the lily family are propagated by bulbs. The basal stem plate to which the layers of fleshy leaves are attached is the area of union between leaves and roots. Tiny bulbs, called *bulblets*, are formed around the base of the mother bulb at the periphery of the stem. As a propagation practice, these bulblets are removed, planted, and grown to larger size before they are set out for blooming. For maximum production of bulblets, the mother bulbs are planted shallow, and seed heads are removed after flowering.

Modifications of bulbs are found in the multiplier onion, tree onion, and garlic. Actually, the multiplier onion and garlic produce bulblets which increase in size without replanting. The onion sets and garlic cloves are separated for propagation, and each produces another clump of bulbs or cloves. Most onions are propagated from seed. The tree onion produces bulblets on a seed stalk instead of seed, or in addition to seed.

Bulbs, like tubers, will remain in a dormant state under good storage conditions.

Growing Plants by Layerage, Cuttage, and Specialized Structures   **175**

FIGURE 7-15   Bulbs and corms. *Left:* Note concentric fleshy leaves of tulip bulb. *Right:* Note solid stem structure of gladiolus corm. (Both photographs courtesy of *Better Homes & Gardens* magazine.)

Corms   These structures are similar to bulbs in appearance but are enlarged modified stems rather than compacted fleshy leaves. The gladiolus and crocus are typical corms. The tiny corms formed around the basal periphery of the mother corm are called cormels. As with bulblets, cormels are separated from their parent corms at harvest time and are planted for increase in size. After a season's growth they are dug and stored for future planting. This type of corm is known as a "high crown" corm because of its fleshy compact upper surface. Gladioli that have flowered produce new corms above the original, but these have sunken areas on the top surface and are called low crown corms. Either type is satisfactory for flower production, but high crown corms are in greater demand, probably because of their plumper appearance.

Rootstocks   Many plants of diverse types are propagated by dividing what are variously called crowns, fasiculated roots, or rootstocks. Examples of these plants are found in the dahlia, rhubarb, horseradish, asparagus, and chrysanthemum. When dividing or sectioning these plants for increase, it is essential that a bud or "eye" be present on each piece of rootstock. Because of the fleshy nature of the roots, they will often stand considerable rough handling. For this reason many plants grown from rootstocks have become widely disseminated.

Suckers   Plants arising from roots horizontal to the soil surface are called suckers. This is sometimes an undesirable habit of otherwise desirable ornamental trees, plums, apple rootstocks, other fruit trees, and shrubs.

Skills and Practices in Horticulture   **176**

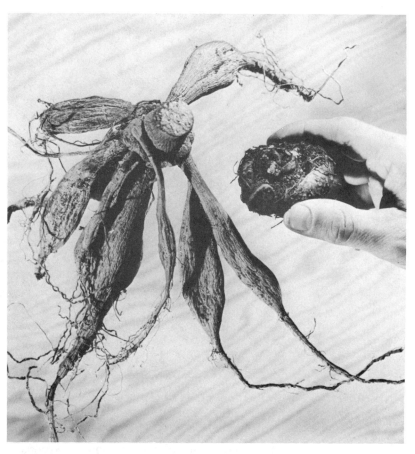

FIGURE 7-16

Dahlia on left is a rootstock. Fingers are holding tuberous begonia. (Photograph courtesy of _Better Homes & Gardens_ magazine.)

It does constitute an important means of propagation of some plants. Notable among this group is the red raspberry. Under favorable conditions the red raspberry has a very extensive root system with many vigorous roots growing horizontally at a depth of 3 to 5 in. ($7\frac{1}{2}$–$12\frac{1}{2}$ cm). Shoots develop from buds differentiated on these horizontal roots and produce what is commonly called sucker growth. The production of suckers is one means of replacing old canes with new in red raspberries and, likewise, it is also their principal method of propagation. Suckering can be further stimulated by severe top pruning of dormant canes.

STUDY QUESTIONS

1. Why are some plants propagated asexually? List five reasons.
2. What are the essential environmental conditions for successful asexual propagation?
3. Layerage is often used in propagating plants. What are the basic reasons for propagating plants by layerage?

FIGURE 7-17   **Suckers arising from a horizontal root of red raspberry.**

4. Describe five different methods of layerage.
5. How does cuttage differ from layerage?
6. What are the advantages of the following types of media for propagating cuttings? Water, sand, peat, vermiculite, volcanic ash (e.g., Perlite), and loam soil. Also what are the disadvantages of each?
7. Distinguish between evergreen and herbaceous cuttings.
8. Why does it usually take longer to propagate woody cuttings than herbaceous cuttings?
9. Distinguish between leaf cuttings and leaf-bud cuttings.
10. Can all plants be propagated from root cuttings? What is a requisite for propagation from root cuttings?
11. List 10 kinds of plants that self-propagate from specialized plant parts.
12. How can each of the 10 kinds in question 11 be forced to increase its rate of propagation?

SELECTED REFERENCES

1. Brooklyn Botanic Garden. 1976. *Handbook on Propagation.* Brooklyn, New York.
2. Carlson, R. F. 1964. Dwarf fruit trees. *Mich. Agr. Ext. Serv. Bull. 432.*

3. Denisen, E. L., and H. E. Nichols. 1962. *Laboratory Manual in Horticulture*, 4th ed. Iowa State University Press, Ames, Iowa.
4. Hall, C. V. 1960. *Horticulture Laboratory Exercises*. Burgess Publishing Co., Minneapolis, Minn.
5. Hartmann, H. T., and D. E. Kester. 1975. *Plant Propagation*, 3rd ed. Prentice-Hall, Englewood Cliffs, N.J. A very complete reference source.
6. Mahlstede, J. P., and E. S. Haber. 1957. *Plant Propagation*. Wiley, New York.
7. Welch, H. J. 1970. *Mist Propagation and Automatic Watering*. Faber and Faber, London.
8. Wells, J. S. 1955. *Plant Propagation Practices*. Macmillan, New York.
9. Wright, R. C. M. 1975. *The Complete Book on Plant Propagation*. Macmillan, New York.

8 Grafting and Budding

THE ART OF GRAFTAGE

Graftage has been practiced by propagators and fruits growers for several centuries, yet the art of graftage is considered mysterious, or unique, by many gardeners. This art does involve the mysteries of life and growth, but the techniques and applications can be mastered by those who earnestly desire such skills. There are a number of reasons why horticulturists have contrived methods of placing twigs or "sticks" of one type or cultivar on another type or cultivar. A primary purpose of graftage is (1) for asexual propagation, the reasons for which were outlined in the previous chapter. The aerial parts of a plant must have a root system to be maintained. Every time a graft is accomplished a stem is being united or reunited with a root, be it adjacent or remote. (2) Plants are also grafted for dwarfing effect, since some roots or trunks exert a dwarfing effect on the tops. Dwarf trees and shrubs are in great demand from home owners, gardeners, fruit growers, and nurseries because of the greater ease of spraying, harvesting, pruning, and general care and maintenance. (3) Trees are grafted to increase hardiness. Greater hardiness of root and trunk of a fruit tree is obtained by grafting the desired cultivar on the top of the hardy stock. (4) Plants are sometimes grafted to increase their resistance to diseases and insects. European type grapes are grafted to roots of certain American type grapes that are resistant to phylloxera, an insect pest which attacks roots of the European grape. (5) Structural strength of trees can be increased by grafting the desired cultivar on strong, wide-angle branches of a vigorous stem or trunk. (6) The form of a plant may be changed by grafting. A notable example is the tree rose. A cultivar of shrub rose is placed on the trunk of a tree type rose to change the form of growth.

Principles of Graftage The terms used in identifying the parts of a graft are *scion* and *stock*. The scion is the part grafted to the root or growing plant. The stock is the root, branch, or trunk upon which the scion is placed. The stock–scion relationships are important to the success of a graft. There must be cambial contact of stock and scion for union and growth of each. This is greatly aided by smooth cuts with a sharp knife. Following insertion of a scion in a stock, callus tissue is produced from the cambium of each. Callus tissue consists primarily of meristematic cells, and as these cells fuse, union is truly established. Although the objective in matching stock to scion is to have immediate contact of cambia, true contact comes only as the callus fuses in intimate cellular proximity.

Another important principle in graftage is compatibility of stock and scion. If the stock and scion cannot form a union, or the union becomes constricted or enlarged, the graft is not considered compatible. One cultivar may not be compatible with another. When two pieces of stem tissue fuse together in growth from the meristematic area, there is a line, or point, or plane at which the adjacent cells originated from two different sources. The xylem and phloem tissues on either side of this line should be similar enough that they will permit passage of food, nutrients, and water through the graft union. Incompatible grafts will sometimes fuse and grow, but normal development is restricted in some manner. In some dwarfing stocks, it may be just such a partial incompatibility which causes one stock to dwarf the top placed upon it.

Associated with compatibility of cultivars or stocks are the limitations of graftage with regard to different species and differnt kinds of plants. In general, only similar or related types of plants will intergraft and many of these only with difficulty. Most cultivars of the pome fruits, apple, pear, and quince, are compatible. Many of the stone fruits, peach, plum, cherry, and apricot, will intergraft if certain kinds or methods are used. Among the conifers, there is compatibility between many of the species. Many types of citrus, orange, lemon, lime, and grapefruit, will intergraft.

Another important principle of graftage is to maintain the moisture content of stock and scion. No matter how intimate the contact of cut surfaces, they must be protected from drying out or there will be very little chance of tissue union. In the aerial parts of a tree, water loss is prevented by using a grafting compound or rubber tape. In stem-root grafts, setting the union below the surface of moist soil or sand is effective. Placing in an atmosphere of 100 percent relative humidity prevents water loss from grafted herbaceous plants.

GRAFTING METHODS

There are numerous methods of grafting. Each has its own purposes, advantages, and disadvantages. Some plants form a compatible union more quickly and effectively by one method than by another. Some methods of grafting will work best on one-year-old wood; others are better for grafting on more mature trees. The several methods described herein are commonly used by horticulturists under various conditions and circumstances.

Whip or Tongue Grafts These grafts are so called because of the back-cut producing a whip or tongue on both scion and stock. Nearly equal size scion and stock should be used. A long slanting cut is made on the lower end of the scion and upper end of the stock. This slanting cut should be about three times as long as the diameter of scion or stock. On most twigs and roots, the length of cut surface will be 1 to 2 in. (2–5 cm). The whip or tongue is

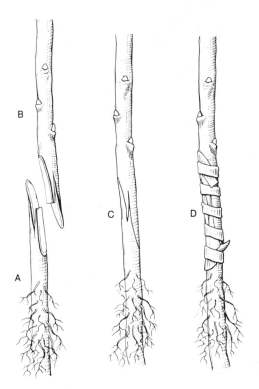

FIGURE 8-1   The whip or tongue graft used in uniting scion to root for a piece-root graft. *A*, root section; *B*, prepared scion; *C*, fitted graft; *D*, graft bound with tape.

then made by inserting the knife blade about one third the length of the cut surface as measured from the pointed end. The blade is drawn slowly into the wood and directed toward the central axis until it reaches a depth of $\frac{1}{2}$ to $\frac{3}{4}$ in. (1–2 cm). Avoid cutting with the grain to reduce the chance of splitting. After both stock and scion have been prepared in this manner, they are inserted one in the other. By pressing the tips of the cut surfaces together the tongues will protrude, and they can then be fitted together. It is important, as with all grafting methods, that cambium of the stock be in contact with cambium of the scion. The possible area of contact is greatly increased because of the interlocking whips or tongues. The stock and scion are also held more firmly in place by this interlocking force. The completed graft is tied with string or bound with adhesive tape.

This graft is used extensively in root grafting for propagation of the apple and pear. In the piece-root graft a 3 to 6 in. ($7\frac{1}{2}$–15 cm) piece of root is used as the stock, and the scion consists of a section of dormant twig containing three or four buds. For a whole-root graft an equal size scion is inserted on the entire root of a seedling. The usual source of roots for both types of root grafts in applies is from one- or two-year-old seedlings. Roots are collected in the fall and stored in moist sand at temperatures just above freezing. Dormant twigs are collected in the winter and stored under similar conditions. The root grafting may then be done during the winter months indoors. Because of this manner of handling and working, piece-root grafting is often referred to as "bench" grafting. After taping or binding, the completed piece-root grafts may be returned to storage and planted in the spring. When planting out-of-doors it is essential that the graft be placed below the surface of the soil to avoid drying. Any shoots which arise below the graft union should be cut off as soon as they appear. If allowed to grow they will compete severely with the shoots on the scion since they are located at a more favored position to the root. All but the strongest shoot on the scion should be eliminated so a single strong trunk will develop.

Aerial (scion to twig) grafts may also be made with the whip or tongue method. Apple, pear, and quince are sometimes grafted in this manner to insert another variety. It is important that the scion be dormant. Grafting wax or rubber tape should be used to thoroughly cover all cut and exposed tissues in aerial grafts.

Modifications of the whip graft include the double-whip and saddle grafts. The double-whip has two incisions and protrusions on both stock and scion. It has the advantage of greater cambial contact but makes fitting of stock and scion more difficult. The saddle graft has a pointed stock and the scion is wedged or split to fit over the stock like a saddle. Cambial contact is not as extensive nor is the interlocking force as great as with the whip or tongue graft.

Cleft Grafts   This method is used mostly on larger trees when it is desirable to change the cultivar. The operation is most effectively performed on branches ranging from $\frac{3}{4}$ to 2 in. (2–5 cm) in diameter. The branch to be cleft grated is sawed off perpendicular to its central axis at least 12 in. (25 cm) from the trunk or other branches. A special cleft grafting cleaver is used to split the stub and produce the cleft. The cleaver is placed on the sawed surface of the stub at the desired position of the cleft and is given a sharp blow with a mallet. The special wedge at the tip of the cleaver is pounded into the stub to hold the cleft open while the scions are inserted.

Two scions are used for each graft, one at each side of the cleft. This increases the probability of producing an effective union and aids healing. The basal end of each scion is tapered to a point from both sides for maximum cambial contact. The side of the scion which will be used for matching with the cambium of the stock is slightly thicker than the side toward the center of the branch. This insures closer contact as the wedge is released and the cleft closes with a vise-like clamp. The scions are tilted outward slightly at the top to make certain of cambial contact at the point of intersection. Any buds near the point of insertion are rubbed out. All exposed tissue, including the end of the stub and tips of the scions, and all openings of the cleft are sealed with grafting wax.

Growth from each scion should be confined to one or two shoots. Excess shoots can be rubbed out after the graft "takes" and growth begins. Only one of the two scions is retained for replacing the former growth. However, if both scions make good unions, they are left for

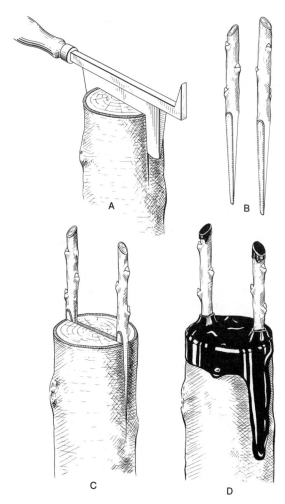

FIGURE 8-2   Cleft graft. *A*, cleaver being used to split stub; *B*, prepared scions; *C*, scions fitted in place; *D*, completed graft sealed with wax.

about two years to hasten healing of the end of the stub. One may be subordinated to the other by occasional pruning. After two years this scion is removed entirely by making a tapered cut outward from the center of the stub. This wound is similarly coated with grafting compound. Growth in length and diameter is usually rapid, since the stored food which was formerly available for the entire branch is now at the disposal of the scions and their selected shoots. In three to five years, the diameter of the inserted wood is usually equal to the stock diameter at the point of insertion.

Related types of grafts which involve removal of limbs or branches prior to insertion include notch, bark, and crown grafting. For the notch graft, an angular cut or notch is made one or more places on the stub and the scion fitted and nailed in place. With the bark graft, the scion is inserted between the bark and wood in an opening made with a knife

FIGURE 8-3   A cleft graft two years after insertion of scion. Note fused tissues and large size of scion.

blade, chisel, or awl. Crown grafts are made near the main trunk on the primary branches of a large tree. Techniques of either the cleft, notch, or bark graft are used in crown grafting. A rapid water sprout type of growth is likely to follow removal of a major portion of the branches for crown grafting.

Side Grafts   The side graft is a quick method for inserting a scion into a branch too small for a cleft graft and too large for a whip graft. A single angular cut is made with a knife blade into the side of a branch. The cut is opened by pressure from the opposite side for inserting the double-tapered scion. Releasing the pressure then tightens the grip of the stock on the scion. Cambial contact is assured by the several points of pressure. On large stems, the scions can be held in place by tacking or nailing. If only one graft is desired per stem, the remainder of the stem or branch can be removed immediately above the insertion.

It is possible with the side graft to actually insert branches on a trunk or primary branch. All existing twigs or secondary branches are removed and scions of 3 to 4 buds each are inserted at desired intervals, directions, and angles. It is one of the most rapid methods of changing over the cultivar on a tree without changing its framework and greatly delaying production. It is important that twigs, spurs, and water sprouts arising from the original branches be kept pruned out if a complete changeover of cultivar is desired. Grafting wax should fill the incision and cover exposed tissues.

Grafting and Budding   **185**

FIGURE 8-4   **Side grafts. Four scions were inserted to completely change over this branch.**

Bridging and   Occasionally, rabbits or other rodents will do considerable damage to
Inarching   young trees by gnawing through the bark and cambium during the
winter months. Since this "girdling" severs the phloem tissue, destroys
the cambium in that region, and causes drying of xylem tissues, the trees
so injured will usually die unless remedial measures are taken. Either of
two grafting methods, bridging or inarching, can be used to save the
tree.

Bridging or bridge grafting utilizes the scion to "bridge" the girdled
portion of the stem. The wound is first smoothed, cleaned, and trimmed
with a knife to promote even healing. Any strips of bark that maintain
contact above and below the wound should be left, since they provide a
channel for supplying food to the roots and are a source of wound tissue
for healing around the trunk. The bark is loosened and scions inserted

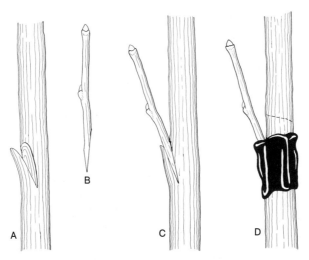

FIGURE 8-5   Steps in side grafting: *A*, cut made in stock; *B*, tapered scion; *C*, inserted scion; *D*, completed graft waxed and upper portion cut off if desired.

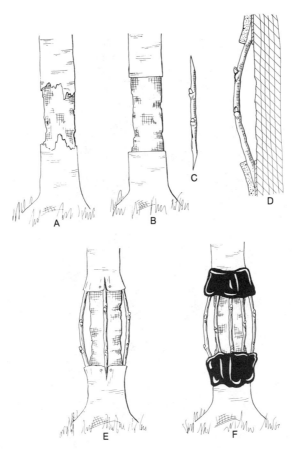

FIGURE 8-6   Bridge grafting: *A*, girdled tree; *B*, bark cut even; *C*, scion tapered at both ends; *D*, scion inserted between bark and wood; *E*, additional scions inserted and tacked in place; *F*, grafting wax applied.

above and below the wound. The scions are double-tapered at each end and placed with a cut surface next to the girdled trunk. The bark may be slit to accommodate the pointed scions. The scions are usually nailed in place. The entire wound area should be coated with grafting wax or other water repellent wound treatments to prevent decay and drying and to promote healing. It is important that the scions be oriented with buds pointed upward as with all grafts. However, any buds which show growth should be removed. The scions are not implanted for any other purpose than to save the tree by providing a channel for food movement between the top and roots.

Inarching, or approach grafting, is another method used on girdled trees but more often on trees with damaged roots. Instead of providing a bridge for the damaged plant, the inarch utilizes another plant or a sucker to by-pass the wound and provide another root. If no sucker plants are present, a seedling may be planted adjacent to the injured tree and grafted to the trunk. A notch is made into the bark of the older tree to receive the inarch. The seedling tree is tapered at the point and placed in the notch flush against the side of the older tree. Cambial contact should occur between the edges of the cut on the

FIGURE 8-7   The "H" tree. These American elms were probably inarched while saplings. Their estimated age is sixty years.

seedling and the notch of bark on the tree. As with bridge grafting, all buds of the seedling should be rubbed out as they appear. The function of seedlings in approach grafting is to replace or supplement the present root system.

BUDDING METHODS

Budding is really a form of grafting. It differs from the grafting methods previously discussed in that single buds rather than scions, are implanted in the stock. It is the principal means of propagating varieties of stone fruits, citrus fruits, avocados, and roses. The stocks used in budding must be under two years of age for greatest likelihood of bud union. The requisites for budding are bark that slips on the stock plant and buds that are resting or dormant for insertion.

Shield or T-budding. This method is generally used during the growing season after buds have formed on current season's growth and while the bark is slipping. Budsticks (slightly hardened current season's growth) are collected from the cultivar to be budded, the leaf blades are removed, and the petioles are left attached for future use as "handles" for the buds. For cutting out buds, a knife is inserted about $\frac{1}{2}$ in. (1 cm) below the bud, and the cut is made into the budstick and upward in the wood to a point about $\frac{1}{2}$ in. (1 cm) above the bud. The cut has thus been made under the bud. The bark, but not the wood, is then severed at the point $\frac{1}{2}$ in. (1 cm) above

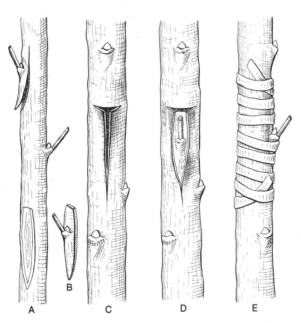

FIGURE 8-8   Shield or T-budding. *A*, budstick; *B*, bud and shield; *C*, T-cut in stock; *D*, inserted bud; *E*, completed bud graft tied with rubber band.

the bud. A slight pressure and twist with the fingers will free the cut bark and bud from the budstick if the bark is slipping. The excised bud with surrounding tissue has a characteristic shield shape.

The stock is prepared for insertion of the bud by making a T-cut in the bark. The longitudinal cut, about 1 to $1\frac{1}{2}$ in. (2–4 cm) is made through the bark and the horizontal cut is made near the top of the first incision. Following the second cut, the bark is opened and the bud is inserted from the horizontal slit. It is pressed into position by using the petiole stub as a handle. When properly placed, the top of the shield is below the horizontal cut and the bud is visible through the vertical opening. Following insertion, the stock and bud are wrapped and tied to avoid drying out. Rubber bands are most satisfactory for wrapping and tying as they stretch when the tree grows and will later disintegrate and fall away. If string or adhesive tape is used, it should be cut after the wound has callused to prevent girdling.

Buds inserted during late summer do not usually start growth until the following spring. After noting that the bud has survived the winter in its new location, the twig or trunk is pruned back. The bud then takes over as the terminal growing point.

FIGURE 8-9   **Tree rose. Buds of desired cultivar were inserted in vertical stem to give tree effect.**

Budding by the shield or T-bud method is not always done during the summer with current season's buds. Dormant budsticks are sometimes used in the spring of the year. It is more difficult to insert the buds since they have no attached petiole. The bark of the stock is more likely to slip if budding is delayed until growth of the stock portion has started. Budsticks are also more likely to slip if kept at warm temperatures for a few days until some buds break dormancy. The buds which have started growth are not used, but those that remain dormant slightly longer will separate from the wood and can be used. These buds will callus and start growth almost immediately, so the remainder of the branch or twig is removed following the budding operation. The principal difference in bud activity between late summer and spring budding is due to the rest period. It is not essential that the section of wood adjacent to the bud be removed; however, it greatly facilitates insertion of the bud and makes a smoother union.

Patch Budding

A less common but often useful method of budding involves relocating a patch of bark which contains the bud. This method is especially adapted for a thick bark as found in nut trees. It is done before growth starts in the spring. The patch must be precisely matched with the peeled area of the stock, especially above and below the bud. A good tool for making equal-length cuts on both budstick and stock can be made by fastening razor blades parallel on either side of a 1 in. (2 cm) block of wood. The horizontal cuts, on both budstick and stock, are made with the parallel blades and the vertical cuts with a budding knife.

The patch from the budstick is slipped off and placed on the wood of the stock with careful matching of top and basal ends to the stock. After implanting, the patch is wrapped with grafting tape or waxed cloth to hold it in place and prevent drying. Since the bark of the stock does not envelop the bud patch, as with the shield bud, drying is more likely to occur if string or rubber bands are used. Grafting tape, string, and

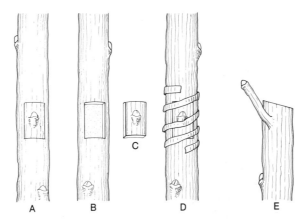

A   B   D   E

FIGURE 8-10   **Patch budding: *A*, cut around bud; *B*, cut made to receive bud; *C*, bud ready for insertion; *D*, bud inserted and taped; *E*, shoot growth from bud about a year later.**

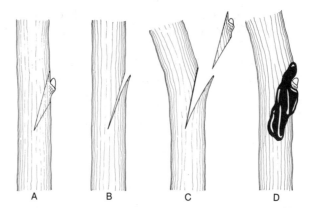

FIGURE 8-11

Chip bud: *A*, bud selected and cut above and below; *B*, stem cut for insertion; *C*, stem flexed and bud inserted; *D*, completed chip bud graft.

waxed cloth should be cut as soon as the patch has healed in place. Shoot growth from the patch bud will occur soon after it has healed. The trunk or branch should then be cut back to within $\frac{1}{2}$ in. (1 cm) of the inserted patch. If the stub is over $\frac{1}{2}$ in. (1 cm) diameter it should be coated with a grafting wax or wound dressing.

Chip Budding　　This method is used primarily for grapes. The buds are taken from matured canes, but the stocks should be in actively growing condition during late summer or early fall. The bud is removed from the budstick by cutting inward and slightly downward immediately below the bud. A second cut is made from above the bud, following down behind, and completing the chip at the basal end. A notch is made near the base of the stock and the bud is inserted. The bud is bound to the stock with a strip of budding rubber and covered with soil. The following spring, the stock is cut back to within $\frac{1}{2}$ in. (1 cm) of the bud union.

TOPWORKING

Grafting or budding to modify the stems or tops of trees is called *topworking*. In addition to replacing the top with other growing points for various reasons, top working is also a means of utilizing stocks as a base, foundation, or support for the tree.

Selecting the Stock　　If grafting is used solely as a method of propagation, then any compatible root may be grafted to a scion. This practice has actually been followed. Seedling roots have been used as rootstocks for most cultivars of tree fruits. This satisfies the need for a root system, but may result in considerable variation in size, vigor, hardiness, and other factors. Considerable attention is now being focused on the use of desirable stocks for fruit trees and ornamental trees and shrubs.

Some cultivars of fruit trees are actually on their own roots or mostly on their own roots. This occurs following the planting of piece-root grafts. Since the union is placed beneath the surface of the soil, conditions may be favorable for root development on the lower part of the scion. If strong roots develop from scion tissue, the grafted root may become increasingly less vital to the tree and may actually be starved for food. Such a tree may be placed entirely on its own roots at transplanting time by cutting off the grafted root at the point of union. Trees on their own roots have no problem of compatibility and no construction in the translocation of food or nutrients. Not all plants are better off on their own roots, however. In many cases there are definite advantages of a known, selected rootstock. Items to consider are depth of rooting, anchorage, spread of roots, and resistance or susceptibility to insect and disease pests.

FIGURE 8-12   **Interstem of Clark dwarf apple stock inserted by whip grafts. Note overgrowth of Clark interstem.**

One of the principal advantages of having all trees of one cultivar on the same rootstock is for uniformity of growth. This factor is important for efficient land use in an orchard. A single rootstock for all trees will give uniform height and spread so that all trees are of near equal productivity. The East Malling Research Station in England has produced several apple stocks which exert various influences on the top. One of the most obvious effects is the uniformity of tree size and shape in an orchard where all the trees are a single type of stock. Some of the East Malling stocks are being used successfully in the United States; however, many of them lack hardiness for the colder regions of the northern states.

When grafting for hardiness, the stem or trunk of a tree is fully as important as the root for cold resistance. A technique of topworking known as "double working" is used for this purpose. A hardy intermediate stock is grafted to the root, the stock grows and develops branches, and the desired cultivar is grafted to the branches. The tree thus formed has trunk, crotches, and primary branches of the hardy intermediate stock. Hibernal, a stock of Russian origin, and Virginia crab are two hardy stocks which have been used as intermediate stocks on apple trees in northern United States and Canada. The search continues for more and better hardy stocks for deciduous fruit trees.

Dwarfing stocks may be of two general types, rootstocks and intermediate stocks. The exact cause of dwarfing is not known. As long as 200 years ago, horticulturists had noted a reduced amount of growth with certain kinds of stocks. It is likely that the varying degrees of compatibility between stock and scion produce varying degrees of dwarfness. Perhaps differences in cell size, permeability of the cell wall, self-perpetuated scar tissue at the union, or other physiological or anatomical factors reduce the upward flow of water and nutrients or the downward movement of manufactured food or both. The effect of a true dwarf is a smaller tree with normal size fruit, early bearing, and the ability to withstand fairly heavy pruning without excess vegetative growth.

Methods of Topworking   Nearly all grafting and budding methods are used in topworking. However, some are much more adaptable for certain plants and for certain purposes. When topworking the desired cultivar on hardy intermediate apple stocks, budding with the shield or T-bud is most commonly used. For young trees about five or six buds will completely topwork the main branches and leaders. Budding has the advantages of being done during the summer and without the use of grafting wax. With experience one can topwork with buds much faster than with whip grafts or side grafts. Another advantage of budding is the second chance for changing over a branch if the bud fails to heal in place. The following spring, the same branch can be whip grafted. Budding is also used extensively in topworking other types of fruits.

Sometimes the crown of the tree is allowed to grow to considerable size before it gets topworked. In such case, budding is impractical since it is best adapted to shoots and twigs. It would take a tremendous

FIGURE 8-13　　　**A topworked tree of Virginia crab. Circles indicate graft or bud unions.**

amount of time and buds to change over the entire tree. Cleft grafting is resorted to instead. Since a large amount of growth is removed, the development of water sprouts from latent buds is common. For this reason, it is advisble to topwork only one third to one half the tree in any one season with the cleft graft. Modified versions of the cleft graft such as notch grafting and bark grafting are also useful when topworking older trees. The side graft and whip graft are useful on trees just past most efficient budding stage, i.e., about four to five years old.

In all topworking, the progress of the buds or grafts should be checked periodically. During the first year it is important that a strong union be developed. This can be facilitated by pruning and disbudding of the stock growing points in favor of shoot growth of the buds or scions. The base of the primary branches should be kept free of water sprouts that will compete with the topworked cultivar. Branches should not be permitted to arise at or near the point of union as they tend to weaken the union.

A topworked tree will usually show the points of union during the remainder of its life. Differences in color of bark, a bulging ring of growth or a persistent scar on the periphery of a limb near the base are criteria of location of the union. These should be noted during pruning operations.

FIGURE 8-14   **Large limb on right has "robbed" the cleft graft of adequate food. This can be remedied by severe pruning of competing limb.**

Bracing   A type of topworking that strengthens the crown is accomplished by bracing. Adjacent branches can be secured together by grafting twigs or small branches so they will fuse into a joining branch. This can be done by twisting the twigs together and tying. Fusion of the twigs is facilitated by scarring the sides of contact so cambium and callus tissues will unite. Bracing can be very practical in areas of extreme wind, or on crowns which are inclined to have narrow crotch angles that split easily.

GRAFTING HERBACEOUS PLANTS

It is reasonable to assume that herbaceous plants can also be grafted when considering their similarities in stem structure to woody plants. However, it is more difficult to graft herbaceous stems and the limitations are greater. The stems are more succulent, are without a rest or dormant period, and have a less definite ring of cambium. Water loss from the scion is one of the major problems in grafting nerbaceous plants.

FIGURE 8-15   Bracing graft between main branches of an apple tree.

FIGURE 8-16   Herbaceous saddle graft on geranium. A piece of paper has been inserted in graft to show its outline. (Courtesy C. H. Sherwood, Iowa State University.)

Purposes   Grafting herbaceous plants is not commercially practical as is grafting of woody plants. The principal reasons for uniting herbaceous stems are for transmitting systemic diseases, indexing for virus diseases, inducing flowering, for scientific study, and for novelty. A potato scion may be grafted on a stem of virus-infected stock to show resistance or susceptibility to the disease. By means of runner grafts, strawberry plants are found to be either virus-infected or virus-free as determined by symptoms or normal growth of the indicator plant. Sweet potato scions are grafted on morning glory stock to induce production of flowers on the sweet potato for breeding purposes. The translocation of synthesized food through a graft union has been studied in tomato-tobacco grafts. A tomato scion on a potato stem makes an interesting and novel plant which has tomato fruits on the tops and potato tubers in the soil.

Techniques   The principal methods used in herbaceous grafts are cleft, saddle, side grafting, and inarching. It is difficult to match cambia of stock and scion, so fusion of callus tissues following insertion is the most reliable observation indicating a good union. Of extreme importance with any of the grafting methods is maintaining high humidity. This reduces water loss and subsequent drying of tissues and wilting of leaves.

The lower leaves or a lower branch should be left on the stock as a stimulus to the upward movement of water. Some of the leaves should be removed from the scion to reduce water loss. The scion is generally of young tissue, often the tip of a stem. However, with the saddle or inarch graft, the scion should be approximately the same diameter as the stock. After insertion of the scion, it should be tied and bound in place. It may then be coated with a grafting wax which can be molded without heating. Certain types of rubber tape which stretch and will stick only to itself are useful for holding the graft in position and excluding pathogens. One of the best ways to maintain high humidity is to place a bell-jar over the plant. Light can enter the jar and transpiration keeps the relative humidity at near 100 percent. After the stock and scion tissues have fused, the stock plant can be pruned of its remaining branches and the scion will take over as the sole top.

STUDY QUESTIONS

1. Suggest several instances in which bridge grafting can be used other than for saving the life of the tree.
2. How can buds be used for budding operations without the use of grafting wax and still provide a high rate of union? What is the comparative amount of water in the tissues of a bud compared to a scion?
3. Draft a rule for judging the compatibility of different plant materials in grafting or budding.
4. When is the inarch method of grafting superior to actual "transplant" of buds or scions?

5. Where does food for new growth come from in a piece-root graft? in a cleft graft? in budding roses? in a bridge graft? in an approach graft?
6. How does "dry budding" compare to budding during the growing season in ease of operation? likelihood of success? ability of bark to slip?
7. Why are some trees of the same cultivar more vigorous than others?
8. What is the mechanism for hardiness in topworking on hardy understocks?
9. How can dwarfing be induced by grafting?
10. What are the reasons for making grafts with herbaceous plants? Is it practical from a production standpoint?

SELECTED REFERENCES

1. Carlson, R. F. 1964. Dwarf fruit trees. *Mich. Agr. Ext. Serv. Bull. 432.*
2. Edmond, J. B., T. L. Senn, F. S. Andrews, and R. G. Halfacre. 1975. *Fundamentals of Horticulture*, 4th ed. Chapter 9: Propagation. McGraw-Hill, New York.
3. Hartmann, H. T., and D. E. Kester. 1975. *Plant Propagation*, 2nd ed. Prentice-Hall, Englewood Cliffs, N.J.
4. Janick, Jules. 1974. *Horticultural Science*, 2nd ed. Chapter 9: Mechanisms of propagation. W. H. Freeman and Co., San Francisco, Calif.
5. Mahlstede, J. P., and E. S. Haber. 1957. *Plant Propagation*. Wiley, New York.

9 Pruning Principles

Pruning is the judicious removal of limbs, branches, twigs, shoots, or roots to increase the usefulness of plants. Many people cut and saw but do not always prune. A good rule to follow in pruning is to have a logical reason or purpose for making each cut. Allied with effective pruning is the follow-up observation of plant response, to note whether the desired effect is being accomplished. Thus, pruning is a skill acquired through knowledge, practice, and observation.

Training of plants is shaping or adapting them to specific forms so they can function more efficiently or effectively. Training includes tying and bracing; however, the principal means of training is by pruning. How this is accomplished may be seen by becoming acquainted with the underlying principles and objectives.

PRINCIPLES OF TRAINING AND PRUNING

Modification of Apical Dominance

The natural forms of plants vary in relation to the extent of their apical dominance. A spruce tree has a characteristic cone shape. Apical meristem of its leader is dominant over the lateral growth of buds and branches. The natural hormones or auxins are produced in stem tips and are inhibitory to growth. As they are translocated down the stem they inhibit or reduce branching and growth from lateral buds. If the spruce is decapitated, i.e., the terminal growing point is removed, the production and flow of auxins to lateral buds is stopped. Lateral growth of branches will occur at an increased rate. Each branch also has apical dominance, and if one of the branches is trained upwards it will assume the leadership and dominance over other branches. Food for growth is translocated in greatest quantity to areas of greatest dominance, areas which produce and translocate considerable quantities of auxin for lateral inhibition. A plant less inclined to high vertical growth, such as the apple, has a more limited degree of apical dominance. Removal of the growing point of a twig or whip (a one-year unbranched tree) when done with the apple will result in several strong laterals. The black raspberry produces strong laterals following removal of the terminal growing point.

Balance of Roots and Tops

The ratio of roots to tops, or vice versa, has considerable influence on the growth, flowering, and fruiting of plants. When top pruning is done during the growing season, the leaf area is reduced; consequently, there

200

FIGURE 9-1   **Training by bracing. This technique is used by fruit growers to increase the angle of branching, open up the tree for increased light and aeration, and bring the tree into earlier fruiting. (Courtesy P. A. Domoto, Department of Horticulture, Iowa State University, Ames, Iowa.)**

is less photosynthesis. The root area has remained the same, however, so the capacity for absorption of moisture and nutrients is maintained. There is no immediate need for more root growth, and the sugars and other carbohydrates are diverted to shoot growth. During the dormant season, top pruning reduces the number of buds on the plant. Stored sugars and other carbohydrates in the plant are reduced only slightly, since they are stored principally in the older parts of the tree (trunk and large branches). With a reduction in growing points (buds) of the top, no decrease in root area, and only a slight reduction of stored food, strong shoot growth will result. Leaf area may be reduced only temporarily, since strong shoot growth is usually accompanied by more and larger leaves per shoot. Large leaf area means more transpiration, more photosynthesis, which increases demands on the root. Food is then channeled to the root for further root growth. Root pruning has a different

FIGURE 9-2   "Pinching" back the terminal growth of a black raspberry shoot results in strong lateral growth due to loss of apical dominance.

effect for it reduces the absorbing area of the plant slowing down top growth. Stored food is then used for root replacement. Resumption of top growth does not immediately follow root replacement since manufactured food must first be restored to the stem.

Altering the
Phases of Growth   As previously discussed (Chapter 2) early growth of a plant is characterized by vegetativeness. Carbohydrates are utilized in growth until the point is attained when growth slows down and carbohydrates accumulate in storage organs. The period of vegetativeness, or the juvenile phase is much longer for woody plants such as trees than it is for herbaceous plants like the tomato. Rapid growth of shoots in the early part of the growing season followed by the development of large leaves and a consequent high rate of photosynthesis are characteristic of the vegetative stage of woody plants. Since this is the period of training, removal of undesirable or unnecessary branches and reducing the number of buds results in continued vegetative growth. If heavy annual pruning were continued as the tree increased in age, flowering and fruiting would be delayed. The carbohydrates would be continually utilized in growth of new shoots to replace those removed. Thus, heavy annual pruning of a young fruit tree is a deterrent to early production. It is stimulatory to shoot growth and is a common practice in the production of scion wood for grafting.

As the woody plant progresses through the transition period from the juvenile to productive state, pruning is held to a minimum. When full production is reached, the carbohydrate supply accumulates in storage organs, fruits, seeds, stems, and roots. Food accumulating in fruits and

FIGURE 9-3   **A scion orchard. Heavy annual dormant pruning stimulates heavy shoot growth for budstick and scionwood production.**

seeds is lost to the plant, but the sugars and other carbohydrates stored in the stems and roots are available for continued growth of roots and further twig growth. For maximum flower and fruit production, a plant should show good annual shoot growth. Leaf area is greater on a plant with strong shoots than on one with weak shoot growth. Thus, if annual shoot growth slows down as in old trees, pruning will stimulate or invigorate more growth, and production is consequently increased. The same general effect of increased shoot growth can be obtained on old trees by addition of a nitrogen fertilizer. Both fertilizing and pruning can be utilized in maintaining good flower and fruit production. An excess of either practice can throw the plant into a resulting vegetative state. Nitrogen as a nutrient, gives rapid and significant response when applied to plants. The first effect is increased growth which utilizes carbohydrates. The second effect is increased photosynthesis from the larger leaf area. The final effect is an increased carbohydrate supply, flowering, fruiting, and moderate shoot growth.

Relating to Environmental Factors   Just as various factors of environment influence the well-being of a plant, their role can be modified by training and pruning practices. The amount of *light* reaching the inner leaves is influenced by the density of branching. Photosynthesis is thus affected and so is color development of fruits. Heavily shaded fruits are poorly colored, have less flavor, and are usually smaller. The inner portion of a clipped hedge becomes denuded of leaves. Thus, frequent clipping and maintaining good form are important to good lighting for producing a dense foliage. *Air movement* is another factor influenced by pruning. A tree or shrub trained to an open structure will be less inclined to harbor fungi than a

Pruning Principles   **203**

dense, compact type. Also a plant which has good air movement permits better spray penetration for insect and disease control.

The *nutrient* supply of the plant plays an important role in pruning and it is influenced by the root-top ratio and the vegetative-productive phases of growth. In general, moderate applications of nitrogen fertilizer accompany moderate pruning practices. Excess fertilizing or excess pruning, or both are likely to cause the plant to be extremely vegetative. The *water* supply is related to pruning in much the same way as are other environmental conditions. An excess of moisture often tends to produce water sprouts on the trunks and primary branches; therefore, if severe pruning precedes or accompanies excessive moisture, considerable wasted growth is lost from water sprout production. Because of its effect in reducing transpiration, pruning can be beneficial in periods of drouth. *Temperature* also has its effect on pruning, since the soft, succulent growth resulting from over-pruning or late summer pruning will be more susceptible to winter injury. There has been less opportunity for hardening of tissues and food storage before cold weather begins. Training trees to a low-headed shape reduces intensity of light reaching the trunk and aids in preventing sunscald of the bark. Crown and crotch injury from cold is more apt to occur on trees which have not been trained or pruned to desirable angles of branching.

OBJECTIVES OF TRAINING AND PRUNING

To Control the Direction of Growth   Plants have a natural form characteristic of their varieties and species. If the natural form is desired, training is necessary only in the event of a broken or misshaped leader or rank growth from previously crowded conditions. If modification of the natural form is desired, generally upright types can be trained to branch more profusely and become spreading; low branching types can be trained to branch higher; branches growing toward utility wires or buildings can be diverted. It is important to follow-up with further training and pruning if the natural form of the plant is changed. The tree or shrub may otherwise tend to revert to its natural direction of growth as determined by its degree of apical dominance.

To Develop a Strong Framework   The trunk and primary branches are the principal structures of support for the aerial portion of the tree. The strength of the trunk and primary or scaffold branches is determined principally by crotch angles and junctures. Some fruit and ornamental trees have naturally narrow crotch angles (less than 40° from vertical) and have resultingly higher loss of limbs from windstorms and heavy loads of fruit. By proper selection of scaffold branches in the early training of a tree, many such problems can be prevented. The strongest crotches are formed when branches arise from the trunk at angles ranging from 40° to 90°. Scaffold branches should be well distributed around the tree; each branch should be at least 90° from adjacent branches. No scaffold branch should be located directly above or below its closest neighbor. They should be

FIGURE 9-4   Tree on left has never been pruned following its loss of apical dominance. Tree on right has been trained to the modified leader system. Note stronger crotches and less crowded structure.

sufficiently far apart on the trunk so as the tree grows, they will not arise at nearly the same point.

There are several training systems for trees, each of which may be most practical or give the strongest framework for a specific kind of tree. The *central leader* more nearly approaches the natural type of growth of most trees. The apical growth of the tree is favored. It provides a strong trunk with well-spaced, well-distributed branches and good crotch angles. This system of training will result in a taller tree than other types.

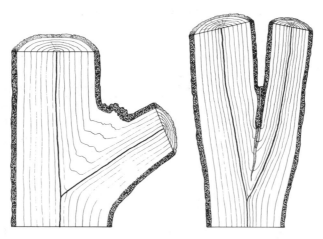

FIGURE 9-5   A wide-angle crotch makes scaffold branches stronger. Note the lack of continuous cambium and squeezed-off bark in the narrow angle on the right. (Adapted from Cornell University Agricultural Experiment Station.)

Pruning Principles   **205**

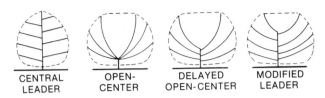

CENTRAL LEADER OPEN-CENTER DELAYED OPEN-CENTER MODIFIED LEADER

FIGURE 9-6   **Training systems for fruit trees.**

The *modified leader* system lowers the height of the tree by reducing the length of main trunk and encouraging the scaffold branches to become larger and grow to greater length. Proper selection of strong scaffolds with good distribution, spacing, and crotch junctures produces a relatively low-headed tree with a strong framework. The third system of training is the *open center*. This shape of tree is less satisfactory for some fruits as with the apple and pear but may be more practical for peaches and plums. The main advantage of this system is to admit light into the center of the tree. Its usefulness is mainly dependent on the ability of the tree to provide strong crotches, for the scaffolds all arise from a nearly central point. If the crotch angles are narrow, breakage is more likely to occur, and a tree can actually "split in half" because of poor support.

To Control the Amount of Growth

Pruning can be correctly termed both a dwarfing process and an invigorating process. The objective, a reduced size or an increased size, determines the type, manner, and time of pruning. Grafting a dwarfing interstock, as described previously, has a pronounced effect in holding down tree size. Pruning in addition to the interstock or natural dwarf is usually needed for maximum dwarfing of woody plants. A combination of both dormant and summer pruning is conducive to dwarfness. Frequent clipping, as with a sheared hedge, is likewise a dwarfing process. The shrub privet will normally grow to a height of 8 to 10 feet ($2\frac{1}{2}$–3 meters) but can be kept as a hedge of 10 to 12 in. (25–30 cm) by frequent clipping. Even trees, e.g., Chinese elm and Russian olive, can be maintained as low hedges by clipping three or four times during the growing season. Frequency of clipping or pruning rather than severity is most effective in maintaining dwarfness.

Increased vigor by pruning is obtained by dormant pruning of older wood. The total growth following dormant pruning is actually less with the reduced number of buds and removal of some stored food. However, the increased growth of the remaining shoots is all in the desired places and total breadth and height can actually be increased.

To Influence Productiveness

Whether the product of a plant be in fruits, flowers, foliage, shoots, or twigs, its yield can be increased or decreased by pruning. The manner of pruning for production is dependent on the flowering and fruiting habits or the vegetative response of that species. Decreasing the number of fruit buds will probably result in fewer but larger fruits and may actually increase the yield, at least of the most desirable fruits. If the objective of

FIGURE 9-7   *Left*: **This normally large-growing Ponderosa pine is maintained as a shrub by annual pruning of the tips before new shoot growth occurs in the spring.** *Right*: **Immediately after pruning.**

growing trees is to produce twigs for scion wood, then severe pruning to stimulate excessive vegetative growth is practiced. In such cases the flowering and fruiting habits are suppressed.

To Improve Quality of Product

Quality is often an intangible entity of description. It usually implies various features of color, texture, and flavor, but many also refer to size, uniformity, maturity, and freedom from disease and insect pests. Fruits are better colored and develop more flavor when they receive adequate light. Removing thin unproductive wood in the center part of an apple tree admits more light, eliminates small poorly nourished fruits, and improves the general quality of the total harvest. Grapes and black raspberries produce larger and more uniform fruits when considerable portions of the dormant canes are removed. The everbearing strawberry produces larger, more uniform, and better colored berries when runners are pruned than when the berries are forced to compete with runners for the food supply. Foliage size and texture of some house plants like philodendron, coleus, and peperomia can be improved by moderate pruning. Improving the quality of the product is one of the principal objectives of annual dormant pruning of many woody plants.

To Utilize Space Efficiently

Under crowded conditions both herbaceous and woody plants are often trellised or staked. The trellis or stake serves as a support for the plant, but also establishes a limit of plant expanse. As a consequence these plants must also be trained and pruned. Training, staking, and

FIGURE 9-8   This Golden Delicious apple tree has been "opened up" by removing excess wood for production of larger, better-quality fruits.

pruning are considered synonymously when tomatoes, rambler roses, and blackberries are grown for most efficient utilization of space. Pruning is also a means of facilitating cultural operations. Low-hanging branches which prevent passage of equipment, extremely high tops on fruit trees which make harvesting hazardous, and dense growth which prevents good spray penetration are sound reasons for removal of either large limbs or small branches. It should be remembered, however, that correct and timely training could have prevented these situations from arising.

To Increase the Usefulness of Plants   Since all horicultural plants are grown for a reason, it should be a goal of the grower to get maximum usefulness from them. Pruning, since it modifies growth, is very helpful in attaining that objective. A shade tree can be made to cast a heavier shade by encouraging dense growth of the top. A hedge can be made practically impenetrable by proper training and frequent clipping. The beauty of plants can be enhanced and maintained by judicious use of the proper cutting implement. Fruit trees produce better fruits, roses grow larger blooms, vines develop more concealing foliage, trees and shrubs provide more screening and wind protection, while large and spreading plants are confined to small spaces by proper training, pruning, observation, and follow-up.

RESULTS OF TRAINING AND PRUNING

Heading Back   When a twig or branch is headed back, approximately one third to one half its terminal end is removed. This reduction in length leaves fewer buds for the twig and the plant and reduces the amount of stored food to a slight degree. However, since most of the food is stored in the older wood of the plant, heading back does not usually result in a great loss of stored food.

Apical dominance has been destroyed by heading back the twig, branch, or young whip. If growth conditions are favorable, this is followed by an almost immediate stimulation of lateral buds. Several shoots will usually develop laterally from a twig which has been headed back. In the case of a fruit tree whip, the heading back results in strong laterals from which the scaffold branches can be selected. In an older tree or shrub a branch may be headed back to one of its own lateral branches. This results in nearly equal distribution of food and nutrients to the various laterals below the cut. Trees or shrubs which are getting too tall can be treated in this manner without appearing to have been "butchered" down to size.

When heading back is practiced on current season's growth or herbaceous stems, it is called "pinching" or "pinching back." It is commonly done on black raspberry, chrysanthemum, and overgrown potted plants. The effect is the same, lateral bud growth and production of strong lateral branches. If only one or two buds are removed with the tip, however, the full effect may not be realized. Often the auxin content of the tip section is so high that it continues to exert considerable apical dominance, the outermost bud takes over as the leader, and laterals remain suppressed.

The natural form of a plant has considerable influence on the extent of lateral branching that results from heading back. The spruces, tree pines, and firs do not usually benefit from heading back of the leader unless special circumstances such as breakage, disease, fire, or other damage demands it.

Subordination

Sometimes a tree or shrub has two or more nearly equal leaders. It is often necessary to select one as the main leader and "subordinate" the others. This is accomplished by severe heading back, that is, removing the upper two thirds or more of each competing leader. The effect of this operation is to so reduce the rate of growth of the subordinated arms that they become branches off the main leader. They are subordinated in location as well as in size. Their demands on the food supply are much less and they allow the selected leader to grow more upright and larger.

A tree with twin leaders is often in danger of splitting at the crotch because of the great weight on either side. By subordinating one to the other, the crotch is strengthened. The more rapid growth of the selected leader tends to envelop the base of the subordinated branch and make a stronger union. Also by slowing the rate of growth of the subordinated limb, there is less likelihood of bark being squeezed and pinched between the cambia on the inner sides of a narrow angle crotch.

Thinning Out

The method of wood removal known as "thinning out" is often contrasted with heading back. They both are useful techniques in pruning for each has its appropriate conditions for usefulness, based

FIGURE 9-9   **Removal of one leader from a twin-leader Douglas fir.**

primarily on the desired or expected results. Heading back is employed for stimulating lateral bud and twig growth by destroying apical dominance. Thinning out aims at reducing laterals and maintaining apical dominance. In a shrub it tends to produce greater height from encouraged terminal growth and opens up the plant for better light penetration and aeration. In a young tree thinning out increases the height and breadth by directing the growth in the desired directions. In an old tree, thinning out of weak growth in the lower and inner parts of the tree is

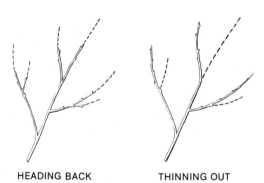

HEADING BACK   THINNING OUT

FIGURE 9-10   **Contrast of heading back and thinning out. Heading back reduces apical dominance. Thinning out maintains apical dominance of remaining branches.**

Skills and Practices in Horticulture

an important means of improving fruit quality. The fruit produced on this weak and thin wood is generally smaller, poorer flavored, and poorer colored. By removal of this type of wood, less poor quality fruit is produced and the food can be diverted to the periphery where more light, better aeration, and more protective sprays are available.

Thinning out is an important means of rejuvenating an old fruit tree. As a tree becomes aged or senile, it has more fruit buds and growing points than it can adequately support under existing environmental conditions. Moderate thinning of branches diverts food to the remaining growing points and fruit buds. As the growing point, or shoot buds, increase in growth, they develop more leaves and the leaves grow to larger size. They carry on photosynthesis on a larger scale. More carbohydrates are produced. There are fewer fruits since some fruit buds were removed in the thinning operation. Those that remain get more of the increased available food. The result, a senile tree brought back to production by judicious thinning. Nutrition of old trees is not to be forgotten as a means of continued productiveness. Pruning and fertilizing can supplement each other in growth and production.

Fruit Thinning   All pruning operations which remove fruit buds can be considered methods of thinning. However, blossoms, tiny fruits, buds, and spurs can be removed with little or no accompanying wood. Various means are used to accomplish this type of fruit thinning. Prior to bloom, spur pruning on some fruits, e.g., the apple, pear, and cherry, will decrease the over-abundance of fruits. More shoot growth occurs, more carbohydrates are manufactured, and the remaining fruits get larger. Also because of the abundance of food produced, assuming adequate nitrogen is available from the soil, a good supply of fruit buds are formed for the following year's production. Thus, fruit thinning in years of heavy production of fruit buds will help to prevent biennial bearing, the alternate year production of fruit. It may also be used to force a biennial bearer to become annual.

In addition to spur pruning, some blossom clusters may be removed during bloom and prebloom or a certain amount of young fruits may actually be eliminated after fruit set. Various techniques are used in thinning at these periods. Blossoms or tiny fruits which have set can be cut out or rubbed off. With peaches, the force of a stream of water from a hose will knock off many of the fruits. The most recent innovation in fruit thinning is with chemical or "hormone" sprays. These fruit-thinning sprays are used commercially on apples, peaches, and apricots. Time of application is extremely important. For most fruits best results are obtained after fruit set but before much growth has occurred.

An extremely important precaution with all fruit thinning, whether it be mechanical or chemical, is to remove the excess fruit while it is yet very small. After they are more than $\frac{1}{2}$ in. (1 cm) in diameter, their removal means considerable loss of food which otherwise would have gone to the remaining fruits. If thinned after 1 in. (2 cm) in diameter, the pattern of food use has been well established, and there may be little beneficial effect on the current crop.

FIGURE 9-11 **Disbudding the rose by pinching of axillary shoots.**

Disbudding This is a thinning process which involves bud removal. It is the method used with roses, peonies, single-bloom chrysanthemum stems, and other ornamentals. Here, as with fruit thinning, timeliness is very important. The excess shoots or blossom buds should be pinched out soon after they appear. This reduces wasted effort and wasted food in the discarded buds. The favored buds respond sooner and more vigorously following early removal of competing buds.

In flowering ornamentals the apical or "king" bud is usually retained because of its advantage of position. The king bloom of a peony or a rose will be decidedly larger if its competitor buds are removed. The giant "mums" frequently used for single bloom corsages have become huge as a result of disbudding.

Disbudding is also practiced in grafting and budding operations. It is desirable for a single shoot to arise from an inserted scion which may have from three to five buds. Often the buds are permitted to show growth and then the more distal or apical bud is retained and the others rubbed out. After an inserted bud has "taken" and begins growth, all buds preceding it on the the stem should be removed to permit all stored food and absorbed nutrients to serve the implanted bud. Immediate and rapid growth of the inserted bud follows the removal of the native buds, for food and nutrients are then channeled to that bud. Disbudding reduces the outlets for translocation of food much as the express train with its single destination is contrasted with the local train that has its cargo distributed at many points.

Summer Pruning Most pruning of deciduous plants is done during the dormant season or before growth starts in the spring. There are instances, however, in

FIGURE 9-12   *Top*: **Disbudding the peony.** *Bottom*: **Note stronger stem and larger bloom of disbudded peony on right.**

which summer pruning is practiced and accomplishes certain results. Actually this practice could more appropriately be termed pruning during the growing season, since it is often done during or following the initial burst of shoot growth in the spring. Pruning of evergreens also can be included in this category for they are never completely dormant.

Summer or spring pruning of new growth has a dwarfing effect on the tree or shrub and is frequently employed for this purpose. The early growth is largely from the stored food reserves of the stem. Removal of this early growth results in depleted stores of food and reduced leaf area with a consequent decrease in total food manufactured. Because fewer carbohydrates are synthesized, less is translocated to the root and root growth is slowed. The dwarfing effect of summer pruning is even more pronounced if a second and third pruning of new growth follows at later intervals. The dwarfing of trees or shrubs in a frequently clipped hedge is an excellent example of the greatly reduced rate of growth due to summer pruning. Such plants cannot remain in the uniform, compact, dwarf form by pruning only during the dormant season. If pruned only while dormant, more wood is removed, a sudden spurt of growth follows when spring comes, and the necessary pruning to limit size becomes more severe each succeeding year.

Summer pruning is also used to accentuate the effect of dwarfing stocks on fruit trees. Removal of some of the current season's growth will reduce the amount of carbohydrates produced, less is available for the root, and there is less increase in absorbing area. The constriction of the graft union, aided by the summer pruning, produces further dwarfness. As the dwarf fruit tree comes into production, summer pruning is reduced so the leaves on the new growth can supply the enlarging fruit with carbohydrates.

Most summer pruning of new growth should be done before midsummer; however, when clipping is frequent, it may be continued until late summer. The plants should be given an opportunity to harden tissues and build up reserves before winter.

Summer pruning of old growth is not generally to be recommended. The older growth of a woody plant is the storage place of reserve carbohydrates. By removing the stored food and the storage place during the growing season, much plant effort is wasted. Also there is very little opportunity for healing the wounds before the dormant season begins. Summer pruning of old growth also increases the danger

FIGURE 9-13   **Dolgo ornamental crab.** *Left*: **With the lush shoot growth of early spring before summer pruning.** *Right*: **After summer pruning of new shoots. One bud per shoot was retained. This tree has increased in height and breadth about 12 in. (30 cm) in eight years, yet produces fruit annually.**

of winter injury to new shoots that are forced to develop late in the season.

Severe Pruning   There are various reasons why severe pruning is done. It may be from lack of a knowledge of pruning, delay in getting a pruning job done, placing large shrubs in small locations, cutting out limbs and tops for power lines, for scion wood production, for topworking old trees, and for production of large attractive twigs as with red and yellow stemmed dogwoods. Severe pruning means that excessive amounts of wood are removed usually during the dormant season. The effect is one of "over-stimulation" or "excessive invigoration." Water sprouts arise from latent buds on the trunk, branches, or main stem following the dormant pruning operation. The size of water sprouts is closely related to the food reserve of the plant and the absorptive capacity of the roots. In old trees they may attain lengths of as much as 6 to 12 feet (2–4 meters) if large limbs are cut back in their entirety. Water sprouts are the product of vegetative conditions. Allied to extreme vegetativeness is non-fruitfulness of fruit trees and non-flowering of ornamental plants.

One of the most hazardous effects of severe pruning, particularly in northern areas, is reduced hardiness. Elongation of shoots continues longer during the growing season, leaving less time for maturation of tissues. As a result these succulent tissues enter winter in an inadequately hardened condition. Die-back of twigs or crown and crotch injury may be quite serious depending on the severity of the winter.

Root Pruning   This type of pruning is generally practiced less and understood less than other kinds. Other than placing roots at the time of transplanting, the locations of main roots, branch roots, and feeder roots are not definitely known. Thus when pruning roots of growing plants one is actually guessing at their location and hoping to make proper and sufficient cuts.

The principal effects of root pruning are (1) reduced absorption, (2) more branched main roots and feeder roots, and (3) reduced top growth. As previously mentioned, reduced absorption favors dwarfness. Root pruning may be used on cabbage plants to prevent cracking of the heads following an excess of moisture. A spade inserted to a depth of about 6 in. (15 cm) around the periphery of the leaf area will sever many of the feeder roots. A similar method can be used for trees and shrubs to reduce absorption and reduce top growth. The spade will need to be thrust considerably deeper than for the cabbage. It may be more effective to dig a trench around the tree far enough from the trunk so that only the periphery of feeder roots is removed. Where continued dwarfing by root pruning is desired, it may be advisable to install a trench, fill it with straw or other mulch packing, and open it periodically to sever new roots.

Nurseries prefer plants with shorter and more compact root systems for ease in handling, shipping, and replanting. It is common practice to root prune young deciduous trees and shrubs by undercutting with a U-shaped cutting blade pulled under the rows of nursery stock. The severed roots branch near the cut end and toward the trunk to form a more fibrous root system of young feeder roots. Undercutting is also effective in making plants with typical tap roots more fibrous rooted.

STUDY QUESTIONS

1. Distinguish between training and pruning.
2. Evaluate this statement: "Anyone can cut, but not all people prune."
3. Explain the term "apical dominance."
4. What is meant by vegetativeness in a woody plant? How is it influenced by pruning?
5. What are the objectives of training and pruning? List them.
6. Why is a wide-angle crotch generally stronger than a narrow-angle crotch on a tree?
7. What is the advantage of a delayed open-center system of training versus the open-center?
8. How can pruning improve the quality of the product in fruit production?
9. Explain the difference in results from heading back and thinning out. Does one result in a smaller tree?
10. Under what conditions is summer pruning recommended?
11. What are the effects of severe pruning of a fruit tree?
12. When can root pruning be used advantageously? What is undercutting in a nursery?

SELECTED REFERENCES

1. Brooklyn Botanic Garden, Plants and Gardens Handbooks, Brooklyn, New York:
   1976 No. 65, *Tree and Shrub Forms—Their Landscape Use*, 64 pp.
   1976. No. 36, *Trained and Sculptured Plants*, 65 pp.
   1975. No. 28, *Pruning Handbook*, 80 pp.
2. Chandler, W. H. 1950. *Evergreen Orchards*. Lea and Febiger, Philadelphia, Pa.
3. Christopher, E. P. 1954. *The Pruning Manual*. Macmillan, New York.
4. Denisen, E. L., and H. E. Nichols, 1962. *Laboratory Manual in Horticulture*, 4th ed. Iowa State University Press, Ames, Iowa. Covers specifics of pruning apples, peaches, woody ornamentals, grapes, brambles, and other small fruits.
5. Weaver, R. J. 1976. *Grape Growing*. Wiley, New York. Thoroughly covers training and pruning of European- and American-type grapes.

10 Pruning Methods

Because of the many kinds and types of plants which require pruning, the methods vary considerably. To attempt to give complete and detailed instructions on how to prune all kinds and all varieties of plants, for all purposes, occasions and uses, and for all environmental conditions would be a noble effort, but far from feasible. The best help can be obtained by: (1) learning the growth processes in plants; (2) understanding the principles, objectives, and effects of pruning; and (3) studying and observing the methods used on certain common ornamental, fruit, and vegetable plants. The first two of these aids refer to earlier chapters. The examples and methods cited in this chapter offer considerable opportunity for acquiring a background for pruning. With such general knowledge and specific examples, it should be possible to predict the effects on other plants or under other circumstances without making serious errors.

GENERAL PRUNING PROCEDURES

Certain preparations, operations, and post-operative procedures apply to pruning of woody plants in general. It is well to become acquainted with these procedures and to be prepared for doing the job completely and correctly.

Selecting Equipment

SHEARS. One of the most useful and versatile shears is the hand pruner. A sturdy shears with a strong spring and a close bite will cut branches up to about $\frac{3}{4}$ in. (2 cm) in diameter, but is most useful for the $\frac{1}{2}$ in. (1 cm) size. The lopper is a shears with a large bite and long handles. It is capable of cutting branches up to 1 in. (2.5 cm) in diameter. The long sturdy handles give it excellent leverage for cutting the hard tissues of matured or dead branches. The long-bladed hedge shears is used largely for shaping shrubs, clipping hedges, and shearing specimen plants. The pruning hook, or pole pruner, consists of a shears or cutting head mounted on the end of a long pole. Some pole pruners are capable of cutting through $1\frac{1}{2}$ in. (4 cm) branches located 12 to 16 feet ($3\frac{1}{2}$–5 meters) above the operator. Power pruners greatly reduce the work load of pruning. They have a strong durable cutting head operated hydraulically.

SAWS. Pruning saws differ from carpenter saws primarily in the "set" of the blade. A wide set is used for pruning saws to give a wider cut and reduce pinching or binding of the saw blade by the living wood and moist sawdust. The tapered carpenter-type saw is very satisfactory for straight sawing. A meatcutter-type saw with a narrow blade and supporting D-shaped frame is especially useful where a shift in direction is made while sawing. A curved-blade saw is adapted for getting into constricted areas of adjacent limbs arising from almost the same point.

CLIPPERS. These instruments differ from shears in that they have oscillating blades on a straight rigid frame. The electric shrub clipper is widely used for hedges and other clipped ornamentals.

Making Cuts   ALONG THE STEM OR TRUNK. When removing an entire twig or branch, the cut should be nearly flush with the stem or trunk. Leaving a minimum of stub promotes more rapid healing of the wound. This does not mean the cut should be made into the trunk or that every bit of branch tissue need be removed. The flare or enlarged portion at the base

FIGURE 10-1   **A sheared pyramidal juniper behind an unsheared Pfitzer juniper.**

of a twig or branch has considerable meristematic activity. A cut at the juncture of this flare and the twig or branch leaves a smaller wound and more rapid production of wound tissue. If the flare itself is removed, a smoother cut is usually obtained, but the wound area is greatly increased and production of wound tissue may be slower. It is important to avoid scratching or "feathering" of the adjacent cambial area. A clean wound heals smoother and more rapidly.

When heading back a branch or leader, it should be cut back to a lateral branch for most rapid healing. An unhealed stub is subject to decay that may spread to adjacent tissue. A slow healing wound may thus eventually envelop a considerable area of decayed tissue. Limb or trunk breakage in later years can often be traced to slow healing as a result of a protruding stub.

BETWEEN NODES. When heading back the twigs of shrubs and small trees and the canes of vines and brambles, pruning cuts are made between nodes. On twigs the cut is made about midway between buds. The direction of growth of the uppermost bud can be influenced by cutting back to a bud located where a branch is desired. The small portion of the twig beyond the bud will dry out, become shriveled, and eventually disintegrate. The wound will then heal over usually without any further treatment.

Canes of vines and brambles usually have much longer internodes than twigs. They are also more succulent and have a much greater central core of pith. Consequently, when pruned the stub dries out more rapidly than with most twigs. The cut is also made about midway between buds. The tissue dies back to the uppermost bud; however, if the cut is made close to the bud, the cane often dies back to the second bud. On the perennial canes of vines the dead stub eventually disintegrates, is shed, and the wound heals. Most brambles have biennial canes and the stubs usually heal over when pruned back during the first year, but do not heal when pruned at the beginning of the second year. After fruiting, the old canes are removed leaving the new canes for the next year's crop.

SHEARING OR CLIPPING. This is done during the growing season and, for most effective results, is done at frequent intervals. A hedge or individual plant is sheared or clipped to give a trim, formal, precise effect. Clipping frequently prevents shagginess due to long protruding shoots and also avoids severe shock to the plant. Only current season's growth should be removed. No specific place of cut is designated other than staying beyond the old wood. No large wounds are produced. It is a heading back process and continued branching and shoot growth will result.

Treating Wounds

The purpose of treating pruning wounds is threefold: (1) to avoid infection by rot-inducing organisms, (2) to prevent drying of the tissues, and (3) to promote healing. Of prime consideration in determining the need for treatment is the size of the wound. It is generally assumed that those under 1 in. (2 cm) in diameter need not be treated. This size wound will usually heal over during the course of a season with a smooth cut that has a minimum of stub and an active cambium. There is very little danger from drying and decay during this brief period and the possibility of cambial injury from a wound treatment compound is eliminated. Wounds greater than 1 in. (2 cm) in diameter should be treated to protect the wound and inner tissues during the longer healing period.

There are several types of wound treating compounds. The most commonly used material is paint. Most outside paints containing white lead or titanium are satisfactory. They are durable and are not injurious to the cambium unless applied heavily in that area. Since interior paints contain other ingredients, some of which may be deleterious to growing tissues, they are not generally used for wound treatment. However, if no other wound-treating materials are available, it is better to apply a protective coat to the wood portion of the wound, using care not to paint the bark and cambium. The periphery of a wound heals rather soon but it is sometimes several years before the central area is sealed over with new growth. Varnish and shellac are also satisfactory but are less durable than paint.

Asphalt compounds in special wound treating preparations are among the best materials available. It is doubtful that healing is actually promoted by their use, except that prevention of drying and infection aids growth of wound tissue. Fungicides incorporated in wound treatment materials are helpful in reducing the incidence of rot. Bordeaux paste, containing copper sulfate and lime, is useful as a fungicidal wound preparation but needs renewing at least annually because of its solubility.

PRUNING WOODY ORNAMENTAL PLANTS

Trees   *Deciduous* ornamental trees are pruned according to their intended uses, desired forms, and natural growth habits (Fig. 17-2, pages 366 and 367). The Lombardy poplar is often used for background plantings or for accent points. Its height is many times its breadth, and its branches are nearly upright. With such extreme apical dominance and natural growth tendencies, it is impractical to attempt drastic changes in its form. Heading back its leader only results in one of the uppermost upright branches assuming leadership. It soon regains its natural shape with very little change of silhouette. Corrective pruning, consisting primarily of removal of dead and diseased branches, is sufficient for the Lombardy and Bolleana poplars and other trees with strong, dominant growth tendencies.

The American elm is characteristically vase-shaped with large limbs arising from the trunk with very narrow crotch angles. The narrow crotch angles cannot be corrected in a practical manner. However, subordination of the lower branches of a young tree reduces their weight and lessens the likelihood of crotch splitting, a common fate of the American elm.

The American linden, or basswood, is slightly taller than it is wide and has a central leader. It can be adapted to modified forms because of its response to training and pruning. When its lower branches are removed its broad leaves and heavy foliage make it desirable as a shade tree. If the lower branches are retained, it is effective for screening and windbreaks. If height is a factor, or a round-headed tree is desired, the leader can be headed back to produce a lower, wider, more oval silhouette. It will generally be necessary to remove some of the lower primary or secondary branches at annual or biennial intervals to maintain the same height of branching. The lower branches of the basswood have a tendency to grow downward and outward for more light. The growth and pruning of the green ash and the pin oak is similar to the linden. The white oak and burr oak require only occasional pruning. They are very slow growing and produce beautifully shaped trees with the training of nature and self-pruning.

Evergreen trees of the narrowleafed type, the conifers, are predominantly cone-shaped in their natural form. They exhibit a high degree of apical dominance from their straight single central leader. If the leader becomes injured or destroyed one or more of the uppermost branches tend to grow upright and compete for leadership. To correct this situation, one branch should be selected, forced to an upright position, and tied or braced. This gives the branch undisputed leadership and the tree will, in time, regain its normal form. When twin leaders are present, one should be subordinated to the other. Many conifers can be made more dense and compact by heading back or shearing the terminals of all branches. This induces further branching. When done while trees are small, full and well-shaped trees are produced. Clipping should be continued if specific forms and sizes are desired. The lower branches of conifers are sometimes removed for various reasons. Unless absolutely necessary, this is not a good practice for it detracts from their natural beauty. The bare trunk of most conifers is not attractive, and some types exude offensive pitch. Little is gained by removing the lower branches to admit light for lawn grass since the fallen needles are not conducive to good grass production.

Broadleafed evergreens are pruned in much the same manner as broadleafed deciduous trees. Their growth habits are similar. It is best to prune in late winter or spring to allow the new succulent growth to harden sufficiently for the succeeding winter. Holly is often pruned in late fall or early winter so that the removed leaves and branches can be used for Christmas decorations. This practice, while not generally recommended, is relatively safe for the tree if not carried to excess. In general, little pruning is needed for holly once the tree form is established.

Tree surgery is occasionally needed particularly for the larger growing types of ornamental trees. It involves the removal of large limbs, cleaning and treating wounds resulting from decay and storm damage, and bracing weak trunks and crotches. Large limbs present a problem in their removal because of the possible danger to neighboring buildings and adjacent properties. Also because of their great weight, care must be taken to avoid skinning the bark from the trunk when the cut is near completion. When removing a large limb from above or near buildings, it is well to use a block and tackle arrangement to gradually lower the limb to the ground. To avoid trunk skinning when a block and tackle are not needed, a series of cuts can be used. The first cut is made from the underside about 12 to 18 in. (30–45 cm) from the trunk and continues until the saw begins to bind from the pressure of the limb. The next cut is from above and about 2 in. (5 cm) outward from the first cut. The limb will fall without tearing away the bark and wood from the underside of the crotch and the trunk. Then the stub is removed near the trunk and the wound is treated.

Cleaning and treating wounds resulting from decay first involves removal of the decayed wood. The cavity is sprayed or coated with a fungicide to prevent further decay. It can then be painted, or in some cases where additional strengthening is needed, it may be filled with concrete and smoothed for even healing. Wounds resulting from storm damage usually need considerable preparation to promote even healing. Broken limbs and torn crotches and trunks should be trimmed as smoothly as possible with a minimum of further injury to the bark and cambium. Thorough coverage with a durable paint or other wound treatment compound is essential for large gaping wounds which require considerable time for healing.

Bracing of weak crotches is best accomplished with large bolts or screws inserted through the base of the limb and the trunk. Wires and cables are sometimes used but the danger of girdling makes them generally impractical.

Shrubs *Spring blooming* deciduous shrubs differentiate their flower buds during the previous growing season. Pruning during the dormant season almost invariably results in a reduction of spring bloom, since some of the flower buds are removed with the prunings. Therefore, spring pruning immediately after flowering is advocated for early spring blooming types. Thinning out of older branches is the usual procedure for most shrubs to renew the shrub, hold it to the same size, and to preserve its natural form. Heading back gives the shrub a "cropped" appearance and often results in succulent vegetative growth. The spring pruning following bloom should not be delayed or the plant loses increasingly more growth and energy. Typical examples of shrubs which should be pruned in this manner are spirea, lilac, forsythia, honeysuckle, and mock orange. Of these, the spirea, lilac and forsythia bear their bloom directly from one-year-old wood. Honeysuckle and mock orange bloom on current season's growth arising for one-year-old wood.

Summer blooming types of shrubs are spring pruned to produce strong vigorous current season's shoots which bear the flowers. Hydrangea (snowhill), roses, althea, and Japanese rose (*Kerria*) are typical examples. Snowberry is also summer blooming, although its blossoms are less conspicuous. It is given moderate spring pruning for best production of its conspicuous berries. Both heading back and thinning out methods are used depending on desired effect.

FIGURE 10-2   **Dormant pruning of "everblooming" rose.**

Because of their varied and diverse types, roses generally require and receive special attention in pruning. The "everblooming" types are given both dormant and summer pruning. Both are heading back processes to stimulate new shoot development, for the rose produces its bloom terminally on current season's growth. Since the rose usually is budded on its stock and does not continue branching in the basal area, thinning out is not practical. Dormant pruning is usually done just before growth begins in the spring. This is especially advisable in areas where die-back of stem growth may occur during the winter. All dead portions are thus removed and danger of winter injury to the cut surfaces is alleviated. The amount of pruning will vary for the types of roses and the effect desired. Long-stemmed roses with large blooms are produced as a result of heading back to strong buds on vigorous canes. With some cultivars of hybrid tea roses more than one bud is produced per shoot, so disbudding of the axillaries is sometimes necessary. If mass effect of smaller blooms is desired, the pruning should be less severe, more shoots will be produced, and the flowers will be more numerous and smaller. Pruning after bloom should be done immediately after petal fall so seed heads do not develop and use the plant's food.

Rambler or climbing roses are pruned less severely since long trailing canes are desired. Many of these types bloom but once during the season, so summer pruning is directed to production of strong canes for the succeeding year. Here as with everblooming types, removal of the seed head should follow immediately after petal fall. Rugosa and floribunda roses are pruned in much the same manner as hybrid teas except they are not disbudded and spring pruning is not as severe. Mass effect of the smaller type bloom is the principal objective of these roses. Thinning out of old growth is sometimes desirable with the rugosa rose to encourage new shoots arising from the crown.

Evergreen shrubs are pruned primarily to reduce their size and prevent overgrowth. Spreading and prostrate junipers and shrub pines can be cut back considerably without a "cropped" effect. This is accomplished by heading back the longest and tallest branches to strong laterals. Since the angle of branching is usually narrow, the wound is not very visible, nor for long, because the feather-like needles are quite concealing. Thinning out of very dense evergreen shrubs is also desirable to prevent dropping of leaves at the base. An extremely dense evergreen shrub is usually green only at the outer periphery where light is abundant. Broadleafed evergreen shrubs can be pruned in much the same manner as the conifers. However, they will withstand regenerative pruning back to old wood much better than conifers. Apparently latent buds in the broadleafed types are more readily stimulated to growth.

Hedges, whether of evergreen or deciduous type, require constant vigilance to maintain their beauty, symmetry, and good health. It has been emphasized previously that clipped hedges benefit from frequent clipping. Even occasional root pruning helps to avoid sudden spurts of growth which may make the clipped hedge unsightly for a time. Hedge form is an important factor in its maintenance (Fig. 17-11, page 377). In order to have foliage extending to the base of the hedge, adequate light and aeration is essential. For this reason tapered sides will usually result in a fuller, more even, and healthier hedge. The tapered form should be maintained at each clipping. The hedge should not be allowed to get taller and wider each year. If it has been neglected for a time, it should be cut back slightly below the desired level so that future clippings of new growth will be above and outside the skeleton framework, concealing them from view.

Vines that have accomplished their principal purpose, that of covering or concealing all or part of an area, will need to be pruned to keep them attractive and within bounds. On buildings it may be necessary to head back growing points around windows at intervals during the spring and summer to keep clear views, admit light, and permit air circulation into the building. Vines can add greatly to the beauty of a building but should afford the viewer at least occasional glimpses of the building itself.

It is likewise important to maintain vines growing on lattices, trellises, and fences. Both heading back and thinning out can be used to keep them in check without destroying their effectiveness. Dormant and summer pruning are both useful, and root pruning helps greatly in controlling exceptionally vigorous vines. Where vines are used for ground cover, pruning is usually necessary to prevent a buildup of too many layers of vines. The leaves near the ground should not be concealed by too much heavy growth above them. If the main function of the ground cover is to hold a bank of soil in place, it is better to set out more plants than to let a few plants produce more vine growth. It is the roots which are most effective in holding the soil. Foliage is the cover.

PRUNING WOODY PLANTS FOR FRUIT PRODUCTION

The various fruits differ in plant size, maturation, earliness of fruiting, fruiting habits, and their response to pruning. The fruit grower is also aware of varietal differences which modify the type and extent of his pruning.

Tree Fruits   APPLES AND PEARS.   These trees bear fruit mostly terminally on spurs. Occasionally some fruits are borne laterally on twigs or terminally on twigs. With these fruiting habits, it is apparent that the productive stage of an apple or pear is not reached until spurs are present in considerable numbers. Each spur does not bear fruit every year. When a spur is terminated by a fruit bud, which is plump and oval, it will usually produce about five blossoms. Usually only one of the blossoms per spur sets fruit and the others absciss. After fruit set, growth, and harvest, the spur is terminated by a shoot bud which is more tapered and pointed. The following spring shoot growth occurs and adds some length to the spur. After this growing season is over, the spur may be terminated either by another shoot bud or a fruit bud depending on the carbohydrate and nutritive condition of the tree, branch, and spur. When all or nearly all spurs on a tree get into a cycle of alternate years of fruit production and shoot production, the tree is called a biennial bearer. Spur pruning, fruit thinning, and thinning out branches in the bearing year tend to force a return to annual bearing.

The juvenile or vegetative phase of apple and pear trees is of fairly long duration, usually from six to ten years. It may be as short as two or three years for dwarf types of the same varieties. During this phase of growth the framework of the tree should become well-established through sound training and pruning practices. Proper selection of scaffold branches, crotch angles, and training systems should be made early. Undesirable and unwanted twigs and branches should be removed before they utilize food reserves for wasted growth. Very little pruning should be done during the years following training or fruit production will be delayed.

As growth produces a tree of larger size, the greater leaf area builds up a greater food reserve, the expanse of roots absorbs an abundance of water and nutrients, and the plant goes into a transition stage preparatory to fruit production. Early in this phase spurs are produced, the tree has a light bloom, and a few fruits set. In two to three years the tree has a heavier bloom, and an increased set of fruit. During this period of transition, very little pruning is done or the plant may revert to a vegetative condition. Removal of a rubbing branch, subordinating a twin leader, or very limited corrective pruning are examples of the light pruning which may be done at this time.

As the tree comes into the productive phase it has a balance of carbohydrate utilization for shoot growth and carbohydrate accumulation for fruit production. Good annual shoot growth is essential to provide a large leaf area for food manufacture. Consequently fertilization with nitrogen and other nutrients is helpful because it stimulates

FIGURE 10-3   **Apple fruits are borne mostly terminally on spurs.** *Top*: **Five flowers from one spur. Note center fruit is becoming dominant.** *Bottom*: **A single fruit remains terminally on a spur.**

growth. Increased shoot growth can also be accomplished by pruning, but it must be remembered that pruning involves removal of fruiting wood. Stimulation of shoot growth, increased leaf area, and increased photosynthesis are not helpful to production if the numbers of fruits are drastically decreased. Thus, fertilizing and pruning are practices which supplement each other in fruit production.

If a tree in its productive stage were allowed to produce sufficient shoot growth year after year without pruning, it would soon become so large and tall that cultural operations would be difficult. The most fruitful area would shift to the outer extremities. Light penetration and aeration would become restricted. The general quality and productiveness of fruit would be lowered. Consequently a tree in its productive phase can be maintained in this condition with good cultural practices and moderate pruning. The extent of pruning operations should be governed by such factors as size of tree, amount of the previous year's twig growth, the previous year's fruit production, and the quality of fruit produced.

Both thinning out and heading back pruning cuts are used on apple and pear trees in their productive stage. Thinning out of weak wood, or thin-wood pruning, in the central area of a tree reduces the number of small and poorly colored fruits. Heading back twigs and branches is helpful in controlling height and breadth of trees.

The process of rejuvenation of old trees involves removing many of the growing points by thinning out branches, eliminating high tops, and heading back lateral branches. The most immediate effect on growth is production of longer and stronger shoots. This builds up the carbohydrate reserve, encourages fruit bud differentiation, and results in a return to fruit production.

PEACHES AND APRICOTS. The manner of growth, training systems, and tree size of these two fruits are similar. However, they do vary in fruiting habits. The peach bears its fruit mostly laterally on one-year-old twigs, and to a minor extent laterally on short spur-like twigs. The apricot produces its fruit mostly laterally on spurs and sometimes laterally on long twigs. Both have adjacent fruit and shoot buds in a multiple bud arrangement. Both have a tendency to set too many fruits if not pruned as a regular practice. Fruit thinning is often desirable with the peach and apricot as a means of removing some of the overly abundant fruit buds or small fruits with less removal of wood.

The stone fruits are generally trained to the modified leader or the open center system to keep the trees low headed and to allow good penetration of light and spray materials. The apricot and peach, as a rule, have stronger crotches with broader angles of attachment than do most cultivars of apples and pears. Newly set two-year-old trees are pruned back to stubs of laterals with one to two buds. As growth starts, further pruning by pinching or deshooting is practiced for selecting the scaffold branches. This procedure eliminates much wasted growth and favors development of the selected lateral branches.

The vegetative or juvenile stage is shorter than for the apple or pear. Under good growth conditions a tree of 6 to 8 feet ($2-2\frac{1}{2}$ meters) in height and breadth is obtained in about three years from planting. It is not uncommon for a tree this size to produce a fair crop of fruit. The vegetative and transition phases have considerable overlap, for rapid increase in size is commensurate with moderate fruitfulness.

FIGURE 10-4 **Peach fruits are borne mostly laterally on one-year-old twigs.** *Top*: **Peach twigs in bloom.** *Bottom*: **Same twigs with developing fruits.**

The productive period which follows is typically one of vigorous shoot growth and heavy fruit set. This is the primary reason why these fruits require heavier annual pruning than the apple and pear.

The two principal objectives of pruning are to keep fruiting wood close to the center of the tree, as well as further out, and to reduce the number of fruit buds so larger fruits can be obtained. Either heading

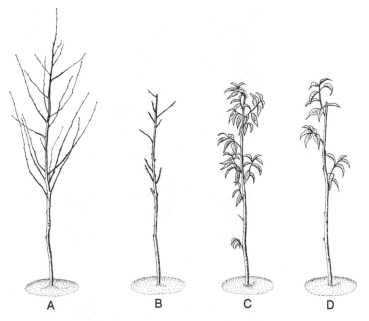

FIGURE 10-5   Training and pruning the peach. *A*, newly planted two-year-old tree; *B*, pruned following planting; *C*, resulting shoot growth; *D*, trained to modified leader by selecting and deshooting.

back of twigs or thinning out of one- and two-year-old branches have been used successfully. The thinning out method is currently in greatest favor since it is less time consuming and gives a more open tree. It also encourages shoot growth from latent buds on scaffold branches near the trunk. For the peach this means increased fruiting area the following year. Thinning out of one- and two-year-old branches with their spur growth is effective in reducing fruit numbers in the apricot.

Winter injury of fruit buds and new growth is common with the peach, and frost injury or killing of blossoms sometimes occurs with both the peach and the apricot. It is because of this possible cold injury that peach and apricot pruning should be delayed until late winter or early spring. In northern peach growing areas removal of winter-killed twigs and branches may be sufficient pruning in itself for maximum size and yield. Trees which have lost their fruit crop due to a late freeze were formerly pruned severely in a "dehorning" process. This was to stimulate latent bud development and water sprout production. This method is being replaced by a more moderate pruning to take advantage of the year of rest to reshape the tree. Thinning out and heading back encourage shoot growth on both old and new wood and increase the fruiting area the succeeding year.

Neither the peach or apricot is as long lived as the apple and pear. Rejuvenation can be accomplished by pruning. However, once the trees become senile, much of the wood has ceased to function. Only a brief comeback will follow rejuvenation, for decay soon results in continued winter injury and drying of tissues.

Pruning Methods   **229**

FIGURE 10-6   **The open center system may also be used for peach scaffolds. Peach crotches are generally of wide angle and strong.**

CHERRIES AND PLUMS.   Some variation of fruiting habit exists between types and cultivars of both the cherry and plum. However, most fruit of the cherry and plum is produced on short spurs arising from wood two years old or older. European and American type plums produce fruit mostly from these spurs but to a limited extent bear fruits laterally on one-year-old twigs. Japanese types often produce the major portion of their crop laterally on twigs. The sweet cherry produces its fruit laterally from spurs on the older wood. The sour cherry also produces some of its fruit in this manner but in addition may have fruit production laterally on one-year-old twigs.

Plums and sweet cherries have a longer life span than most stone fruits. They produce larger and more spreading trees. Scaffold arrangement and attachment are extremely important because of size and spread. The modified leader is the most generally satisfactory training system, although the open center with about three primary branches, or equal leaders, is also used extensively. Sweet cherries tend to grow quite upright. When headed back early in their growth another leader or multiple leaders may develop. The sour cherry is frequently trained to a modified leader early in its development. Then as growth continues in

FIGURE 10-7   **Sour cherry fruits are borne both on spurs and laterally on one-year-old twigs.**

an upright direction, the tree is changed over to a "delayed" open center. The weight of fruit then tends to keep the top more open for light and spray penetration. Wounds on the plum and cherry do not heal as readily as on the fruit trees previously discussed. It is thus very important that there be a minimum of stub and that a protective coat of wound treating compound be applied.

Cherries and plums come into bearing earlier than the apple and pear but, in general, not as early as the peach and apricot. Less wood is removed annually than for the peach due to the spur type of fruiting. Pruning and fertilizing in relative and adequate amounts to promote 6 to 8 in. (15–20 cm) of shoot growth annually is sufficient. As with other fruits, if a crop is lost from frost, that year can be used profitably to reshape the tree. Rejuvenation is practiced to a limited degree. It takes at least two years following severe pruning before a tree becomes productive again. This is because most of the fruit is borne on spurs arising from one-year-old wood.

TREE NUTS.   Among the many and diverse types of deciduous nuts, the fruiting habits differ widely. The English walnut, black walnut, chestnut, pecan, and filbert are monoecious and are produced in clusters on current season's growth arising from spurs or twigs. The almond is perfect-flowered and nuts are produced laterally on one-year-old twigs. Nut trees are usually longer lived than the trees of fleshy fruits. Consequently, training for strong crotch unions and good distribution of scaffolds is important. Since many nuts are terminal bearers, heading back is not usually practiced. Thinning out is used to a very limited

FIGURE 10-8   **Plums are borne mostly on spurs and sometimes laterally on one-year-old twigs.**

extent to admit light, allow for aeration, which is important for wind pollination, and to encourage growth of lateral branches for increased fruiting area. The nuts require relatively little pruning.

EVERGREEN FRUITS.   Summer pruning of deciduous plants results in greater reduction of growth than dormant pruning. Evergreen fruits such as the orange, grapefruit, lemon, lime, avocado, and fig respond to pruning in much the same manner as summer pruning of deciduous fruits. Accompanying the absence of a dormant period is a much shorter rest period for buds, continuous photosynthesis with continuous root growth, lack of a well defined period of carbohydrate storage in the trunk and branches, and among types the coexistence of buds, flowers, immature fruits, and maturing fruits all at one time. Consequently there is little stimulating or invigorating effect from pruning. It is mostly a reduction process used primarily for shaping the tree, keeping it within bounds, and opening a dense head for increased aeration, light penetration, and spraying effectiveness. Pruning of evergreen fruits does not play the extremely important role in fruit production that is accomplished by pruning in most deciduous fruits. The balance of fruitfulness and vegetativeness is regulated principally by such cultural practices as fertilizing, irrigating, and pest control.

Small Fruits   GRAPES.   The grape bears its fruit laterally in clusters near the base of current season's growth. It responds significantly to pruning. The

American type grape is one of the extremes among the fruits in that it thrives on what is considered severe pruning. It has been demonstrated repeatedly that under good growth conditions maximum yields are obtained when about 90 percent of the previous season's growth is removed during dormant pruning.

The most used method of training and pruning the American type grape (*Vitis labrusca*) is the 4-cane Kniffin system. A single trunk is trained to a trellis, and four canes, two on each side, are retained as fruiting wood. It is from these canes that shoots arise. These shoots produce flower clusters on this, the current season's growth. During dormant pruning each year, vigorous canes near the trunk are selected and cut back to 10 or 12 buds in length. All other canes, which amounts to approximately 90 percent of the previous season's growth, are removed. A stub containing about two buds is retained on each of the four arms. These are called renewal or replacement spurs. They are located near the trunk to provide growth for new canes which may possibly be used for fruiting canes the following year. Other training systems for American types include the 6-cane Kniffin with an extra tier of canes, the fan system with several radiating canes arising from a short

FIGURE 10-9   **The grape bears its fruit in clusters near the base of current season's growth. Two shoots are arising from a replacement spur.**

Pruning Methods

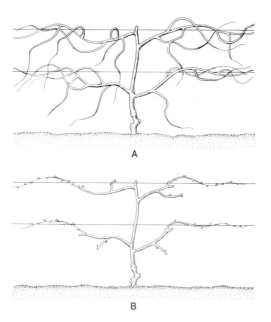

FIGURE 10-10   **The American type grape trained to the 4-cane Kniffin system.**

trunk with several arms, and the umbrella Kniffin with canes arising from the top of the trunk and tied so they taper slightly downward.

The European type grape (*Vitis vinifera*) has a stronger trunk which may even be self-supporting. Buds are closer and for some cultivars the most productive are near the base of the cane. For this reason the spur systems of training and pruning are frequently used. For some cultivars the cane system, much like the umbrella Kniffin, is better adapted. In the spur system, arms or heads on the upright trunk support canes which are pruned back to spurs of about two or three buds. With the cane system, four or five canes arising from a head are selected and trained to a trellis. The European type grape is grown principally in California and Arizona. Various adaptations of the spur and cane systems are used, depending on type of growth and cultivar.

Occasionally grapevines are neglected and not pruned for several years. When such is the case, corrective pruning should not be too drastic or yields will be sacrificed. A good rule to follow in overgrown vines is to remove about 90 percent of the previous season's growth plus some older wood each year until the vine is of the desired size. This may mean as many as 100 to 120 buds per vine instead of the usual 50 to 60 on regularly pruned vines. Grapes on arbors may be pruned in a similar manner. Vegetative growth for cover is the first objective. After shade requirements have been reached, annual removal of 90 percent of cane growth helps fulfill the second objective, production.

BRAMBLE FRUITS. The brambles bear their fruits terminally in clusters on current season's growth which arises from the previous season's canes. Although the roots and crowns are perennial, the canes are biennial. Shoots arise from the crown or roots the first year and build

FIGURE 10-11   **The raspberry fruit is borne terminally on current season's shoots arising from one-year-old canes.**

up food reserves. The following spring they produce the branch shoots which are terminated by berry clusters. After fruiting the canes die. A regular pruning program is essential to good production of the bramble fruits.

The red raspberry is pruned twice during the year. The first is a dormant pruning. The tip $\frac{1}{4}$ to $\frac{1}{3}$ of each cane is removed and weak canes are removed entirely. This eliminates the smaller and weaker buds, provides more stored food for the remainder, and results in larger berries. Thinning is also done at this time. The second pruning consists of removing the old fruiting canes. Their removal facilitates growth of the new shoots by giving them more room. Reducing possible infestation of diseases and insects is another reason for pruning out the old canes.

Black and purple raspberries are pruned three times during the season. Since they do not sucker, additional plants are not formed except by occasional natural tip-layering. Increased fruiting area is obtained by branching of the shoots in response to pruning or "pinching back" (Fig. 9-2 page 202) when they are about 24 to 30 in. (60–75 cm)

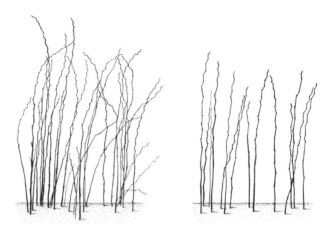

FIGURE 10-12   Pruning and thinning the red raspberry in the hedgerow.

tall. The top 4 to 6 in. (10–15 cm) is removed to adequately discourage any lateral buds from acquiring apical dominance. By the end of the growing season several long lateral branches are produced per cane. In the dormant pruning operation these laterals are cut back 6 to 15 in. (15–35 cm) depending on size and vigor. As with the red type, old canes are removed after fruiting. The juncture of the canes and crown is rather fragile in purple and black raspberries; consequently, trellises are often necessary to prevent breakage from wind and weight of fruit.

The erect blackberry is pruned in much the same manner and frequency as the black raspberry. It does differ in plant habit, however, in that it produces suckers. These suckers are used to fill in the row and replace old canes. Excess suckers should be removed as soon as they appear so the food is channeled to the remaining desired growth. New

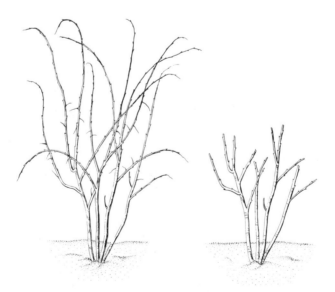

FIGURE 10-13   Dormant pruning of the black raspberry.

shoots are pinched and old canes removed after fruiting. Trellises may be used for support.

The trailing blackberry or dewberry and brambles of similar habit of growth (Boysenberry, Loganberry, etc.) are often trained to trellises of the type commonly used for grapes. They need support for easier handling and cleaner berries. Old canes are removed after fruiting. They are not generally pinched back but are allowed to trail.

BUSH FRUITS. Gooseberries and red and white currants bear their fruit from spurs on two- and three-year-old wood and at the base of one-year-old branches. Annual pruning by thinning out is the most effective means of maintaining uniform production and large clusters. Removal of old wood (over three years old) is of greatest importance for it keeps the plant within bounds, permits adequate renewal by basal shoots, and gives good light penetration.

Highbush blueberries produce fruit in clusters both laterally and terminally on the previous season's growth. Thinning out of weak branches is the principal means of pruning. Moderate pruning will result in larger fruits and better fruit set.

PRUNING HERBACEOUS PLANTS

Although most pruning operations are concerned with woody plants, they are of considerable importance to successful production in several herbaceous types. The balance of productiveness and vegetativeness can be controlled or altered by pruning of many herbaceous plants as well as with the woody types.

Strawberries   Removing runners from everbearing strawberries often results in fantastic increases in yield. Plants are set closer together and by continuous runner pruning develop large crowns. Food manufactured by leaves is diverted to flower and fruit production instead of new plants. Under conditions of no runner pruning, there is competition within the plant between runner growth and fruit production, between carbohydrate utilization and carbohydrate accumulation, between vegetativeness and productiveness. Runner pruning thus swings the balance to production with everbearing strawberries. With springbearing types continuous pruning is generally impractical. However, some cutting back of runners is usually needed to avoid crowded conditions. Certain growth regulators have been found effective for inhibiting runners.

Blossom removal following the setting of new plants is a widely used pruning practice. It enables the newly set plants to utilize their food for root and top growth and results in earlier production of runner plants.

Tomatoes   It is a common practice among many market gardeners and home gardeners to stake tomatoes. This is a tying process and must be accompanied by pruning to keep the plant productive while within

FIGURE 10-14   **Pruning chrysanthemum for large single bloom.** *Left*: **Axillary buds have been removed.** *Right*: **One massive bloom results.**

bounds. After the ultimate height is reached, the plant is decapitated. Axillary shoots are removed to reduce the number of growing points and to favor fruit production.

Chrysanthemum   Two types of pruning are practiced with "mums" depending on the objective of the grower. The cushion and border chrysanthemums are most effective when presenting a mass of blooms. Pinching back of the terminal shoots in early summer destroys apical dominance and promotes branching. Each branch is terminated by blooms when the autumn season arrives and contributes to the mass bloom effect.

With large "mums," size of individual blooms is greatly increased by training to a single stem. Disbudding of all axillary buds results in one large bloom per stem.

Annuals for   Adjusting the ratio of roots and tops is the purpose of pruning for
Transplanting   transplanting. The tender, soft, and succulent roots of annuals like tomatoes, petunias, snapdragons, and peppers are damaged by exposure to air, lifting from the soil, and handling. To compensate for the loss of absorbing area, some of the foliage is removed. This reduces transpiration, and wilting is less severe.

STUDY QUESTIONS

1. Why is the selection of pruning equipment considered important?
2. Contrast the results of shearing or clipping of twigs with complete removal of those twigs.

3. Why is the Pfitzer juniper not generally sheared as may be done with Sabin or Maney juniper?
4. List the advantages and disadvantages of outside house paint and asphalt as wound treatment compounds for pruning.
5. Does severe pruning of "everblooming" roses produce long-stemmed or short-stemmed cut roses? Explain.
6. How can root pruning help to keep a shrub within bounds?
7. Does an apple cultivar grafted on dwarfing or semi-dwarfing stock have the potential for increased production per tree? per unit of area?
8. Why is it important to know and understand the fruiting habits of fruit trees?
9. What are the approximate amounts (in percent) of previous season's growth removed in the annual pruning of apple, peach, raspberry, and grape?
10. Give the procedure for pruning certain kinds of herbaceous plants for increased production.

SELECTED REFERENCES

1. Childers, N. F. 1976. *Modern Fruit Science.* Horticultural Publications, New Brunswick, N.J.
2. Christopher, E. P. 1954. *The Pruning Manual.* Macmillan, New York.
3. Denisen, E. L., and H. E. Nichols. 1962. *Laboratory Manual in Horticulture*, 4th ed. Iowa State University Press, Ames, Iowa.
4. Teskey, B. J., and J. S. Shoemaker. 1972. *Tree Fruit Production.* AVI Publ. Co., Westport, Conn.

11 Pests of Horticultural Plants and Their Control

Mankind has been plagued by the ravages of pests on crop plants since the dawn of history. The Bible makes frequent references to pests of horticultural plants. Hordes of locusts devoured entire crops resulting in famine. The mustard and other weeds smothered out crop plants, robbing them of moisture and light. The fig and olive became barren, suggesting degenerative diseases. Rats and mice consumed stored food, destroying the fruits of toil. Other pages of history also record the impact of plant pests on the destiny of many peoples. The late blight epidemic of potatoes in Ireland caused the great famine of 1849, which resulted in the migration of many Irish to the Western Hemisphere. The introduction of Johnson grass to America for a meadow and pasture crop resulted in its uncontrolled spread. It is now considered a serious weed and much expense and effort are being employed toward its control. The shift in agriculture to intensive cultivation and specialization has increased the problem of the specialized pests. Because an orchard, a flower crop, or a vegetable field provides such an abundance of host plants to a pest, that pest establishes itself and becomes a perennial problem. This has had a tremendous effect on the rapid development of chemical pesticide industries.

In spite of its long history, and our improved techniques of culture, pest control remains one of the principal problems of the horticulturist. The losses due to pests of horticultural crops are in the billions of dollars annually throughout the world. This loss is composed not only of reduced yields of crops ravaged by pests, but also of the added expenditures to combat pests and the extra man-hours of labor required for control. All of these are costs and losses which would not be involved were pests not such a universal problem.

CLASSIFYING HORTICULTURAL PESTS

The pests of horticultural plants can be logically placed in four groups: (1) insects, mites, and nematodes, (2) diseases, (3) rodents, and (4) weeds. The *kind* of pest and its *life history* are important factors in developing effective control practices.

Insects, Mites and Nematodes   This group of pests belongs to the lower and more primitive forms of animals. They have either chewing or sucking mouth parts. This is important from the type of injury and control standpoint. Insects in the

FIGURE 11-1   **Pests of the strawberry leaf.** *Left*: **leaf spot, a fungus disease.** *Right*: **feeding injury of strawberry slug, an insect pest. (Courtesy Iowa State University, Ames, Iowa.)**

adult form have six legs, mites have eight legs, and nematodes, being true eelworms, have no legs.

Insects have a life cycle or metamorphosis, which is described as either complete or incomplete. The complete metamorphosis consists of four stages: egg, larva (or "worm" stage), pupa (or resting stage), and adult. The cabbage worm is an excellent example. The egg hatches into a larva or cabbage worm which feeds voraciously, with chewing mouth parts, on the cabbage leaves. When full grown it forms a cocoon and becomes a pupa. A few weeks later it emerges in the adult form, a butterfly in this instance. Incomplete metamorphoses are not specific in numbers of stages, forms, or even uniformity of life cycle from one generation to the next. The gladiolus thrips has an incomplete metamorphosis. The egg hatches into a nymph which grows to be an adult. The nymph can best be described as a miniature adult. The aphid family has various types of incomplete metamorphoses. Some young of the aphid are born rather than hatched. Some adults are winged and others wingless. Some females lay eggs without being fertilized, a type of asexual reproduction in animals.

Insects vary regarding their destructive stage or stages. The codling moth, or apple worm, is one of the most destructive pests of the apple. It tunnels in the fruit during the larval stage, utilizing its chewing mouth parts. The adult moth is harmless in itself, except as an egg-layer bringing on the next generation. Insect pests with a nymph stage are destructive both while nymphs and while adults. The leafhopper nymph pierces the cells of leaves with its sharp proboscis, or snout, and sucks

FIGURE 11-2 **Result of grasshopper feeding on rhubarb leaves.**

the plant juices into its body. It continues this type of feeding into and throughout its adult life. The grasshopper nymph feeds on foliage with chewing mouth parts. It is a voracious feeder and grows rapidly. It may pass through two nymph stages before it becomes a winged adult.

The wintering stage and location are important to the life history of any insect in temperate climates. It is a period which may determine the ability of an insect to survive and be adapted to an area. The wintering stage of an insect is essentially a hibernation. It occurs over the entire range of stages of the various metamorphoses but is generally specific for each kind of insect. The codling moth overwinters as a larva in a cocoon which is often found on the rough scaly bark of the apple tree or on walls and equipment in the packing shed. The potato beetle overwinters as a pupa in the soil. The plum curculio overwinters in trash and refuse and along fence rows as an adult. Some aphids overwinter as eggs on the bud scales of woody plants.

Mites have an incomplete metamorphosis consisting of egg, nymph, and adult. The red spider mite and the two-spotted mite are extremely small, barely visible to the naked eye. They are wingless and must crawl or be transported to their feeding place on plants. They have sucking mouth parts and destroy leaf cells by removing cell sap. For winter survival they may hibernate in the protected crowns of plants. However, in cold climates the winter mortality is extremely high. Once they become established, they are a serious pest of household and greenhouse plants. They thrive under continued warmth and have no apparent resting stage after they have hatched.

Nematodes or eelworms infect the roots and crowns of plants. They are nearly microscopic in size and may be either ectoparasitic, which is outside the roots, or endoparasitic, inside the roots. Some types, such as the root-knot nematode, form "knots" or galls on the root system. The nematodes feed on living roots and suck or absorb plant juices and food from the plant.

Insects, mites, and nematodes have periods of *vulnerability*. These periods during their life cycle are of major concern to the horticulturist. It is the phase or period during which they are most vulnerable to the control efforts of the grower. For some insects this period is during the larval stage when they are feeding voraciously on foliage. It may be during the flight and egg-laying period of adults which is true of most woody stem borers. It can be during overwintering, when defenseless and immobile, which finds curculio hibernating in trash. It can be during wet weather when the red spider mite is easily washed or shaken from

FIGURE 11-3   **Red spider mite on chrysanthemum. Mites are on web.**

Pests of Horticultural Plants and Their Control   **243**

the leaves. It is prior to planting for most nematodes, for at that time the soil can be fumigated and plant roots can be treated.

Diseases   **Causal agents** which are neccessary for plant diseases to develop can be grouped in four categories: (1) fungi, (2) bacteria, (3) viruses, and (4) certain environmental conditions. *Fungi* are multicellular plants that have no chlorophyll, thus are not capable of manufacturing their own food. Parasitic fungi live off other living things such as plants. They are reproduced by spores which are tiny, usually one-celled, bodies capable of germinating and giving rise to new fungus growths. Apple scab, anthracnose of raspberry, downy mildew of grape, and brown rot of stone fruits are specific diseases caused by fungi. Most fungus diseases are localized on their host plant. They spread by invasion of adjacent tissue but are not generally systemic, that is, transported internally in the plant's vascular system. Since invasion is from without, protective sprays and dusts can often prevent infection.

Bacteria are one-celled organisms. Those types which are parasitic on plants invade various tissues and produce rots, cankers, and blights. They may also get into the vascular system and produce bacterial wilt of certain plants by plugging the xylem vessels. Reproduction in bacteria is by fission, a simple cell division. Sprays and dusts are generally ineffective for controlling bacterial diseases unless they are in contact with the bacteria before infection occurs.

Viruses are not living entities in themselves. They must be in a proper environment, such as a plant cell, before they are capable of regenerating themselves and spreading to other cells. Viruses can be separated from plant tissue by centrifuging. They can be crystallized, then reinoculated into noninfected plants to produce the virus disease again. They are systemic in the plant. Aster yellows, measles of geranium, mosaic of potatoes, and mild streak of raspberries are virus diseases. Viruses

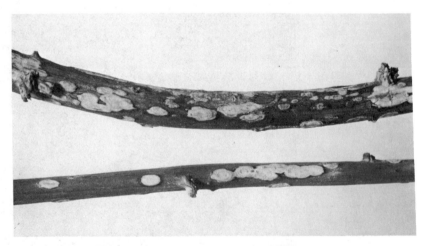

FIGURE 11-4   **Anthracnose lesions on specimen raspberry canes. (Courtesy Iowa State University, Ames, Iowa.)**

FIGURE 11-5 Measles, a virus disease of geranium. (Courtesy C. H. Sherwood, Iowa State University, Ames, Iowa.)

rarely kill the host plant, but they frequently have a devastating effect on production.

Environmental or *physiological* diseases are caused by one or more unfavorable conditions of the environment. They are usually marked by a collapse or breakdown of tissues. Jonathan spot of apples is caused by uneven temperatures during storage or occurs on fruits that are over-mature when harvested. Blossom-end rot of tomatoes is often found during periods of drouth but has also been traced to inadequate phosphorus in the soil. The nutrient deficiency diseases of plants can actually be considered in the category of environmental or physiological diseases.

Means of transmission of the causal agents include (1) soil, (2) wind, (3) water, (4) seed, (5) vegetative propagation, (6) mechanical, and (7) insects. The *soil* is the home of many fungi, especially of many of the lower forms which attack a vast number of plants both dead and living. The molds are typical of this group. Plant refuse in the soil may harbor fungi, bacteria, and also insects which spread disease. Diseases may spread to other host plants of the same or related kinds or, if not specific, to most succeeding crops. Damping-off of seedlings, a disease of the seedbed, is caused by soil-borne bacteria and fungi, which attack many species of plants. Seed decay and root rots are other effects of this soil inhabiting complex of bacteria and fungi.

Wind blows and disseminates the spores of many fungi. The spores of cedar-apple rust are blown between alternate hosts, the apple and cedar, upon their release from their respective fungal bodies. Apple scab fungi will winter over on dead leaves of the past season. During sporulation in the spring, the apple scab spores are released into the atmosphere and further disseminated by winds.

FIGURE 11-6   **Environmental diseases. *Top*: Jonathan spot of apples. *Bottom*: Blossom-end rot of tomatoes. (Top photograph courtesy Iowa State University, Ames, Iowa.)**

Water aids in disease transmission as humidity, rain, and flowing liquid. Many spore-producing bodies need moisture to release their contents. Nearly all fungi need moisture for germination. Bacteria are carried or are mobile in water. Fireblight, a bacterial disease, can be

FIGURE 11-7 **Cedar-apple rust on alternate hosts.** *Left*: **apple leaf with rust lesions.** *Right*: **cedar (juniper) twig with sporulating rust gall. (Photo on right, courtesy Iowa State University, Ames, Iowa.)**

spread by splashing rain or water that is washing down the twigs from infected tips. In a dry season, or under dry conditions during spore production, fungus diseases are relatively less severe than during moist, humid weather.

Seeds can be the means of transmission for some fungi and bacteria. Occasionally bacterial and fungal damping-off organisms are carried on the seed, although the more usual habitat of these agents is the soil. Some fungus diseases actually infect the seed, where they remain in a resting condition, and resume growth when the seed germinates. It is rare for seeds to be infected with viruses.

Vegetative reproductive structures of plants are a frequent means of disseminating viruses and other systemic diseases. Potato tubers from plants with mosaic will also have the disease and in turn will produce foliage that is infected. Strawberry plants with viruses produce runner plants with viruses. Some cultivars of asexually propagated plants seem entirely infected with certain viruses.

Mechanical transmission of diseases may occur during cultivation, pruning, and other cultural operations. It involves not only the transfer of the causal agent but must also provide a channel of infection if the disease is to be spread. Pruning in an apple or pear orchard where fireblight has been active can spread the disease unless proper precautions are taken. Crown gall can be spread by pruning and grafting operations. These are both bacterial diseases on woody plants. Mosaic on beans, potato viruses, and tomato viruses can be spread by cultivation. The cultivating equipment contacts infected foliage, brushes

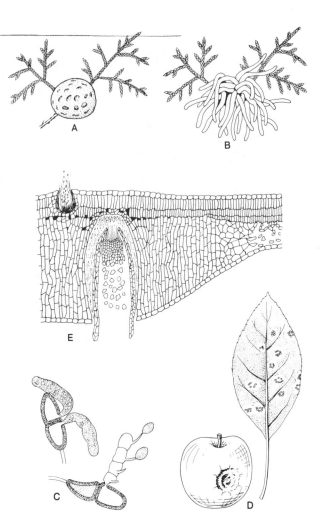

FIGURE 11-8   **Life cycle of cedar-apple rust. Of the two-year cycle, 19 months is spent on the cedar and 5 months on the apple. *A*, gall on cedar; *B*, gall develops tentacles and sporulates; *C*, spores released to apple; *D*, spores develop on leaves and fruit of the apple; *E*, new spores released and return to cedar tree. Total cycle 2 years.**

against it, and transfers the virus to healthy plants. This means of spread is further facilitated by the presence of dew on the plants. Bacterial ring rot of potatoes may be spread by cutting knives while preparing seed pieces for planting.

Insects are an extremely important instrument of disease transmission for many horticultural plants. The 12-spotted or striped cucumber beetle can transfer bacterial wilt of cucurbits from infected plants to healthy plants. The bacteria even overwinter in the digestive tracts of the beetles. Infection of new plants thus occurs after emergence of the beetle from its overwintering stage when young cucumbers, pumpkins, squash, and melons are getting started. Insects which transmit diseases are called *vectors* of that disease. In many cases the insect vectors cause

FIGURE 11-9   Cedar and apple in adjacent positions make control of cedar-apple rust extremely difficult.

more damage and destruction through the disease they transmit than by their own feeding on crop plants. Many kinds of aphids are vectors for virus diseases on a wide range of crops, including strawberries, potatoes, chrysanthemums, and cherries. The bark beetle is a vector for Dutch elm disease, which has destroyed many plants of one of the most widely grown ornamental trees, the American elm. Insects can also be an aid to infection and not actually carry the causal agent. Feeding punctures and egg-laying wounds can be a means of causal agent entrance to the tissues and cells of a plant.

Active and dormant stages of plant diseases influence the control programs. *Destructive stages* of plant diseases occur during active growth of fungi and bacteria and when viruses affect the growth and productivity of plants. The late blight fungus on potatoes and tomatoes infects the leaves. Blighting and loss of considerable leaf area reduces photosynthesis and total production. The fungus also moves down the stem to the tuber. It may continue active growth in the tuber and produce a rot; or it may remain in an inactive state during the early part of the storage period and develop later as a storage rot. Some viruses are present in plants with, seemingly, no detrimental effect. The latent mosaic virus of Irish Cobbler potatoes was not recognized for many years and then caused little concern, since the infected variety continued to be productive. It is capable of spreading to other varieties where it does considerable damage.

Any phase of an environmental or physiological disease can be considered destructive. Jonathan spot of apple and blossom-end rot of tomato not only destroy some of the tissue of the fruits involved, but decrease their market value, consumer appeal, and monetary return.

Overwintering stages of plant diseases may have considerable effect on their adaptibility to an area and their potential for reaching epidemic proportions. Most fungi overwinter in the spore stage which is adapted for tolerating greater extremes in temperature than are the growing fungal bodies. Natural mulches on the soil such as leaves and snow often protect dormant spores. In perennial plants they may stay on the host. Bacteria frequently overwinter in the soil, e.g., soft rot of carrots, on the host plant, e.g., fire blight of apple and pear, and in some cases in the digestive tract of insects, e.g., bacterial wilt of cucurbits. Viruses remain in the host plant, in vegetative organs used for propagation, and in crystalline form at favorable temperatures.

FIGURE 11-10   **Slime mold in strawberry bed. This fungus overwinters in the spore stage on organic matter.**

Vulnerability of the causal agents of plant diseases vary with locality. During germination of spores and early growth of fungi, they are extremely susceptible to injury by chemical toxicants. In the fallen leaves and dropped fruits they are subject to disposal by the clean gardener. They are favored in growth but also are more vulnerable to spray injury during humid weather. Bacteria are most vulnerable while present on the surface of plants before they gain entrance. Systemic bacteria and viruses are quite well protected within the host plant. However, if the plant shows symptoms as with cucumber wilt and tomato mosaic, they are more vulnerable since the plant can be "rogued" and destroyed. Diseases spread by insects are vulnerable in that control of the insect vector controls spread of the disease.

Rodents
Rabbits, rats, mice, gophers, and *moles* are called rodents, a class of mammals which feed by gnawing and nibbling on plants and plant products. *Rabbits* are offensive to gardeners in eating young shoots of tulips, vegetables, and other herbaceous plants. They can be extremely destructive of young fruit trees, brambles, and ornamental shrubs when rabbit food is scarce during winter months. Their gnawing and eating of the bark, through the phloem and cambium, often kills the woody plants. It may result in complete girdling. Root starvation and drying of the xylem make survival very difficult.

Rats and *mice* are especially offensive in storages where they have a haven among food supplies. Mice can be extremely troublesome to trees and shrubs. They burrow in the soil around the base of woody plants and feed on the young and tender feeder roots.

Gophers and *moles* are soil-inhabiting. Gophers feed on the roots and crowns of herbaceous perennials such as strawberries, lawn clovers, and lawn grasses. Moles tunnel in lawns in their search for insects.

Vulnerability of rodents is proportionate to their offensiveness. Rabbits that are damaging horticultural plants are subject to elimination by traps or firearms. The food-eating tendencies of all rodents make them subject to poison baits and repellents. Dogs and cats are natural enemies of rodents. The trails left by moles are indicative of their presence, and the burrows of other rodents make them subject to asphyxiation.

Weeds
Just as the rodents are plant pests of a more advanced form of animal life than the insects, so are weeds a more advanced form of plant life than the fungi. It is through competition for moisture, soil nutrients, and light that certain plants are classed as weeds. Weeds can be grouped as: (1) annuals, (2) winter annuals or biennials, and (3) perennials. Some which are very difficult to control and are extremely serious pests are termed "noxious" weeds.

Annual weeds are propagated by seed. They are easily controlled by cultivation and specific herbicides during their early growth. They become more difficult to eliminate if permitted to grow to larger size. Lambs quarters (*Chenopodium album*), red-root pigweed (*Amaranthus retroflexus*), smartweeds (*Polygonum* spp.), and foxtail grasses (*Setaria* spp.) are typical of the annual weeds. Some are very low growing, others are extremely large, attaining heights of 6 to 8 feet (2–2½ meters) during the growing season. They are generally very productive of seeds and unless controlled will disburse seed many fold for succeeding seasons. Many of the common weeds are responsive to photoperiod and temperature relations. Late germinating pigweeds will often form flowers and seed heads while very small and they mature their seeds before they are killed by fall frosts.

Winter annuals can be considered biennial because they grow during two seasons. However, these weeds generally germinate in late summer or fall, they will winter over, and then resume growth and flower the next spring. Typical examples include shepherd's purse (*Capsella Bursapastoris*), peppergrass (*Lepidium virginicum*), chickweed (*Stellaria media*), and squirrel-tail grass (*Hordeum jubatum*). They are serious weed problems the following spring in strawberries which were protected by a winter mulch.

Perennial weeds pose a serious problem in many horticultural crops, especially perennial crops. Many perennial weeds are capable of propagating themselves both by seed and vegetatively. These very troublesome weeds include field bindweed (*Convolvus arvensis*), Canada thistle (*Cirsium arvense*), quack grass or couch grass (*Agropyron repens*), leafy spurge (*Euphorbia esula*), and Johnson grass (*Sorghum halepense*). Other perennial weeds which are propagated primarily by seed— dandelion (*Taraxicum officinale*), buckhorn plantain (*Plantago lanceolata*), and curled dock (*Rumex crispus*)—can become extremely competitive even as individual plants. The perennial weeds are often deep rooted, very persistent, and withstand many efforts at elimination. In some cases a crop season has to be sacrificed in order to reclaim land infested with perennial weeds.

METHODS OF PEST CONTROL

Physical Methods   *Sanitation* is one of the most common and effective methods of controlling pests of horticultural plants. The raking and burning of dead leaves infected with apple scab, leaf roller, or mildew decreases the likelihood of carry-over to the next crop. The burning or deep burial of plants contaminated with viruses or bacteria prevents their spread. Removal of trash along fence-rows and neglected areas removes the hibernating place for insects and diseases. Sanitation around buildings, equipment, and supplies greatly reduces losses from rodent activity. If no haven is provided for rats and mice they are not likely to become a problem. Prompt removal of fallen fruits in an orchard is a sanitation measure that eliminates many insects which otherwise would be next year's egg-laying adults.

Rotation of crops is especially helpful in preventing a build-up of insect, nematode, and disease populations in short-term crops. Organisms that winter over in the soil or in the refuse of plants have less opportunity to infect their favorite host if the planting plan calls for a change of site. Allied to crop rotation is the elimination of weed hosts of the same species or host range of the crop. Verticillium wilt, a fungus, has many weed hosts, in addition to tomatoes, potatoes, and berry crops. Soil infected with the crowngall bacterium should be given a rest from woody plants for several years. Irish potatoes succeeding themselves on the same land are likely to have more scab on the tubers the second year.

Cultivation should be done only as necessary. Plowing is an effective means of turning under trash. In some cases, late fall plowing may be more helpful in pest control than early fall plowing. With the potato beetle, the larva enters the soil after feeding, goes into the pupal stage to overwinter, and is protected from the extremes of weather by several inches of soil. Plowing after the beetle has settled itself for the winter, places it in an exposed or near surface position where survival is less likely. Cultivating the crop for weed control, not only removes competing plants but also eliminates other hosts of insects, nematodes, mites, and diseases.

Guards of wire, screen, cloth, and paper are effective means of protecting various horticultural plants. Encasing the trunks of small trees with rolls of hardware cloth or wire screens is a precaution against girdling by rabbits and mice. Paper collars around the stems of newly set vegetable and flower transplants guard against cutworms.

Pruning of infected twigs, branches, and shoots helps in controlling some diseases such as fireblight and bacterial canker. It should be used with discretion. Pruning should not be excessive. It may stimulate excess shoot growth which is usually more subject to reinfection. The pruning cut should be made well below the point of visible infection. Even with this precaution, the cutting equipment should be disinfected between each cut to avoid possible spread by pruning. Bichloride of mercury solution is a good disinfectant.

Roguing of plants with serious diseases is a regular practice of progressive vegetable growers, nurseries, seedsmen, berry growers, and amateur gardeners. It consists of uprooting the plant and destroying it, usually by burning. Good sanitation methods should accompany roguing. A rogued plant should not be carried about the field while looking for further infection. It may contact other plants and further distribute the disease. The purpose of roguing is to remove sources of infection.

Seed purification and grading improves the quality of seeds. Purification refers primarily to weed seed removal but also includes elimination of chaff, dirt, and other foreign matter. Grading is done on the basis of seed size and shape. Large seeds, within a cultivar produce more vigorous seedlings than small seeds. Mechanical planters give more uniform distribution of graded seeds than if the seeds are different shapes and sizes. Uniform distribution of vigorous plants helps provide favorable growth conditions. Vigorously growing plants are more tolerant of disease and insect pests than weak and crowded plants.

Trapping of both insects and rodents is effective in reducing pest problems. Plowing a furrow around a field infested with army worms will trap the migrating larva where they can be treated with a suitable toxicant, or insecticide. Trap crops can be used to attract insect pests.

The insect goes to his favorite crop, which may be sacrificed to obtain freedom from the pest on the primary crop. The corn ear worm and the tomato fruit worm are the same insect. Sweet corn is preferred by the insect, and when planted around or near tomato fields, it serves as a trap

FIGURE 11-11   **Wire guard on young fruit tree as a precaution against girdling by rabbits.**

FIGURE 11-12   **Broccoli plants growing in tin cans with the bottoms removed serving as a guard against cutworms. (Photograph by K. A. Denisen.)**

crop. Different planting dates or maturity dates can spread the period of effectiveness of the sweet corn. Small-seeded legumes are favored by leafhoppers and can give considerable protection to bramble fruits and vegetables. The effectiveness of these methods is in large part determined by a plan for extermination of the insect pest when concentrated in the trap crop.

Isolation of the crop is a technique employed to avoid insect and disease pests. The insect vector for most viruses of white potatoes is the aphid. In climates of extremely cold winters and cool summers, aphids are not a pest. Since they are not present, virus-free stock can be easily maintained in a virus-free condition. Isolation is thus a means whereby foundation stock can be maintained and produced. Prince Edward Island, Canada and Fairbanks, Alaska provide conditions favorable to production of foundation stock. Other locations may have certain characteristics of climate, sea, native flora, or natural enemies which render the area isolated from certain pests.

Chemical Methods   Numerous chemical pesticides are available for use as sprays, dusts, paints, soil treatments, and vapors. Many cover a wide range of activity and others are quite specific in their usefulness. All should be treated with respect, and warning cautions on the manufacturer's label should be observed.

Insecticides include (1) stomach poisons, (2) contact insecticides, (3) systemic poisons, and (4) fumigants. *Stomach poisons*, which include the arsenicals, cryolite, and ryania, are effective against the insects with

FIGURE 11-13   **Young plants of spinach dusted with an insecticide-fungicide mixture for late summer pest control. (Photograph by K. A. Denisen.)**

chewing mouth parts. Lead arsenate sprayed on the foliage of elm trees is eaten by the spring canker worm along with the leaves. Ingestion of the poison gives a quick kill. The stomach poisons are generally ineffective against sucking insects. *Contact insecticides* kill by physical contact of the insect and the chemical. The chemical may be in a solid, liquid, or gaseous state at the time of contact. The insecticide is either inhaled or absorbed and need not be eaten. Most organic insecticides are of this type. They vary in their toxicity to mammals and should be handled with caution. Their range of effectiveness as insecticides includes both sucking and chewing insects. Those which are useful in killing insect eggs are called *ovicides.*

Systemic poisons are absorbed by the plant, through the roots or foliage, and translocated to other parts of the plant. Insects feeding on the plant are killed or repelled by the systemic insecticide. Systox and thymet are examples of such materials. Since the chemicals become a part of the plant's system, their use on food crops must be carefully guarded. Only after considerable research on residues and human tolerances can these chemicals be declared safe or unsafe for use on food crops. Reading the label and observing the precautions becomes increasingly important as systemic insecticides become more abundant and widely used. Coordinated efforts of insecticide manufacturers, governmental agencies, and growers are essential to efficient use of systemic insectides.

Fumigants are actually contact insecticides since they kill by entering the respiratory tract of insects. However, certain contact insecticides function only as vapors or fumes, thus the more distinctive term, fumigants. Paradichlorobenzene (PDB) kills woody plant borers by fumigant action. This same material is the active ingredient of moth crystals.

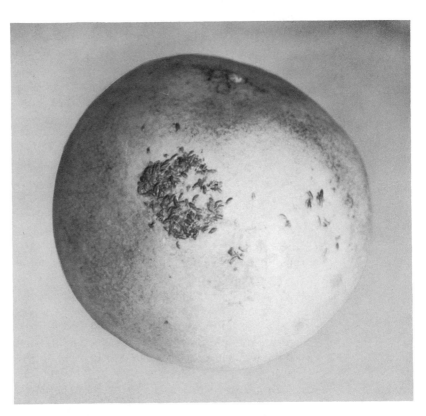

FIGURE 11-14   **Purple scale on grapefruit.**

Cyanide gas (hydrocyanic acid) is a powerful fumigant. It has been used for scale insects on citrus by placing huge gasproof "tents" over the trees. A parathion "bomb" placed in a closed greenhouse is a fumigant for all susceptible insects. All doors should be locked and poison gas signs posted during greenhouse fumigation. The structure should be thoroughly ventilated following the operation.

Miticides are specifically intended for mite control. They are often effective for some insects also. Nicotine sulfate and malathion are effective miticides. The two-spotted mite and red spider mite on conifers can be controlled by either of these materials.

Nematocides are for nematode control. Vapam, methyl bromide, and ethylene dibromide are soil sterilants and effective nematocides. Pre-planting treatment of the soil, followed by planting nematode-free stock, results in vigorous growth. This treatment with perennial crops may give control for several seasons before they become reinfested.

Insects may develop resistance to certain insecticides in time. It is rare to get complete and absolute control of insects with a single insecticide. Those insects which survive may be less subject to injury because of a certain degree of resistance. As they mate with other survivors, more of the progeny will have resistance. After several generations it may seem that the insecticide has lost some of its effectiveness. In reality the insecticide has not changed, but the insects have actually built up a "super-race" which resists the insecticidal action. It thus appears there shall continue to be a need for new and different insecticides.

Fungicides and antibiotics are used in combatting plant diseases. Most fungus diseases are readily controlled at certain stages by *fungicides*. Bordeaux mixture, consisting of copper sulfate and lime, is one of the oldest of the fungicides and modified forms of it are still used for potato late blight and mildew on lilacs. Sulfur is also one of the old, yet widely used, fungicides. More recent developments in fungicides have been organic chemicals such as ferbam, captan, dithane, and tersan. These materials, where effective, have generally given less injury to foliage and fruits than some of the earlier inorganic chemicals. Those fungicides which are effective against bacterial diseases have also been termed *bacteriacides*. Bordeaux mixture will control the blossom blight phase of fire blight in apples and pears, and as such, it is a bacteriacide.

Antibiotics have been used in controlling some plant diseases. Antibiotics are absorbed through the leaves and are residual within the plant for varying periods. They are used principally against bacterial diseases located in the tissues. They are impractical for most fungus diseases which can be controlled more readily with fungicides.

Fungi and bacteria may "mutate" to "superior strains" which are resistant to the chemicals which killed the original types. Consequently, there remains a constant need for new and different fungicides, bacteriacides, and antibiotics.

Rodenticides include *asphyxiants* and *poisons*. A common asphyxiant, cyanide gas, is produced when cyanide pellets are placed in burrows of gophers and moles. The burrows are closed and cyanide gas fumes kill the rodents. Poisoned grain or other food may be placed in the runways and burrows for effective control of rodents. Warfarin, a rodent poison, is excellent for killing rats and mice. With all poisons, extreme care should be observed to avoid ingestion by humans, pets, and livestock. The use of *repellents* is a means of discouraging the visits of rodents without killing them. Dusting sulfur on emerging tulips in early spring reduces rabbit injury of the tender unopened blooms.

Herbicides, or weed killers, may be *contact* or *systemic* in their killing action. The contact herbicides kill only the plant parts which they contact. On small weeds the entire plant dies. On large weeds the foliage dies, but the stem usually remains and new shoots arise later from axillary buds. The dinitro compounds (Dow General, Premerge, and Sinox) are typical of this group. They also have residual characteristics which make them effective over a longer period as *pre-emergence* herbicides. These may be applied before crop emergence but not necessarily before weed emergence. The herbicides which dissipate after application include the petroleum compounds such as Stoddard solvent (a dry-cleaning fluid), fuel oil, and kerosene. They are also used on weeds which are present before crop emergence. On carrots, these petroleum herbicides are selective and can be applied *post emergence* without injury to the carrots. They give complete kill of nearly all annual weed species.

Systemic herbicides do not give immediate kill. They are absorbed and then translocated to other parts of the plant. Plants so killed are

FIGURE 11-15   **2,4-D injury to grape leaves resulting from spray drift. (Courtesy Iowa State University, Ames, Iowa.)**

killed in their entirety, including the root system and storage organs. Dalapon, 2,4-D, MCP, and amino triazole are systemic herbicides while 2,4-D also produces a hormonal action. It may cause proliferated growth and in light doses is sometimes a growth stimulant. On turf grasses, 2,4-D is selective. It kills most broadleafed weeds without apparent injury to the established grass.

All herbicides should be handled carefully to avoid injury to man, animals, and desirable plants. In many cases, spray drift from 2,4-D and other chemicals has caused serious losses of vegetables, shrubs, and flowers. Some herbicides are absorbed through the skin, so prolonged contact should be avoided. It is essential to read and observe the precautions on the label.

Natural *Methods*   Insects, diseases, rodents, and weeds have natural enemies just as they are the natural enemies of the plants on which they prey. Birds destroy many insects and help to keep them under control. Hawks and owls eat mice. Insects parasitize other insects; braconid wasps grow and pupate on the tomato hornworm. Insects eat other insects; lady-bird beetles feed on scale insects. Some fungi and bacteria parasitize other

FIGURE 11-16   **Pupae of the braconid wasp parasitizing the tomato hornworm.**

fungi, or they may produce toxins that are lethal to other fungi and bacteria. The *Chrysolina* beetle parasitizes the Klamath weed and keeps it under control.

Even the environment plays an important role in the natural control of insects. Heat, cold, drouth, floods, low humidity, and bright sunshine, each has its detrimental effects on certain of the pests of horticultural crops. Indeed Nature often restores the balance of productivity following the ravages of serious pests.

Introduction of natural enemies sometimes results in good control of pests. However, before such introduction is made, it should be determined that the natural enemy is not likely to become a pest in itself, once it has destroyed the present pest.

Legal Methods   Laws have been enacted and regulatory control established as aids to better pest control. Methods of legal action include: (1) quarantine, (2) inspection, (3) regulatory enforcement, (4) pesticide tolerances, and (5) approval or clearance of pesticides. The last two are functions of the Food and Drug Administration and Environmental Protection Agency (EPA) of the United States.

Quarantine laws have been established in nearly every country and to some extent between states. They are designed to prevent plant pests of other areas from gaining entrance and becoming serious economic pests. This type of law is frequently annoying to the importer or exporter of plants, yet it is a protection against widespread distribution of serious pests. There are very good reasons which cause customs officials to deny entrance of certain vegetatively propagated plants or seeds.

Inspection is used widely in the nursery and seed trade. It is not only to comply with state or federal laws that nurseries and seedsmen desire inspection. It indicates that their produce has passed inspection and can be so labeled. Inspection houses operate in conjunction with customs officials in enforcing quarantine laws.

Regulatory enforcement is required for those few who ignore laws and regulations or who may be unaware that such laws exist. Weed laws are of special significance under regulatory control. Seed production of noxious weeds should be prevented. Serious weeds, whose seeds are disseminated by wind, should not be permitted to develop to the seed stage. Most horticulturists observe these points for their own and community welfare without prodding by enforcement officials.

Pesticide tolerances are determined on the basis of mammalian toxicity of residues. Other factors such as latest possible date of application and maximum residue in parts per million (ppm) of the edible product are specified. The tolerances established are well within the limits of safety. This is an added precaution by the Food and Drug Administration, for the health of the consuming public.

FIGURE 11-17 **Fruits are among the crops that may have spray residues on the edible product. New spray materials must be approved by the Environmental Protection Agency (EPA) before their use can be recommended. (Photograph courtesy John Bean Division, Food Machinery and Chemical Corporation.)**

Approval or clearance of pesticides is based on ingestion tests on experimental animals, residual quantities on fruits and vegetables, chemical changes and breakdown products of the pesticides, and cumulative effects of continued use on animals. The results of tests by the manufacturer together with other data concerning the pesticide must be submitted to the Food and Drug Administration for their approval. Permission is then granted to print on the pesticide label recommendations for specific crops and uses.

Genetic Methods Some species, cultivars, or strains of plants are resistant to certain disease and insect pests. Under epidemic conditions they may be the only plants which survive. Their importance and value is much amplified by such an occurrence. There may be a shift to these resistant types by growers. If their quality and yield in non-epidemic seasons is of questionable value, these plants may be advantageously used by plant breeders as parents of new cultivars or hybrids.

SPRAY SCHEDULES

In many programs of pest control, more than one application of pesticides is needed. This involves correct timing and selection of proper materials. A schedule can be set up in advance to plan the program, anticipate the pesticide needs, and avoid incompatible combinations of fungicides and insecticides. It is essentially a plan for organized and systematic control of pests. It is planned to prevent insect and disease infestations before they reach epidemic proportions.

The experiment stations and extension services of most states develop and distribute spray schedules that are adaptable to local conditions. These schedules indicate recommended materials, periods or dates of application, quantities of each pesticide to use, and frequency of application. They may also suggest modifications in the event a less common pest problem should develop. These spray schedules are available on request.

STUDY QUESTIONS

1. How does the Environmental Protection Agency (EPA) influence the control of horticultural pests?
2. Is quarantine an effective means of pest control? Give examples. What are its principal weaknesses?
3. What is a predator? What is its potential in pest control?
4. What is a hazard or shortcoming of biological control with predators? with parasites?
5. Can pest control be completely attained under biological control with *natural enemies*? with *pesticides*?

FIGURE 11-18   Spraying potatoes for disease and insect control. Spray schedules indicate materials to use and frequency of application. (Photograph courtesy of John Bean Division, Food Machinery and Chemical Corporation.)

6. Under biological control, what does the predator feed upon after completely eliminating a serious pest of tree fruits? of vegetables? of ornamentals?

7. How are pest-resistant cultivars acquired? Who vouches for their resistance? Is 100 percent control attained?

8. Is it simpler to remove the causal agent than to control directly? If so, how is it done?

9. What is a systemic insecticide? Are there hazards in its use? What precautions must be followed?

10. Can rotation of kinds, cultivars, or strains of vegetables be helpful to an economic extent for pest control?

11. In what instances does the plowing under of diseased or insect infested plants not control the pests involved?

12. Distinguish between stomach poisons, contact poisons, anti-fungals, and systemic poisons.

13. Give three examples of pest control by control of the vector.

14. In what manner can climate aid in pest control?

Pests of Horticultural Plants and Their Control   **263**

SUGGESTED REFERENCES

1. Blakely, Ransom, 1977. PYO liability insurance. *American Fruit Grower (Oct.)*, Willoughby, Ohio.
2. Brooklyn Botanic Garden, 1976. *Garden Pests* (No. 50), and *Biological Control of Plant Pests* (No. 34). Brooklyn, New York.
3. Childers, N. A. 1976. *Modern Fruit Culture*. Horticultural Publications, New Brunswick, N.J.
4. Laurie, A., and D. L. Kiplinger, 1972. *The Florist Industry*. McGraw-Hill, New York.
5. Metcalf, C. L., and W. P. Flint, 1962. *Destructive and Useful Insects*, 4th ed. McGraw-Hill, New York.
6. Shurtleff, M. 1969. *Plant Diseases and Their Control*. Iowa State University Press, Ames, Iowa.
7. Teskey, B. J. E., and J. S. Shoemaker, 1961. *Practical Fruit Growing*. Wiley, New York.
8. Thompson, H. C., and W. H. Kelly, 1969. *Vegetable Production and Marketing*. McGraw-Hill, New York.
9. Yearbook of Agriculture, 1959. *Insects*.
10. Yearbook of Agriculture, 1967. *Plant Diseases*.

12 Harvesting and Storing

The products of toil are the goals of horticulturists. To the amateur gardener this production may vary from the satisfaction he derives from growing even a single plant, to a bountiful harvest of fruits, vegetables, or flowers. To the commercial grower, production is his livelihood, and it is important to his welfare that his crops yield well and give him a good return. In order to have complete fulfillment of their objectives, the *harvest* is vital to all who are engaged in horticultural enterprises and pursuits. *Storing* for temporary or prolonged periods has important aspects for both amateurs and professionals. The time of harvest, among other factors, is determined by maturity and quality. Storage aims to maintain or improve quality. Good quality is thus extremely important to all phases of the harvest and post harvest period. Good quality of fruits and vegetables is a combination of flavor, texture, appearance, and food value which give pleasure or satisfaction to the consumer. Good quality in ornamentals is dependent on appearance and longevity.

HARVESTING

Maturity and Time of Harvest

Maturity can be described as the state of ripeness. However, it may have different meanings. For uniformity of definition, maturity can be considered the stage of development which results in maximum quality of the product. Fruits such as peaches will have a different standard of maturity if picked for immediate consumption than if picked for shipment to a distant market. Peaches for immediate use are fully ripe or "tree-ripened." If the same peaches are shipped to a distant market, even under good refrigeration and handling, their chances of reaching the consumer in good quality are quite remote. They must therefore be harvested under a different standard of maturity for maximum quality at the destination. Therefore, they are harvested at an earlier state of ripeness. Sweet potatoes or yams may be harvested over a relatively long period. They are not marketed immediately since they must be "cured" to make the skin more firm, less easily bruised, and more resistant to invading disease organisms. Flowers for sale or for home bouquets and arrangements are cut before fully opened to prolong their period of beauty. Fruits, vegetables, and ornamentals thus vary widely in their maturity standards at harvest. Separate treatment of each group follows.

Fruit maturity can be determined by firmness, coloration, tasting sample fruits, and by familiar varietal characteristics. Each of these factors will vary with the kind or type of fruit. *Firmness* in a peach or plum can be readily detected by feel and is a reliable guide to maturity. The apple and pear are more difficult to judge by hand pressure, since they are quite firm up to the point of ripeness. A pressure tester is sometimes used to determine maturity of representative fruits. Experienced pickers acquire judgment of the right degree of firmness for proper maturity. *Color* is the sole guide in the harvest of many fruits. Sweet and sour cherries are harvested when fully colored whether sold on the fresh market or processed. Grapes for fresh market are harvested as they acquire complete coloration. However, for juice, jams, jellies, and wine, the sugar content is sampled, and the grapes may be left on the vines for increased sugar content even after full color development. Citrus fruits are harvested for market when there are still traces of green color in the skin. If the green color persists they may be artificially colored before being placed on the retail market. The berry fruits are harvested principally on the basis of color. Strawberries may be picked before they have turned completely red for less spoilage on distant shipments. Both red and black raspberries must be ripe before they can be separated from the torus (receptacle). This factor has limited their shipment to distant markets in the fresh state. Color of seeds is a guide of maturity in pears and certain winter apples. *Tasting* of sample fruits can be used effectively with apples, plums, and white grapes. A tart Jonathan apple or a crisp juicy Delicious apple are equally ready for harvest. They both improve by proper storage when picked at this stage on the basis of their peculiar *cultivar characteristics*. McIntosh and Cortland apple cultivars have distinctive but pleasant odors which are indicative of maturity.

FIGURE 12-1   **Harvesting snap beans is principally on the basis of size and color. Careful handling is essential to good quality. (Photograph by K. A. Denisen.)**

Skills and Practices in Horticulture   **266**

Vegetables may be harvested as mature or immature, depending on which is considered the edible stage. The salad crops depend on rapid growth under favorable conditions for their tender, crisp, and succulent characteristics. The harvest should be done at the point of highest quality commensurate with high yields. Delayed harvest at that point may increase the tonnage yield per acre but reduce the gross and net returns because of deteriorating quality. Cauliflower and broccoli kept in the field or garden beyond the full-head stage will discolor and become more fibrous. The root crops should be harvested when minimum acceptable retail size is reached if an early market is the goal. If high yields of a late crop are desired, they can be left to grow until greater size is attained, provided the quality remains at an acceptable level. White potatoes should be mature when harvested to avoid skinning and bruising. It may be necessary to use vine killers on late varieties to hasten maturity.

Of the vine crops, summer squashes are harvested when immature. Pumpkins and winter squash should mature on the vine for best quality and keeping ability. Muskmelons (cantaloups) are ready for eating at the "full-slip" stage of the stem. For shipping, the "half-slip" stage and refrigeration in transit is the best combination. Watermelons are harvested for shipping when a metallic sound results from thumping. For immediate use, a dull, hollow sound indicates ripeness. Cucumbers for pickling are picked daily to get uniform size throughout the harvest season. For slicing they may be picked less often, since there is a wider range of acceptable sizes. Cucumbers left to ripen on the vine will decrease future production of that vine, consequently, thorough picking should be done.

Sweet corn and peas should be harvested at their peak sugar content and before toughening of the seed coat occurs. The use of an instrument called the tenderometer or puncture tester is a guide to processors.

Ornamentals may be divided into flowers and nursery stock. Since these are not food crops, maturity and quality have different meaning. *Flowers* are usually ready for harvest before the blooms are fully opened. Rose buds with the color showing, but yet in a tight whorl, are ideal for shipping or holding. For immediate use, they may be picked as the bloom expands to show individual petals. They will finish opening in the vase or arrangement and last for several days. Gladioli are cut for long blooming and holding just as the basal buds of the long spike are opening. For home or immediate use, three to five of the lower blooms may be open. The tip blooms will be larger with this delay in harvest. Tulip, peony, and narcissus have large blooms and succulent stems. They should be harvested as they are beginning to open and should be kept moist continuously after being cut. Annual and biennial garden flowers may be picked from beginning bloom to full bloom. They are not generally satisfactory for distant markets or storage.

Nursery stock should be mature in that the tissues are hardened to water loss and shrinkage. It is often desirable to defoliate deciduous trees and shrubs during a late fall to hasten digging operations in a nursery. This can usually be accomplished by chemical defoliants. Withholding water or using competing cover crops also hastens maturation of tissues. For spring digging of nursery stock, earliness is extremely important. Delayed digging means that more stored foods have moved to the buds and root tips and growth may have even started. This represents a loss to the plant, since roots of deciduous plants will dry out and die back to some degree even with the most careful handling, and there will be water loss from opening buds.

Handling the Harvested Products

The manner in which the harvest is conducted and how the products are handled have considerable influence on the efficiency of production and the ultimate quality. During this age of technology, many harvesting operations are done by machine, but many are still done by hand. In either case, care in handling the product is important.

Careful handling will reduce bruising injury for all types of horticultural products. An apple, peach, pear, or any other fruit will have a shorter storage life if bruised. If marketed immediately it may have a soft spot with discolored tissue, or it may produce cork tissue to overcome the invasion of disease organisms through the blemish. The vegetable tuber, root, and bulb crops will take rougher treatment than will most fruits. However, they too can become bruised by bouncing on the chain of a digger or being dropped from high elevators into a bin. The storage life of most vegetables of this type can be lengthened by care in handling. Cut flowers may be bruised by knocking off blooms and buds or breaking stems and receptacles. The market value or aesthetic value for the home is drastically reduced by bruising. There is considerable mechanization in the harvest of nursery stock. For deci-

FIGURE 12-2   **Tomato harvester in operation. Development of uniform ripening cultivars was concurrent with harvester development. (Courtesy W. L. Summers, Horticulture Department, Iowa State University, Ames, Iowa.)**

Skills and Practices in Horticulture

268

FIGURE 12-3   **Carrot harvester with carrot field in background. (Photograph by K. A. Denisen.)**

duous trees and shrubs, undercutting of the roots with a U-blade loosens the soil and prunes the roots for efficient handling. Care in handling results in a minimum of broken twigs, broken roots, and skinned bark. Evergreen trees and shrubs are dug with a ball of soil remaining on the roots. Burlap is wrapped around the ball to hold the soil. These "balled-and-burlapped" evergreens retain their freshness due to careful handling and provision for moisture absorption to compensate for transpiration loss. Machines have been devised for the balling operation.

Proper environment for the harvested products contributes to quality maintenance. Cooling of highly perishable products is often desirable. The salad vegetables, peas, sweet corn, and cut flowers will retain their quality longer if placed in a cool place following their harvest. Balled-and-burlapped evergreens, and the exposed roots and tops of deciduous plants will benefit from shade while waiting to be placed in their permanent location.

Removal of residues may follow the actual harvest for some crops, particularly the fruits. Washing or brushing is used on apples, pears, sweet cherries, plums, citrus, avocados, and tomatoes to remove spray residues and improve their appearance. They should be dried if washed.

STORAGE AND PRESERVATION

Horticultural produce is stored or preserved to increase its usefulness over a longer period. It may also result in improved quality in some

FIGURE 12-4   *Left*: bagged onions in storage. *Right*: bulk carrots in temporary storage. (Photographs by K. A. Denisen.)

cases. Winter cultivars of apples have a more mellow flavor after a few weeks in storage and this improved flavor may last for several months. Sweet potatoes become sweeter after storage. White potatoes make higher quality potato chips when conditioned for one to three weeks at 70°F (20°C) following storage at 40°F (4°C). Enzyme activity, sugar-starch conversions, and respiration are all influenced by storage temperature. Length of storage period and type of storage influence quality. Preservation is also a form of storage which involves processing prior to storage.

Storage of Fresh Produce

Types of storages are temporary, short-term, and long-term. Highly perishable fruits, vegetables, and flowers for immediate market are often placed in *temporary* storage. This may consist of a refrigerator or a refrigerated room which reduces the field heat of the produce. The temporary storage is also used to maintain favorable conditions for very brief periods such as overnight, during weekends, or in the shipping season when larger lots are being accumulated for carload or truck quantities. This type plays an extremely important role for the market gardener, roadside stand, florist, retail store, and shipper of highly perishable products. The refrigerator car is a means of temporary storage under favorable conditions while the produce is in transit. *Short-term* storage may last as long as a month to six weeks. It is designed not only to provide favorable conditions while awaiting sale, but is used to extend the market season. Chinese cabbage, eggplant, green-wrap tomatoes, peaches, and gladiolus can often be stored for a week to six weeks after harvest to avoid a "glut" on the market. They are taken from short-term storage and sold when their quality is still good, and prices have returned or been maintained at a good level. Continuous checking of the products in storage is vital, for a decrease in quality from too long storage or poor storage can be costly. *Long-term* storage is dependent on good storage conditions and produce that is adapted. Economic factors of increased prices due to improved markets later in the season play an important role in this type of storage. The risk involved is intensified by costly storage structures which must be amortized by increased returns. Adaptable crops for long-term storage include white potatoes, sweet potatoes, onions, rutabagas, cabbage, apples, pears, ornamental bulb crops, herbaceous perennials, and woody nursery stock. The various types of nursery stock are stored for early spring and fall shipment.

Requisites of good storage are proper temperature, humidity, and aeration. *Temperature* in the temporary, short-term, or long-term storage is often considered the most vital factor. Perhaps this is because of the great losses experienced when proper temperatures are not maintained. *Humidity* builds up from fresh produce and does not vary appreciably unless there is excess ventilation. *Aeration* is not usually a problem since respiration is at a low ebb if temperatures are sufficiently low. Tables 3, 4, and 5 give favorable storage temperatures and humidity levels for many fruits, vegetables, ornamentals, and planting stock. The holding time commensurate with good quality is also given.

FIGURE 12-5   **Rutabagas stored in pallets. (Photograph by K. A. Denisen.)**

Skills and Practices in Horticulture

TABLE 3 Favorable Temperatures, Relative Humidity, and Approximate Length of Storage Period for the Commercial Storage of Fresh Fruits and Vegetables

Commodity	Temperature (°F.)	(°C.)	Relative humidity (Percent)	Approximate length of storage period
Fresh Fruits				
Apples, winter cultivars	30 to 32	−1 to 0	85 to 90	3 to 8 months
Apricots	31 to 32	−1 to 0	85 to 90	1 to 2 weeks
Avocados	45 to 55	7 to 13	85 to 90	3 to 4 weeks
Bananas, ripe	56 to 60	14 to 16	85 to 90	7 to 10 days
Blackberries	31 to 32	−1 to 0	85 to 90	5 to 7 days
Cherries	31 to 32	−1 to 0	85 to 90	10 to 14 days
Coconuts	32 to 35	0 to 2	80 to 85	1 to 2 months
Cranberries	36 to 40	2 to 4	85 to 90	1 to 3 months
Dewberries	31 to 32	−1 to 0	85 to 90	5 to 7 days
Figs, fresh	31 to 32	−1 to 0	85 to 90	10 days
Grapefruit	40 to 45	4 to 7	85 to 90	4 to 6 weeks
Grapes				
European	30 to 31	−1	85 to 90	3 to 6 months
American	31 to 32	−1 to 0	85 to 90	3 to 4 weeks
Lemons	55	13	85 to 90	1 to 4 months
Limes	48 to 50	9 to 11	85 to 90	6 to 8 weeks
Mangoes	50	11	85 to 90	15 to 20 days
Olives, fresh	45 to 50	8 to 11	85 to 90	4 to 6 weeks
Oranges				
Florida	30 to 32	−1 to 0	85 to 90	2 to 3 months
California	35 to 37	2 to 4	85 to 90	1 to 2 months
Papayas, firm-ripe	45	7	85 to 90	7 to 21 days
Peaches and nectarines	31 to 32	−1 to 0	85 to 90	2 to 4 weeks
Pears				
Bartlett	30 to 31	−1	90 to 95	7 to 9 weeks
Fall and winter cultivars	30 to 31	−1	90 to 95	2 to 7 months
Persimmons, Japanese	30	−1	85 to 90	2 months
Pineapples				
Mature-green	50 to 60	11 to 15	85 to 90	2 to 3 weeks
Ripe	40 to 45	4 to 7	85 to 90	2 to 4 weeks
Plums, including prunes	31 to 32	0	85 to 90	3 to 4 weeks
Pomegranates	34 to 35	2	85 to 90	2 to 4 months
Quinces	31 to 32	0	85 to 90	2 to 3 months
Raspberries	31 to 32	0	85 to 90	5 to 7 days
Strawberries	31 to 32	0	85 to 90	7 to 10 days
Tangerines	31 to 38	0 to 3	90 to 95	2 to 4 weeks
Fresh vegetables				
Artichokes				
Globe	32	0	90 to 95	30 days
Jerusalem	31 to 32	0	90 to 95	2 to 5 months
Asparagus	32	0	85 to 90	3 to 4 weeks
Beans				
Green, or snap	45 to 50	8 to 11	85 to 90	8 to 10 days
Lima				
Shelled	32	0	85 to 90	15 days
	40	4	85 to 90	4 days
Unshelled	32	0	85 to 90	14 to 20 days
	40	4	85 to 90	10 to 14 days

TABLE 3—*continued*

Commodity	Temperature (°F.)	(°C.)	Relative humidity (Percent)	Approximate length of storage period
Beets				
Topped	32	0	90 to 95	1 to 3 months
Bunched	32	0	90 to 95	10 to 14 days
Broccoli (Italian, or				
sprouting)	32	0	90 to 95	7 to 10 days
Brussels sprouts	32	0	90 to 95	3 to 4 weeks
Cabbage				
Early	32	0	90 to 95	3 to 6 weeks
Late	32	0	90 to 95	3 to 4 months
Carrots				
Topped	32	0	90 to 95	4 to 5 months
Bunched	32	0	90 to 95	10 to 14 days
Cauliflower	32	0	85 to 90	2 to 3 weeks
Celeriac	32	0	90 to 95	3 to 4 months
Celery	31 to 32	0	90 to 95	2 to 4 months
Corn, sweet	31 to 32	0	85 to 90	4 to 8 days
Cucumbers	45 to 50	8 to 11	85 to 95	2 to 3 weeks
Eggplants	45 to 50	8 to 11	85 to 90	10 days
Endive, or escarole	32	0	90 to 95	2 to 3 weeks
Garlic, dry	32	0	70 to 75	6 to 8 months
Horseradish	30 to 32	−1 to 0	90 to 95	10 to 12 months
Kohlrabi	32	0	90 to 95	2 to 4 weeks
Leeks, green	32	0	90 to 95	1 to 3 months
Lettuce	32	0	90 to 95	2 to 3 weeks
Melons				
Watermelons	36 to 40	2 to 4	85 to 90	2 to 3 weeks
Cantaloupes (muskmelons)				
Full-slip	40 to 45	4 to 7	85 to 90	4 to 8 days
Half-slip	45 to 50	8 to 11	85 to 90	1 to 2 weeks
Honey Dew	45 to 50	8 to 11	85 to 90	2 to 3 weeks
Casaba	45 to 50	8 to 11	85 to 90	3 to 6 weeks
Mushrooms, cultivated	32	0	85 to 90	3 to 5 days
Okra	50	11	85 to 95	2 weeks
Onions	32	0	70 to 75	6 to 8 months
Parsnips	32	0	90 to 95	2 to 4 months
Peas, green	32	0	85 to 90	1 to 2 weeks
Peppers				
Chili, dry	—	—	65 to 70	6 to 9 months
Sweet	45 to 50	8 to 11	85 to 90	8 to 10 days
Potatoes				
Early-crop	40	4	85 to 90	5 to 8 weeks
Late-crop	40	4	85 to 90	5 to 8 months
Pumpkins	50 to 55	11 to 13	70 to 75	2 to 6 months
Radishes				
Spring, bunched	32	0	90 to 95	10 to 14 days
Winter	32	0	90 to 95	2 to 4 months
Rhubarb	32	0	90 to 95	2 to 3 weeks
Rutabagas	32	0	90 to 95	2 to 4 months
Salsify	32	0	90 to 95	2 to 4 months

TABLE 3—*continued*

Commodity	Temperature (°F.)	(°C.)	Relative humidity (Percent)	Approximate length of storage period
Spinach	32	0	90 to 95	10 to 14 days
Squashes				
Summer	32 to 40	0 to 4	85 to 95	10 to 14 days
Winter	50 to 55	11 to 13	70 to 75	4 to 6 months
Sweet potatoes	55 to 60	13 to 16	85 to 90	4 to 6 months
Tomatoes				
Ripe	50	11	85 to 90	8 to 12 days
Mature-green	55 to 70	13 to 21	85 to 90	2 to 6 weeks
Turnips	32	0	90 to 95	4 to 5 months

Adapted from Agriculture Handbook 66, U.S.D.A.

TABLE 4   **Favorable Temperatures, and Approximate Length of Storage Period for Cut Flowers (80 percent humidity)**

Commodity	Temperature (°F)	(°C)	Approximate length of storage period
Acacia	40	4	3 to 4 days
Anemone	45	8	1 to 2 days
Anthurium	55	13	2 to 3 days
Aster (China)	40	4	1 week
Babysbreath	40	4	1 to 2 days
Bird-of-paradise-flower	45	8	2 to 3 days
Bouvardia, sweet	35	2	1 week
	40	4	4 days
Butterflybush, orange-eye	40	4	1 to 2 days
Calendula (pot marigold)	40	4	3 days
Calla, common and golden	40	4	7 days
Camellia	45	8	3 to 6 days
Candytuft	40	4	3 days
Carnation	33	1	1 week
	40	4	4 days
Chrysanthemum	35	2	2 weeks
Clarkia	40	4	3 days
Columbine	40	4	1 to 2 days
Cornflower	40	4	3 days
Crocus	33	1	1 week
Dahlia	50	11	2 days
Daisy, English	40	4	3 days
Delphinium			
Hardy larkspur	40	4	1 to 2 days
Annual larkspur	40	4	1 to 2 days
Eucharis	45–50	8–11	7 to 10 days
Feverfew	40	4	3 days
Forget-me-not, true	40	4	1 to 2 days
Foxglove, common and common white	40	4	1 to 2 days

Harvesting and Storing

TABLE 4—*continued*

Commodity	Temperature (°F.)	(°C.)	Approximate length of storage period
Freesia	33	1	2 weeks
	36	2	1 week
Gaillardia, common perennial	40	4	3 days
Gardenia	45	8	3 to 6 days
Gerbera	35	2	2 weeks
	40	4	5 to 7 days
Ginger	55	13	3 to 4 days
Gladiolus	35	2	1 week
Godetia	50	11	1 week
Heath	40	4	1 week
Heliconia	55	13	3 to 4 days
Hyacinth	33	1	2 weeks
	36	2	1 week
Iris, Dutch	33	1	1 week
Laceflower, blue	40	4	3 days
Lilac, forced	40	4	4 to 6 days
Lily			
Easter	35	2	2 weeks
Goldband	35	2	2 weeks
Regal	35	2	2 weeks
Speciosum rubrum	35	2	2 weeks
Lily-of-the-valley	35	2	1 week
	40	4	3 days
Lupine	40	4	3 days
Narcissus			
Daffodil	33	1	2 weeks
	36	2	1 week
Paperwhite	40	4	2 to 4 days
Orchid	55	13	2 to 3 days
Peony, Chinese and common			
Tight buds	35	2	6 weeks
Loose buds	35	2	3 to 4 weeks
Phlox, garden	40	4	1 to 2 days
Poinsettia	60	16	2 to 3 days
Primrose, baby	40	4	1 to 2 days
Ranunculus	40	4	2 to 3 days
Rose			
Tight buds	40	4	4 to 5 days
Loose buds	40	4	2 to 3 days
Snapdragon, common	40	4	3 days
Snowdrop	40	4	2 to 4 days
Squill	33	1	2 weeks
	36	2	1 week
Statice (sea-lavender)	35	2	6 weeks
	40	4	3 weeks
Stephanotis	40	4	1 week
Stevia	40	4	3 days
Stock, common	40	4	3 days
Strawflower	35	2	6 weeks
	40	4	3 weeks

TABLE 4—*continued*

Commodity	Temperature (°F.)	(°C.)	Approximate length of storage period
Sweetpea	50	11	1 to 2 days
Sweet-william	45	8	3 to 4 days
Tulip	33	1	1 week
Violet, sweet	40	4	3 days

Adapted from Agriculture Handbook 66, U.S.D.A.

TABLE 5 **Storage Temperature and Period Stored for Rhizomes, Tubers, Roots, Bulbs, Corms, and Nursery Stock**

Commodity	Storage temperature (°F)	(°C)	Period stored
Rhizomes, tubers, roots, bulbs, and corms:			
Amaryllis	40 to 45	4 to 7	Fall and winter
Begonia, tuber	45	7	Fall
Caladium, fancy-leafed	60	15	Fall
Calla	36 to 40	2 to 4	Fall
Canna	40 to 45	4 to 7	Fall and winter
Crocus	55 to 60	13 to 16	Fall
Dahlia	40 to 45	4 to 7	Fall and winter
Freesia	55 to 60	13 to 16	Fall
Gladiolus	40 to 50	4 to 11	Fall and winter
Gloxinia	50	11	Fall
Hyacinth	55 to 60	13 to 16	Fall
Iris			
Dutch	35	2	Fall
Lily			
Lilium auratum	32	0	Winter
L. candidum	32	0	Fall
L. longiflorum (Easter lily)	31 to 35	−1 to 2	Fall
L. regale	32	0	Late fall and winter
L. speciosum rubrum	32	0	Fall
Lily-of-the-valley	25 to 28	−4 to −3	All year
Narcissus			
Daffodil	55 to 60	13 to 16	Fall
Paperwhite	75 to 80	25 to 27	Fall and winter
Oxalis	40 to 45	4 to 7	Fall and winter
Peony	40 to 45	4 to 7	Fall and winter
Tuberose	40 to 45	4 to 7	Fall
Tulip	65 to 70	18 to 21	Early fall
Nursery stock: 85% humidity			
Deciduous fruit trees and shrubs	32 to 35	0 to 2	Fall and winter
Rose plants	30	−1	Fall and winter
Strawberry plants	30 to 32	−1 to 0	Fall and winter

Adapted from Agriculture Handbook 66, U.S.D.A.

The storage temperatures are very influential in determining the length of storage period. Carnations can be kept four days at 40°F (4°C) with good quality maintenance but can be kept one week if the storage temperature is 33°F (0°C). Both the hyacinth and narcissus (daffodil) can be held safely for two weeks at 33°F (0°C) but only one week at 36°F (2°C). Temperature control will vary with the type of storage and facilities. A refrigerated unit with thermostatic control is easily regulated. Uniform temperature may be a problem unless good circulation of air is maintained, which can be done with the help of fans. With "common" storage, temperature regulation is dependent on outside weather and the insulating effect of soil. These storages have no refrigeration unit but are dependent on slow response to winter temperature. The principal problem with common storages is too high a temperature at the beginning of the storage period in the fall. Unless the temperature of many stored products can be reduced to near optimum in a relatively short period of time, some breakdown or loss in quality will occur early in the storage period. Common storages are consequently better adapted to those products not stored at or near the freezing point. Sweet potatoes, pumpkins, and winter squash will usually do well in the early period of common storage since they prefer temperatures above 50°F (11°C). The temperature at which fresh commodities are stored also affects the extent of respiration and the

FIGURE 12-6   **A common storage built into a hillside. The headhouse services four storage units. Ventilators aid in temperature and humidity control.**

amount of heat generated. Since these products are living, they must carry on respiration. At high temperatures more oxygen is consumed, more carbon dioxide is released, more heat is evolved, and in general, the storage life of the product is shorter than at low temperatures. Winesap apples release nearly twice as much heat when stored at 40°F (5°C) and about seven times as much at 60°F (15°C) compared to storage at 32°F (0°C). Lettuce will give off about four times as much heat at 60°F (15°C) as at 32°F (0°C). However, lettuce at 32°F (0°C) evolves about 35 times more heat per unit of weight than Winesap apples do at the same temperature. Adequate cooling of fruits and vegetables before storage reduces heat loss and increases storage life.

Some fruits and vegetables are ripened in storage. Tomatoes picked at the mature-green stage for distant markets will ripen most rapidly at 70°F (21°C). They will ripen slowly but can be held longer if stored at 55°F (13°C). Tomatoes in home gardens are often picked at this stage in the fall when a killing frost is imminent. The fresh tomato season is, in this way, considerably prolonged. Pineapples are also harvested at the mature-green stage for shipping, while bananas are harvested green, then ripened in storage at 62° to 70°F (16–21°C) to develop the best texture and flavors. Plant-ripened bananas are poor in quality. Humidity for ripening fruits and vegetables is important and should be maintained above 85 percent to reduce shrinkage.

Preservation of Fruits and Vegetables

Canning, freezing, drying, and pickling are methods of preserving horticultural food commodities. The food preservation industry has become extremely large, and it makes a great contribution to a better nourished population. Food preservation in the home also plays an important role in food conservation and family health. The advent of home freezers, pressure cookers, and pectin additives for jellies and jams has increased the ease of home processing of foods for preservation. Important to the processing of all foods is the inactivation of enzymes or reduction of enzyme activity. In canning this is done by heat. In freezing it may, in some cases, occur with low temperatures alone as with berries and rhubarb. However, sweet corn, snap beans, lima beans, asparagus, and many other vegetables must be subjected to high temperatures for a brief period to inactivate the enzymes. This process is known as "blanching." Dehydration in itself inactivates the enzymes in dried foods, while the acid media or low pH of pickled foods is the retarding factor in that method.

Quality of preserved fruits and vegetables is determined in large measure by the quality before processing. For high quality products, high quality foods should be used.

Canning and *freezing* on a commercial scale are closely allied. Many processors use both methods as a means of preservation. Much of the equipment for preparing foods for canning and freezing is the same. The high temperature retorts or pressure cookers used for sterilizing cans and food can be used for blanching vegetables for freezing. The quick-freeze method and deep-freeze storage are the principal additions to a canning factory for its conversion to frozen foods processing. High

quality produce is desirable for both products. However, there may be different standards of quality for the different processes. Peas and sweet corn for canning should have a tender pericarp or seed coat for highest palatibility. Peas and sweet corn for freezing should have a seed coat which will not break down with freezing, thawing, and cooking. For this reason, canned peas and sweet corn are usually more tender than the frozen products. Different cultivars are adapted to each processing method. The introduction of the frozen concentrate process in fruit juices has had a great impact on the fruit industry. Fruits can be juiced at or near their source of production, greatly reduced in volume by

FIGURE 12-7 **Commercial storage and facilities.** *Left*: grader and elevator. *Right*: carrots being loaded following washing and sizing for transport to the processor. (Photographs by K. A. Denisen.)

concentration, and shipped in a small fraction of the space occupied by fresh fruits. The resulting juice is frequently more palatable to the consumer than canned juice and is often lower in cost than juice from the fresh fruits.

Dehydrated fruits and vegetables are increasing in usefulness since flavors now are better preserved. Onions are one of the most successful of the dehydrated vegetables. However, for certain products, such as dehydrated soup mixtures, carrots, tomatoes, radishes, parsley, celery, and other vegetables are also satisfactory. Dehydration without excessive heat is essential to maintain good quality. Rapid drying in vacuum results in dehydrated products that have the original flavor of freshness when properly prepared with water. Prunes, raisins, figs, dates, and apricots are among the fruits that were first dried for shipment and storage. For many centuries they have been sundried, and in some countries are staple foods in the nation's economy. They were an important source of vitamins during the winter months for early pioneers, the crews of sailing vessels, and explorers. After drying, these fruits are still moist because of the high concentration of sugar in the fruit pulp and their great affinity for the remaining traces of water. Sulfur dioxide fumes are used as a bleaching agent for the lighter colored, dried fruits, e.g., apricots.

Pickling is a process by which bacteria, fungi, and enzymes are prevented from breaking down the tissues of the preserved product. It involves the use of vinegar, citric acid, or other organic acids that are capable of resisting breakdown by organisms. Nearly any type of fruit or vegetable can be pickled by placing it in a suitable media.

Harvesting and Storing

FIGURE 12-8   **High-quality foods are best for processing. Removal of leaf blades from rhubarb stalks at harvest reduces transpiration maintaining quality.**

Food preservatives or *anti-oxidants* are used to prolong the edible period of fruits and vegetable products. Ascorbic acid (vitamin C), when added to the syrup used on peaches for freezing, prevents oxidation or discoloration of the fruit. Sodium benzoate is a preservative effective in reducing deterioration in purees and catsup. Sorbic acid, a mold inhibitor, prolongs the storage life of many foods.

STUDY QUESTIONS

1. When does it become feasible to harvest a horticultural crop mechanically? When is it not feasible?
2. Which horticultural crops are not yet harvested mechanically? Why?
3. What are the essentials for mechanical harvest of a certain vegetable or small fruit?
4. How does mechanical harvest of sweet corn compare to mechanical harvest of field corn? Can they be used interchangeably?

5. In what manner and to what degree does mechanically harvested produce affect storage of that commodity?
6. Explain "controlled atmosphere" storage, common storage, long-term storage.
7. Why is the temperature of produce in storage generally higher than air temperature in the storage room?
8. Uniformity of characteristics is often mentioned in both mechanical harvesting and storage. Why is it considered highly important for each?
9. Plastic is used considerably in the handling and storage of horticultural produce. Why? What are its hazards?
10. Is sun-drying of fruits practiced in your community? Why or why not?
11. What is the potential for machine harvesting of the following crops: celery, radishes, blueberries, apples, strawberries, peaches?
12. What role does perishability play in mechanical harvest, storage, and processing?
13. Why is high humidity considered a requisite of storage?
14. How important is size of storage facility for short-term storage, long-term storage, and the general feasibility of storage?

SELECTED REFERENCES

1. Cargill, B. F., and G. E. Rossmiller, 1969. *Fruit and Vegetable Harvest Mechanization, Technological Implications.* Rural Manpower Center, Michigan State University, East Lansing, Mich. A landmark treatise on mechanical harvesting for the horticultural industry.
2. Cargill, B. F., and G. E. Rossmiller, 1969. *Fruits and Vegetable Harvest Mechanization, Manpower Implications.* Rural Manpower Center, Michigan State University, East Lansing, Mich.
3. Childers, N. F. 1976. *Modern Fruit Science*, 5th ed. Horticultural Publications, Rutgers University, New Brunswick, N.J.
4. Hardenburg, R. E. 1971. Effect of in-package environment on keeping quality of fruits and vegetables. *HortSci*, **6**:198–201.
5. Smock, R. M. 1949. Controlled atmosphere storage of apples. *N.Y. (Cornell) Agr. Ext. Serv. Bull. 759.*
6. Thompson, H. C., and W. C. Kelly, 1957. *Vegetable Crops*, 5th ed. McGraw-Hill, New York.
7. Wright, R. C., *et al.* 1954. The commercial storage of fruits and vegetables and florist and nursery stocks. *U.S. Dept. Agr. Handbook 66.* This handbook is especially valuable for its recommended storage temperatures and humidities.

13 Marketing Horticultural Products

The market and market prices are vital to the horticulturist who is dependent on the products of his profession for his livelihood. The products may be fruits or vegetables for food, flowers, and house plants for beautification, nursery plants for landscaping or food production, or horticultural services. All of these products or services have economic value and fit into the scheme of supply and demand.

Other factors being equal it can be stated that those products or services of highest quality will provide the greatest return. A goal of good quality is a reasonable endeavor for horticulturists. As described in Chapter 1, the Basic Four Food Groups with recommended daily food allowances has delineated food needs. Where incorporated into daily food intake, this plan has virtually wiped out deficiency diseases. Food consumption has increased. Interest in fruits and vegetables has increased, probably due to publicity on the nutrient and vitamin value of these products.

SUPPLY AND DEMAND

Marketing is buying and selling. It is a human process subject to the economic laws of supply and demand. To buy or want to buy indicates a demand. To sell or want to sell indicates a supply. If the supply of an item is greater than the demand for that item, the price trend is generally down. If the supply is short, i.e., not great enough to meet the demand, the price trend is upward. Many factors tend to modify the relationships of supply and demand of horticultural products. Among them are: (1) distances between areas of production and areas of consumption; (2) buying power of consumers; (3) competition with other products; (4) relationship of luxury or necessity items; (5) perishability of commodities; (6) possible alternatives to selling; (7) volume of sales and margin of profit; (8) degree of salesmanship and amount of advertising; (9) regulation by government; and last, but not of least consideration, (10) quality of produce.

Distances between producing and consuming areas influence prices. The cost of transportation must be included in the ultimate selling price. A great supply of oranges in California, grapefruit in Texas, or tangerines in Florida may lower the price slightly in New York or Chicago. However, so many of the costs are fixed that the retail price does not fluctuate widely. If the retail price is not adequate, the grower or shipper may not risk shipping to a distant market because of rela-

FIGURE 13-1   An attractive assortment of vegetables and fruits from distant and local
sources of supply whets the appetite of the consumer. (Photograph by K. A.
Denisen.)

tively high fixed costs of transportation, commissions, containers, and
retailing. If in short supply, the fixed costs remain nearly the same and
prices rise somewhat. The spread in price between surplus and short
supply is much less at the distant market, however, than at markets near
the producing area.

Buying power of consumers influences prices by modifying the relationships of supply and demand. This does not mean that wealthy
people are necessarily influential in price rises. General prosperity, large
factory payrolls, and full employment contribute to the buying power of
the majority of families. With an increase in buying power the demand
for staple vegetables by a family may shift from potatoes in large
containers at a low price per unit of weight to potatoes and several other
vegetables in smaller containers, each at a higher price per unit. The
increased buying power changed the demand to several vegetables
instead of one, and quantity purchasing of that vegetable to get it at a
lower price was no longer considered essential. Increased buying power
gave freedom for broader diet and a change in demand for the various
products.

Competition with other products may change food habits and alter
price relationships. High carbohydrate vegetables such as potatoes,
sweet potatoes, beans, peas, and rutabagas compete with the high-
carbohydrate cereal grains. Fruits compete with candy. A complete
landscaping of the home grounds may compete with decorating the
interior of the house. Flowers compete with synthetic products for
centerpieces and arrangements. Fads, fancies, and fashions may designate the next area of competition for various horticultural commodities.

FIGURE 13-2   **Ornamental plants are sometimes considered luxury items in the home grounds. (Photograph by K. A. Denisen.)**

Luxury or necessity plays an important role in priority of purchases in many homes. Some home owners consider the planning and planting of the home grounds a luxury that can wait. Others point out that a well-planned and planted home is not a luxury, but is essential to family development. To others, the increased resale value of a planted home justifies immediate development. Prices of new and different fruits and vegetables are also influenced by luxury-necessity decisions.

Perishability of commodities will determine how soon a supply has to be distributed before there is loss due to spoilage. A large supply of peaches at or near a consuming area must be disposed of rather rapidly because they are highly perishable. The price goes down; consequently, they will sell faster. The demand increases until they are sold out. The demand may be present the next week, but the supply may then be short. Perishability has caused fluctuations in price through altered supply and demand.

Possible alternatives to selling when the supply is large include storage, other markets, processing, and dumping. A grower or shipper may find any one of these outlets preferable to taking a very low price.

Volume of sales and margin of profit can mean the difference between profit or loss during low prices. Efficiency of production and size of the operation, even at extremely small profit per unit, can often help the producer through a crisis of over-supply.

FIGURE 13-3   Careful handling, occasional use of leaves, icing of produce and plastic wraps contribute appeal in a display. (Photograph by K. A. Denisen.)

Degree of salesmanship and amount of advertising can readily alter the demand for a certain commodity at a particular time. National promotion programs, illustrative posters, suggested recipes, and nutrition drives are some of the techniques employed by advertising to play on the emotions and desires of the public as consumers.

FIGURE 13-4   Evergreen container stock is a convenience item at the local nursery. (Photograph by K. A. Denisen.)

Marketing Horticultural Products   **287**

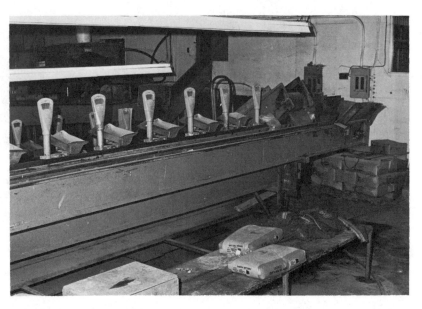

FIGURE 13-5   *Left*: **production line weighing and packaging facility.** *Right*: **preparation of green onions. (Both photographs by K. A. Denisen.)**

Regulation by government agencies include price supports, marketing agreements, food stamps, and school-lunch programs. These affect both supply and demand but may counteract the usual effects of the economic law, i.e., increasing the supply merely because of a guaranteed price when the supply is already great.

Quality of produce is a principal factor that a horticulturist relies on to alter the law of supply and demand. When supplies are great, greater emphasis is placed on quality of produce. High quality products are in demand and command a higher price than low-quality products. A producer who emphasizes high quality in his fruits, vegetables, or ornamentals is likely to find a market whether the supply is abundant or short. Highest quality has greatest demand especially during periods of surplus.

QUALITY AND QUALITY MAINTENANCE

The principles of good harvesting techniques and good storage practices are elementary to good quality and quality maintenance. Even more basic is the production of good quality horticultural commodities. Adapted cultivars, pest control, proper irrigation, and good cultural practices contribute to both maximum yield and quality.

Careful handling to avoid bruising is important to quality maintenance. Not only do bruised products appear less attractive and result in varying amounts of waste, but also increase the likelihood of loss by spoilage even in the display counter. Some commodities are not as fragile or easily bruised as others. Potatoes can take more vigorous

treatment than apples, but nevertheless can be damaged by dropping on hard surfaces or great distances. Cut flowers are generally quite fragile. A broken stem or damaged buds on a gladiolus, rose, snapdragon, or other cut flowers can render them practically valueless.

Grading on the basis of size and quality, favors the better grades. It gives the customer a choice of both quality and price. A single item of inferior grade lowers the quality of the entire lot. Such individual items should be removed to improve the grade and general appearance. Uniformity is important to customer satisfaction. A very large fruit or vegetable should not be left with smaller ones. The size comparison dwarfs the others in the lot even though they may be of top quality and in the same grade class. Grading also eliminates defects due to blemishes, disease, and malformation.

FIGURE 13-6   Proper environment during merchandising is essential to good quality nursery stock. Balled and burlapped evergreens (*left*) and balled and burlapped deciduous shrubs (*right*) all need watering and careful attention. (Photographs by K. A. Denisen.)

Proper environment during merchandising is vital to quality maintenance. This may mean refrigerated display cases for the retailer. It may require shading from sunlight or adding chopped ice or placing in the refrigerator by the wholesaler or shipper. It actually begins in the grower's field, orchard, or greenhouse where immediate transfer to favorable conditions starts the program of quality control.

WHAT TO LOOK FOR IN MARKETING

Whether buyer or seller, whether consumer or producer, factors to consider in evaluating various items of produce are the same. The intermediate objectives will vary, but the ultimate goal is satisfaction of the consumer. If the consumer is satisfied, he is more willing to pay the price, the retailer is satisfied because he is making a profit and is serving the community, the jobber or wholesaler is content because he can move his produce and increase his volume of business, and the basic producer of horticultural commodities reaps his monetary reward. His best efforts have resulted in good products that give him a profit and the satisfaction of helping to feed the world.

Vegetables should be clean, fresh, and free of disease and insect damage. The *salad* crops should be crisp and firm. Avoid wilted leaves on celery, lettuce, Chinese cabbage, endive, and watercress. Soft or water-soaked leaves indicate prolonged or unfavorable storage conditions. *Cole* crops should be firm, solid, and of good color. Avoid soft or soggy leaves on cabbage and Brussel sprouts, dark, open, ricey heads of cauliflower, and blooming flowers on broccoli. Offensive odors indicate mold or decay. *Greens* should be crisp and well colored. If held too long they become discolored and wilted. The "*fruit*" vegetables should be firm and free of blemishes and water-soaked areas. Tomatoes should be ripe and well colored. Eggplant that is heavy and glossy, cucumbers that are green and unshriveled, and peppers that are bright and smooth are to be preferred. Sunken areas in the skin of these vegetables often indicate fungus activity. *Root crops* that are smooth, solid, and free of large secondary roots are best. Slow growth because of unfavorable growing conditions produces fibrous, tough tissues. Roots of rutabaga and turnips may be small because of stunting and therefore are frequently more fibrous than those making rapid and continuous growth. Radishes with hairy side roots are more apt to be pithy and strong than the smoother, plumper ones. Both *sweet potatoes* and *Irish potatoes* should be dry, firm, and smooth. Sweet potatoes referred to as "yams" are moist fleshed, and the others are generally dry fleshed or of intermediate type. Irish, or white, potatoes are classed as cooking or baking types. The Burbank Russet variety is typical of the baker type and is characterized by mealiness after baking. *Summer squash* is edible

FIGURE 13-7   **Use of cellophane, plastic wraps and chipped ice all suggest freshness to these vegetables. Note the radishes have their tops, a practice now revived to indicate freshness. (Photograph by K. A. Denisen.)**

Marketing Horticultural Products   **291**

when immature. If the fingernail can penetrate the skin of the squash it has not passed the edible stage. *Winter squash* and *pumpkins* should have hardened skin before they are ready for use. *Peas* and *lima beans* with green, well-filled pods are preferred over wilted, discolored or leathery pods. *Snap beans* and *okra* should be crisp and green except for the yellow-podded bean varieties. *Sweet corn* that has soft and unshriveled kernels is preferred to that which is firm and beginning to dent.

Fruits with mature coloring are desired. *Citrus* fruits that are smooth and firm, yielding slightly to pressure, are sufficiently ripened. Discolored, water-soaked areas on the fruit indicate partial breakdown. Unlike many fruits, this gives an off-flavor to the entire fruit even after removal of the infected area. Many citrus fruits, especially oranges, are artificially colored. This is primarily for uniformity of appearance rather than to hasten the ripening process. *Apples and pears* should be firm and crisp. Soft fruits indicate that their storage period is at an end and they should be used immediately. The *stone fruits* are easily bruised. Firm, but not hard, peaches and apricots with yellow ground color are preferred over soft fruits, especially if not used immediately. Completely ripened fruits may be desired for home canning. However, very ripe fruits will not retain their shape as well when canned. Plums and cherries should be free of blemishes and also firm, but not hard. *Grapes* that are plump, firm, and in clusters are of good quality. When they become overripe the stem ends of individual grapes break down and grapes shatter from the cluster. At this stage mold organisms invade the tissues and distintegration and spoilage is rapid. *Tropical and subtropical* fruits, such as the banana, avocado, pineapple, and papaya, are harvested when green or mature-green for shipment to distant markets. When ready for use they should be firm but yielding to slight pressure of the hand or fingers. *Strawberries* that are shipped from great distances complete their ripening in transit, in storage, or on the display counter. Local berries are picked when fully colored. In either case, the berries should be firm, uniformly colored, and bright. A wilted or dried calyx may indicate that the berries have not been kept cool or were not shaded after picking. *Cranberries* are commonly sold in cellophane bags. Bright clean berries that are free of blemishes, soft areas, or mold indicate good handling and storage. The *bush berries* include raspberries, blackberries, Boysenberries, gooseberries, currants, and blueberries. They should be fresh and not overripe. It is wise to avoid soft berries for they will deteriorate rapidly and may cause others in the container to start molding or breaking down.

Ornamentals should also be carefully scrutinized before being bought or sold. *Cut flowers* should have vitality for potential and continued bloom. Fully opened blossoms are at their maximum showiness and can only proceed in the direction of deterioration. Opening flowers and unopened buds will give longer usefulness and greater satisfaction for most purposes. *Potted house plants* that have sufficient soil and adequate sized pots for a considerable period are more satisfactory than those which are pot-bound and will soon need repotting. They should have clean foliage which is free of insects and diseases and good in color.

FIGURE 13-8   A neat, orderly display should be varied and rearranged at intervals to add interest for the shopper. (Photographs by K. A. Denisen.)

Deciduous nursery stock, which is dormant, should have the roots in moist material or encased, bare root, in polyethylene wraps. Desiccated roots or tops may result in poor growth or death of the plants. *Evergreens or growing deciduous plants* are "balled-and-burlapped" or "canned" for increased livability. Instructions for planting balled-and-burlapped stock and for removing the can from canned stock should be followed. The purpose of the can or the ball of soil is to reduce root injury and drying. Carelessness in transplanting can defeat its purpose.

Transplants for the vegetable garden and flower border should be healthy, vigorous, and stocky. The gardener and supplier should both avoid leggy, spindly plants or stunted, unthrifty plants.

CONSUMER APPEAL AND SATISFACTION

The merchandiser should make every effort to have his produce look appealing and show to best advantage. Neat, orderly display tables and counters that have high quality produce are a drastic contrast to cluttered, unkempt stacks of produce that are not sorted and contain decaying specimens. Clean floors, clean tables, clean containers, clean produce, and clean employees give consumers confidence in a retail store. Sorting out and disposing of solitary spoiled items in a lot will improve

FIGURE 13-9   **A merchandiser should make every effort to have plant materials look appealing and show to best advantage. (Photograph by K. A. Denisen.)**

Skills and Practices in Horticulture   **294**

the general appeal of that commodity. Some retailers follow the practice of not displaying poor-quality merchandise. Many try to dispose of it at a low price just to get what they can out of it. Others do not handle it for sale at all, since they want a reputation for handling only good quality merchandise.

The neat, orderly display should be varied and rearranged at intervals to add interest for the shopper. Signs should be changed occasionally to avoid monotony. All items should have prices clearly indicated to avoid any suggestion of overcharging. Instructions for use of the products, recipes, handling instructions, prepared pamphlets, and the like may acquaint customers with new and different items that contribute to their future pleasures.

The factor of courtesy should never be forgotten in any phase of marketing, whether it be at a roadside stand, nursery yard, floral shop, supermarket, jobbing house, or shipping point.

STUDY QUESTIONS

1. What is marketing? How much handling is involved in marketing?
2. How does marketing affect the horticulture producer? the consumer? the processor?
3. How is marketing associated with storage? It is practical to store produce when the market will accept it?
4. How is marketing affected by or controlled by the economy?
5. What are the advantages in grading produce? to the producer? to the consumer? to the merchandiser?
6. Of what value is promotion of certain products?
7. What is quality? How does it affect price, holding ability, sales volume, satisfied customers?
8. What is the future for PYO (pick-your-own) horticultural operations? On what do you base this judgment?
9. As a consumer, what do you look for besides price in 5 fruits and 5 vegetables of your choice in a supermarket?
10. Why are consumer appeal and satisfaction generally considered important by a supermarket merchandiser?
11. List several factors which give consumer confidence in a retail fruit, vegetable, and flower operation.
12. What are market check-offs? Are they generally popular with producers?

SELECTED REFERENCES

1. Brunk, Max E. 1960. How the apple industry can strengthen its selling. *Cornell Univ. Agr. Econ. Mimeo Report, M.B.60 June.*
2. Desrosier, N. W. 1970. *The Technology of Food Preservation*, 3rd ed. Avi Publication Co., Westport, Conn.
3. Shepard, G. S., and G. Futrell. 1970. *Marketing Farm Products.* Iowa State University Press, Ames, Iowa.

14 Improvement of Horticulture Plants by Breeding and Selection

A goal of many horticulturists, especially those who are plant breeders, is to develop new and improved varieties or cultivars. Improvement through breeding has resulted in a tremendous increase of new kinds, types, selections, lines, and parental stock. Although plant breeders have been making crosses, growing seedlings, and making selections for several centuries, the science of genetics is of relatively recent origin. Genetics seeks to study and determine the role of heredity in breeding procedures. Some plant breeders, with a long-term view, have devoted great energy to establishing a "bank of germplasm" or parental material on which they may draw for fitting together the building blocks or genes for new cultivars. It is indeed an important endeavor for the long-term improvement of Horticulture. In addition to having an understanding of genetics the horticultural plant breeder knows parental material, the objectives of the breeding programs, and how to proceed in fitting together the bits of information for the emergence of a new cultivar. Either the anticipated objective was attained or a new and unexpected product resulted from the cross or self. It is one of the uncertainties of plant breeding which makes it interesting, fascinating, and exciting. The appearance of certain features in a population sometimes determines the direction of a breeding program, essentially providing some of the needed characters for inserting into the genetic makeup of a potential cultivar.

GENES AND HEREDITY

Many breeders of both animals and plants have used the techniques of crossing and selection for improvement of breeds and varieties. A noted early plant breeder was Luther Burbank who made many crosses and produced numerous cultivars, some of which exist today. When one considers that the science of genetics had not yet been born, his was a great accomplishment of plant breeding. Although Burbank has been attacked and criticized for his lack of complete records, vague breeding objectives, and his presumed desire for publicity, the fact remains that he contributed much to the introduction and distribution of new and adapted cultivars in California and other western states.

The work and reports of the Austrian monk, Gregor Mendel form the basis for the science of genetics. Mendel worked with garden peas and reported the results and inheritance patterns in one of the journals of his time. Although his work went unrecognized for several decades, when it

FIGURE 14-1   **Breeding results are evident. From the single blossom of the wild rose to the many multi-petaled modern cultivars, rose breeders have made great advances. (Photographs by K. A. Denisen.)**

was found it was re-reported somewhat later by three investigators working separately in three different institutions; later, it became widely known and recognized. It was indeed a landmark in our knowledge of genetics.

Many other geneticists and breeders have expanded on the work of Burbank, Mendel, and other early workers. Because of this we have great advancements in plant breeding and plant-breeding technology. Numerous cultivars are available both for the horticultural industry and as parental material for plant breeders to combine for new and more useful cultivars throughout the array of fruit, vegetable, and ornamental plants. Today our cultivars are the result of either chance seedlings, selections from a population, or determined efforts through carefully planned mating programs using the materials at hand. The search continues for means of improving techniques of plant breeding and for new or previously unknown characters or "genes" for transfer to the particular cultivar via breeding.

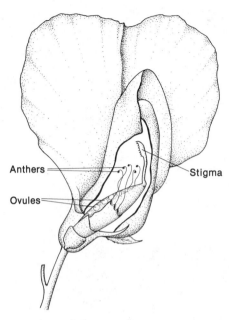

Anthers

Stigma

Ovules

FIGURE 14-2 **Pea blossom. This is the plant utilized by Mendel in his genetic studies. (Adapted from *Invitation to Biology*, Helena Curtis, Worth Publishing, New York.)**

BASIC PROCEDURES IN PLANT BREEDING

To describe the techniques involved requires some delineation of terms. The symbols for male (♂) and female (♀) are widely used by plant breeders to designate individual parents in a cross. Since one of the female parts of the flower, the ovule, develops into a seed after fertilization, that particular flower or line is called the *seed parent*. Likewise the parent which serves as the male is called the *pollen parent*. If both male and female *gametes* come from the same flower or the same genetic line, we have a *self-pollinated* plant (e.g., the tomato). If they come from different cultivars or different flowers (e.g., sweet corn), we refer to them as *cross-pollinated* plants. A plant which is naturally self-pollinated or one which is of controlled self-pollination and has consistent inheritance is called a *pure line*.

Those plants which are naturally cross-pollinated sometimes show a loss of vigor under conditions of continued selfing. This is referred to as *inbreeding depression*. As these inbreds are mated with other inbreds there is often a great increase of vigor in the progeny. This has been given the term *hybrid vigor* or *heterosis* and is a commonly observed result among crosses within many vegetable and flower species. It is not as frequent an occurrence among the fruit species and other asexually propagated plants. This may be an important reason for asexual propagation among the fruit species. When an important advance is made with a new progeny, it need not be repeated for additional seeds to

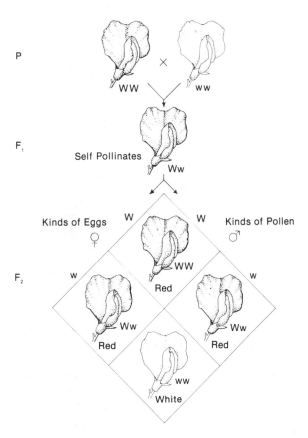

P

WW × ww

F₁

Self Pollinates

Ww

Kinds of Eggs ♀

Kinds of Pollen ♂

F₂

W   W

w   w

WW
Red

Ww
Red

Ww
Red

ww
White

FIGURE 14-3   Mendel crossed red (WW) and white (ww) blossom types of peas and found red was dominant. Fortunately he did not stop at that point but grew more generations, and found a 3:1 ratio of red to white. (Adapted from *Invitation to Biology*, Helena Curtis, Worth Publishing, New York.)

produce the cultivar, but rather another means of increase such as grafting or use of specialized plant parts can be utilized for production of similar or "identical" plants.

Plants which are naturally self-pollinated usually can be increased by seed. Those which are naturally cross-pollinated but also can be increased by seed need to be grown under very controlled conditions to maintain a pure line and to avoid mixtures due to blowing pollen or insect pollination. Thus a very important criteria for increasing seed-propagated cross-pollinated plants is to maintain strict isolation of the new cultivar to avoid foreign pollen contaminating the seed stock. It is also an important reason for a grower to use new seed each year from a controlled program rather than saving seed from current plantings. If the seed producer maintains vigilence in his operation, the likelihood of contamination is greatly reduced. Because of the great diversity of plant materials, their characteristics, and the potential for mixtures, it is imperative that seedsmen be totally aware of the possible contamination dangers that accompany seed increase.

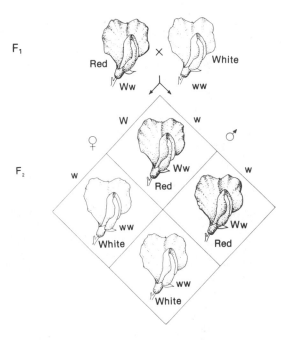

FIGURE 14-4   A homozygous white (ww) blossom when crossed with a heterozygous red (Ww) showed a 1:1 segregation of characters. (Adapted from Helena Curtis, *Invitation to Biology*, Worth Publishing, New York.)

SELECTING PARENTAL MATERIAL

A plant breeder must first of all be a plantsman in his own right. He must recognize differences among plants even within species and cultivars. This awareness of plant characteristics is a trait that helps him observe differences, slight as they might be, between cultivars or even between lines within a cultivar. If he is breeding for disease resistance, he must be able to recognize the disease if it appears among his seedlings. If he is breeding for improved culinary quality in a vegetable or fruit, he must be able to evaluate that quality for processing or fresh consumption—or its ability to tolerate handling as for quality in market channels. Much of plant breeding is in selection, whether it be selection of parents or selection among the progeny. Selection is generally done by making judgments, i.e., making decisions with regard to other individual plants or plant products. Selection can be aided by instruments, yields, epidemics, or other adverse environment conditions. More often, the breeder is faced with deciding if certain of the progeny are deserving of being selected for further evaluation or as a parent for incorporation into the breeding program. As one reads about the origin of new cultivars, generally in articles written by plant breeders, it is noted they have made numerous judgments, backed by supporting data obtained from trials. A significant amount of that information relates to the parent. Either it acquired certain desirable characteristics or

FIGURE 14-5   **Genetic diversity of a population of pinks evidenced by color variation. (Photograph by K. A. Denisen.)**

tolerance of or resistance to serious maladies that had previously plagued one of the parental cultivars or selections.

The genetic *diversity* of a population helps determine the extent to which it may be valuable in providing parental material. It is sometimes quite rewarding to utilize native types which may have survived epidemics of diseases or insects, or those which survived drouth conditions, or those annuals which survived frosts, or those perennials which tolerated unduly severe winter stresses. When utilized as parental material, some of these native species might be invaluable as sources of tolerance of the varied stresses.

A plant breeder develops a great sense of awareness of plant characteristics and utilizes his judgment in making his decisions of selection. He must also be aware of the breeding potential of the parental material and be in control of crossing plans and procedures.

MAKING CROSSES

Of foremost concern is what to expect when making crosses and growing out the seedlings. This should be related to the objectives of the breeding program. In other words, there must be a logical reason for making the cross. This assumes we have a fundamental belief we can

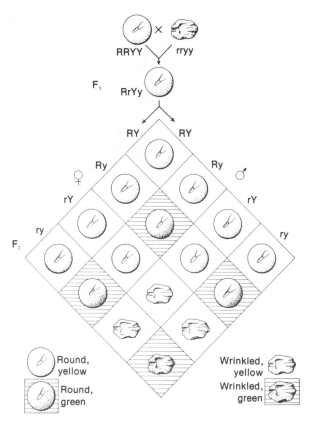

FIGURE 14-6   When Mendel crossed round and wrinkled seeded peas that were green and yellow he found a more complex inheritance which segregated in the F_2 in a 9:3:3:1 ratio.

transfer characters from parent to offspring. Mendel showed we have inheritance of characteristics from one generation to the next. He did his experiments with garden peas. It is fortunate he worked with peas because of their relatively specific scheme of inheritance. It laid the base or groundwork on which the science of genetics was founded. When Mendel crossed wrinkled and smooth seeded peas, all the progeny were smooth seeded. Thus we have the origin of the term, *dominance* among characters (genes) of plants. When the progeny (all smooth seeded because smooth is dominant over wrinkled) were planted and grew to maturity, they exhibited segregation of component characters in that about three fourths were smooth seeded (dominant) and one fourth were wrinkle seeded (recessive). This was one of the basic discoveries in genetics and formed the groundwork for that science of the heredity of living organisms. In a single character cross (monohybrid, e.g., smooth seed vs. wrinkled seed), the first generation of the progeny (F_1 for first filial generation) showed dominance of the superior gene. The second generation, or F_2 progeny resulted in segregation in a 3 to 1 ratio of smooth to wrinkled.

Genes are usually designated by letters in genetic literature. Thus the dominant smooth seeded character may be given the designation *S* and the recessive wrinkled seeded type the designation *s*. The monohybrid is thus *Ss*. It is *heterozygous* as contrasted with its two parents *SS* and *ss*, which are each *homozygous*, one a dominant and the other a recessive. The sum of visible characters in a progeny is called its *phenotype*, as contrasted to its *genotype*, which is its genetic constitution and may not be visible to the breeder or observer. The genotype may be determined by further crossing or selfing since the segregation of characters will follow, as Mendel found when he grew the F_2 seeds of his pea crosses. The chart in Fig. 14-6 shows the manner of segregation following selfing of the monohybrid seed types. As more genes or characters are involved in each parent, the inheritance may become more complex. Two genes on different chromosomes are inherited independently and the progeny is termed a *dihybrid*. When more than one gene is located on a chromosome the inheritance plan becomes more complicated. This is known as *linkage*, since the two or more genes are linked together on the same chromosome and generally are inherited together. This phenomenon of linked genes on the same chromosome caused confusion on the part of some early plant bredders, since it did not provide the clear ratios as reported by Mendel and other early geneticists. It was later demonstrated that a procedure of *crossing-over* was occurring that resulted in abnormal ratios. This happened at the time of chromosome "pairing" and "splitting" and actually occurred at a frequency which was often at a predicted rate, at least to the extent that characters which did not usually appear together, occasionally did occur in the same progeny. Crossing-over, in chromosomes, is one of the phenomena which has given greater diversity to much of the germplasm in horticultural plants.

FIGURE 14-7   **Iris breeding. The breeder is making a cross. Tagged blooms indicate parentage. Records are helpful to the breeder.**

TECHNIQUES OF CROSSING

Since the flower is the base for making crosses, the plant breeder must be thoroughly acquainted with the structure of the flower of the plant material with which he is working. Also certain flowers have characteristics which may place restrictions on techniques or time of day when pollen will shed or stigmas will be receptive of pollen. Usually the plant breeder will become so closely associated with his plant material that, by close observation, he will learn how and when to handle these

FIGURE 14-8   **Looking down the "throat" of a petunia blossom. The stigma is large among the sex organs of the flower. Anthers are shedding pollen. (Photograph by K. A. Denisen.)**

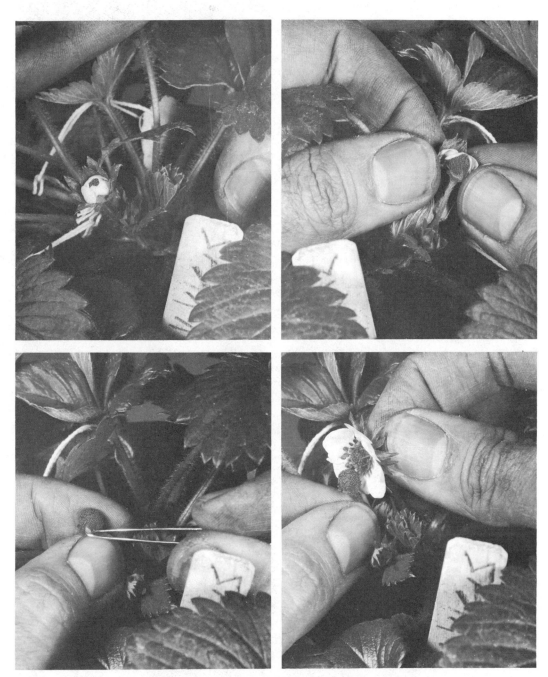

FIGURE 14-9   A series in crossing in strawberry breeding: *Upper left*, blossom just beginning to open and ready for emasculation; *Upper right*, emasculation using the thumb-nail in a cutting action. Calyx and anthers are removed in one operation; *Lower left*, a tweezers is being used to remove an anther that did not come off with the calyx; *Lower right*, pollen from a fresh blossom of the male parent is rubbed on the pistil.

Improvement of Horticulture Plants by Breeding and Selection   **305**

FIGURE 14-10   A honey bee visiting a bloom to collect nectar. While there he gets pollen on
                     his body, then transfers it to the stigma of the next flower visited.
                     (Photograph by K. A. Denisen.)

FIGURE 14-11   Female flowers of squash. When making crosses the tip of the blossom must
                     be tied shut so insects cannot enter and make pollinations before the plant
                     breeder. (Photograph by K. A. Denisen.)

Skills and Practices in Horticulture   **306**

FIGURE 14-12   **Breeder hybridizing cabbage. (Photograph courtesy Iver Jorgensen, Northrup King and Co., Minneapolis, Minn.)**

limitations. Plant breeders can have a mutual exchange of information to the benefit of all individuals. This often happens at scientific meetings.

Depending on species, several means of collecting, extracting, and transferring pollen are available. If pollinations are made in a greenhouse the possibility of contamination by wind, insects, and proximity to related plants can be greatly reduced. Often a watch glass is useful for collecting and transporting pollen. The watch glass is small, pollen is visible from above and below, and it is easily maneuvered into position for administering to the stigmatic surface. It is widely used in crossing many vegetables and ornamental plants. In some species it may be practical to pick a bloom which is shedding pollen and make the transfer direct to the stigma. This can be done with the pumpkin, strawberry, raspberry, woody ornamental shrubs, and certain other plants. Naturally the blossom is sacrificed and cannot be used again.

Insect-pollinated plants need special precautions in a breeding program to avoid pollination prematurely or unknowingly. A useful and practical technique to avoid these problems is to use a paper or glassine bag over the blossom or blossom cluster. Again, depending on species and flowering habit, one will need to give careful attention to the stage of development of the inflorescence. If there are shedding anthers and the blossom cluster is perfect-flowered, it may be necessary to discard all exposed stigmas, since they could very likely already be pollinated.

Wind or gravity pollinated plants likewise also have vulnerable periods during which they might be naturally self-pollinated when covered with a paper or plastic bag. Constant vigilance is the best means of avoiding unwanted self- or cross-pollination in a breeding program. It was the vigilance of Gregor Mendel which resulted in his good records and keen observations. Likewise the vigilance of plant breeders in making crosses, keeping parental and progeny records, and making judgments based on observation and performance can be the solid base for improvement of horticultural cultivars and a more efficient horticultural industry.

Inbreeding is a widely used technique for concentrating or intensifying certain characters. It is extremely useful but can also lead to sterility of the offspring, so it is somewhat limited in usefulness for some species. Some test selfs and crosses can help determine the usefulness of inbreeding or selfing.

GROWING SEEDLINGS

The plant breeder is very much concerned in the potential of certain crosses. One can see the results of these crosses by growing out the seeds obtained from cross-breeding. Depending on the nature of the plant material a large number of seedlings is considered desirable to learn the likelihood of improvement of the progeny compared to its parents. Often the best means of learning the potential of a cross is to grow a relatively large number of seedling progeny. This may be rather readily determined depending on the number of seeds produced in a crossing operation, e.g., a tomato produces a large number of seeds from a single flower compared to the peach or plum, which generally produces one seed. In order to explore more fully the potential of the peach or plum cross, pollinations of numerous flowers must be made.

For highly heterozygous material, it is desirable to grow a large population of progeny to determine the potential of a certain cross. This is because of its diversity—a hundred seedlings will generally give a broader spectrum of possibility than ten seedlings. By the same token a selfed progeny may also have a highly segregated population and selfing may help to concentrate certain characters for future use as parents in the breeding program when accompanied by vigilant selection.

THE SELECTION PROCESS

Prior to the era of plant breeders, new cultivars originated either by chance seedlings or selection from a population. The person making the selection depended on a set of determined criteria and made judgments from among the seedlings. Selection continues to be an important means of improvement even with a breeding program. Many judgments are made by the plant breeder as evaluations of the population of progeny are determined. The selection process actually starts as the plant breeder selects parental stock for the breeding program. It is the goal of every plant breeder to combine desirable features of one parent with those of another parent to produce a more ideal progeny in compliance with the objectives of the breeding program. For example, a vegetable breeder may attempt to incorporate disease resistance into a selection or new cultivar. As the seedlings are grown, it will be relatively easy to discard susceptible progeny if they are subjected to pathogens which generate the disease. Those seedlings which survive and have relatively good characters are selected either for consideration for becoming cultivars or to utilize as parental material for further breeding. Thus a breeding program becomes a compilation of selected parents and progeny. It is cumulative in effect and composition. The plant breeder is challenged by the potential for improvement of progeny by careful selection of parental material and making noteworthy selections among the progeny.

In establishing criteria for selection, it is helpful to be well acquainted with current cultivars and their characteristics. Also any native species with certain features can be very useful as parents since there is good likelihood of broadening the germplasm base for breeding. Since parental material is such an important part of any breeding program, careful consideration should be given to establishing a "germplasm bank" with desired characteristics or traits that can be used as parents. This type of plant repository is useful to plant breeders, whether it contains asexually propagated plants as is true of most fruit plants and ornamentals, or if the plants are seed propagated, as is true of most vegetables and flowers. The repository should be a ready source of germplasm to the plant breeder. Care of the repository is important to avoid plants acquiring and becoming contaminated with diseases of a systemic nature. A national program of plant repositories is contemplated for the United States.

EVALUATION PROGRAM

Most plant breeders conduct cultivar trials of the species in which they are involved. These trials are invaluable in that they acquaint the plant breeder with characteristics of available cultivars and provide opportunity for detailed comparisons. Selections from a breeding program fit nicely into cultivar trials, since the trial provides the conditions of environment, soils, and types of culture characteristics of the specific kinds of plants being tested. Of specific interest to both growers and

FIGURE 14-13   **Greenhouse production of hybrid cabbage seed. (Photograph courtesy Iver Jorgensen, Northrup King and Co., Minneapolis, Minn.)**

plant breeders are yield and quality comparisons. Since both yield and quality are also controlled by cultural and environmental conditions, it is advantageous that they have proximity of location. It gives greater confidence of data.

In addition to production records, notes on characteristics and unusual features are very helpful in making decisions of release and introduction. Plant breeders will often exchange selections with coworkers at other locations and institutions for testing purposes. This procedure can provide additional information on adaptability to other locations, soils, and epiphytotic conditions all of which can be beneficial to both investigators. Information from these trials is useful at the time of naming and introduction of a new cultivar in addition to providing more information on current cultivars. Many growers are eager to receive information on cultivar trials from year to year for comparisons with their own favorite cultivar and to note the progress being made on new selections and cultivars they may have seen or heard about from other growers or at field days. Some plant breeders have a constituency of growers who supply very reliable comments and judgments on potential cultivars. This can be a valuable source of information for the breeder because he does need public acceptance for the new cultivar if it is to replace or be added to some of the current cultivars.

NAMING AND INCREASING CULTIVARS FOR DISTRIBUTION

Following the long and intense period of breeding, selection and thorough evaluation, the plant breeder makes the decision to name and distribute the new cultivar. It is indeed a sense of accomplishment to a program thay may extend over a period of several years, and with some species over a lifetime. Thoroughness of records, photographs and descriptions become vital to the introduction of the new cultivar.

The name of a new cultivar can determine its success to a considerable degree. There are many kinds of names that may be used: some are descriptive, e.g., 'Delicious'; some are named after members of the family, e.g., 'Sharon'; some are named after North American Indians, e.g., 'Chippewa'; some are named after other horticulturists, e.g., 'Darrow', etc. In general, single names are more satisfactory than double names, short and simple names are preferred to complex and easily misspelled names, and repetitive names are undesirable from the confusion they create among growers or gardeners. Use of the word "new" in a name is not desirable because eventually it will not be "new".

Because of the time required for increase of stock of a new cultivar added to the years involved in breeding, growing, selection, and testing, it is often many years from time of crossing to final release. Thus it is not uncommon for 10, 20, or even 30 years to elapse before a cultivar is widely available to the general public. This means one must have another source of income, usually a primary profession, before one devotes considerable effort to plant breeding. It is an important reason why plant breeding is useful as a hobby or avocation or why much of plant breeding is done by public supported institutions.

PHILOSOPHIES OF PLANT IMPROVEMENT

There are many reasons why plant breeders are interested in development of new cultivars. It may be the creative impulse, the desire for originality. It could be a great need for resistance to a serious disease pest, and the breeder has some ideas on the sources of resistance that have not been tried in the breeder's locality or under his conditions. It could also be the needs of a processor of vegetables who wants more consistency in the canned product that could be supplied with a new cultivar selected for that specific quality. It could be the desires of a rose breeder to combine various colors in the petals of the rose bloom, as has been accomplished in some current new cultivars. In all breeding programs, one must consider the challenges of solving problems, creativity, increased beauty, increased hardiness, and special desires, *ad infinitum*.

Generally, profit is not one of the motives of a plant breeder, but in a sense it always is a motive. Even the motive of satisfaction is a profit to the breeder though it may not be expressed in a monetary sense. History of cultivars tells us that great profits do not generally accompany production of new cultivars. Even Luther Burbank with his hundreds of named cultivars did not become wealthy because of it. Sometimes the breeder can "cash in" on his creativity. Perhaps that is what is responsible for more frequent patenting of cultivars under the recently enacted plant patent programs.

PATENTING NEW CULTIVARS

As in the patenting of inventions, new creations of horticultural plants can now for the most part be patented. There are a few exceptions, and anyone desiring to patent a new plant creation should proceed by contacting the Plant Variety Protection Office, U.S. Department of Agriculture, Beltsville, Maryland. The originator of the new cultivar will then proceed to describe the new cultivar, indicating especially how it differs from previously introduced cultivars, and how it can be recognized or identified. The process of claiming a new cultivar and establishing its identity can be greatly simplified by engaging a plant patent attorney who is experienced in filing applications, acquiring the necessary data, and supplying appropriate descriptions for the specific kinds of fruits, vegetables and ornamental plants. Once a patent is obtained, the originator can license growers and distributors to produce the plant cultivar for sale, generally with a nominal return from each plant to the controller of the patent.

STUDY QUESTIONS

1. Why is plant breeding important to a horticulturist?
2. How is the science of genetics related to plant breeding?
3. Among the many species of plants, Mendel might have worked with, why is it fortunate for the science of genetics that he worked with peas?
4. Actually there seem to be more differences than similarities in the personalities and objectives of Mendel and Burbank as plant breeders. Why then do we relate them as pioneers in plant breeding?
5. What is heterosis? Is it an end in itself?
6. How do cross-pollination and self-pollination enter into the concept of maintaining new cultivars?
7. Why is selection considered such an important factor in cross-pollinated crops, self-pollinated crops, seed-propagated crops, asexually propagated crops?
8. How can one prevent pollination of a self-pollinated plant so it will be available for crossing when it is producing pollen?
9. How does a plant breeder determine the criteria for selection among the progeny of a cross?
10. What is meant by "broadening the germplasm base" in a collection of the germplasm bank?
11. Why are some cultivars patented? Is it usually profitable?
12. Are all new cultivars superior to those previously in existence?
13. Why does plant breeding lend itself readily to being a hobby?
14. Experiment stations in the past have not patented new cultivars to a large extent. Why is this true?

SELECTED REFERENCES

1. Brooklyn Botanic Garden 1976. *Breeding Plants for Home & Garden* (No. 75). Brooklyn, New York.
2. Cargill, B. F., and G. E. Rossmiller 1969. *Fruit and Vegetable Harvest Mechanization, Technological Implications.* Rural Manpower Center, Michigan State University, East Lansing, Mich. This publication emphasizes breeding for adaptability to mechanical harvest in vegetables and fruits.
3. Crane, M. B., and W. J. C. Lawrence 1962. *The Genetics of Garden Plants.* Macmillan, London.
4. Curtis, Helena A. 1972. *Invitation to Biology*, Section 3, *Genetics.* Worth Publishers, New York.
5. Frey, K. J. 1966. *Plant Breeding* (A Symposium). The Iowa State University Press, Ames, Iowa.
6. Hayes, H. H., and F. R. Immer 1949. *Methods of Plant Breeding.* McGraw-Hill, New York.
7. Howard, W. L. 1945. *Luther Burbank's Plant Contributions.* Bull. 691. Univ. of Calif., College of Agr., Agricultural Exp. Station, Berkeley.
8. Janick, Jules, and J. N. Moore. 1975. *Advances in Fruit Breeding.* Purdue University Press, West Lafayette, Indiana.
9. Mendel, G. J. 1866. *Experiments in Plant Hybridization.* Harvard University Press, Cambridge, Mass. English translation of his original paper.
10. Merrell, D. J. 1975. *Introduction to Genetics.* Norton, New York.
11. U.S. Department of Agriculture Yearbook 1937. Washington, D.C. Over a 40-year span this has remained an outstanding source of information on breeding techniques and related information on horticulture crops including apples and pears, stone fruits, nuts, subtropical fruits, small fruits, vegetable crops, potatoes, and flowers.

Horticulture
for the Home

15 Planning the Home Grounds

The home grounds are an integral part of family life. It is here that the family can live together, work together, and play together. It is here that they and their friends can enjoy the great outdoors. It is a place for the development of ideas, plans, and arrangements that are a reflection of individual tastes and family compromises. It lends itself to family projects, nature study, hobbies, outdoor cookery, recreation, fruit and vegetable production, relaxation, and aesthetic beautification. It can be the all-purpose area for summer living. How it develops and the purpose to which it is put is determined by its occupants and their desires. Whatever the goal, it requires a plan.

PURPOSE OF THE PLAN

Organized Thinking for the Future

In any planning, the ideas and thoughts should be accumulated, recorded, and arranged. It is then often desirable to let the ideas and plans remain dormant for a time. This absence of action often gives a better opportunity for reevaluation or reappraisal.

Determining individual preferences or tastes is a good place to begin in developing a plan for the home grounds or in changing an existing arrangement of the physical features. Some people prefer a rustic atmosphere, some prefer strictly functional surroundings, others like formal balance, geometric lines, and architectural grandeur, and still others would rather blend these motifs to varying degrees. Individual preferences flavored with originality and imagination are good guides for the home owner and home planner in developing his home grounds.

The *evaluation of ideas* is a practical approach to the problem at hand. It may involve discussions with neighbors, friends, or perhaps persons experienced in plant materials and design. Certainly a clear view of objectives is elementary to evaluation. A critical study of the factors under discussion will help in the ultimate appraisal of the organized plan.

An *estimate of costs* is vital to most home plans and their development. In a new home or a newly purchased home, most people have invested considerable capital in the house, which precedes development of the grounds. The result is an attempt to accomplish with as low cost as possible the planning, developing, planting, and completing of the outdoor surroundings. For this reason, the outdoor planning should be considered along with the purchasing or completing the permanent features of the urban home or farmstead. Any investment is judged by

317

FIGURE 15-1   **Individual preferences flavored with originality and imagination are guides for planning the home grounds.**

its present value. A home that includes planned, developed, and planted home grounds has a much higher present value than the same home without developed grounds. Yet, it may entail but little more additional investment.

There are two major categories of expense to the new home owner, immediate needs and future needs. The immediate needs, or essentials, require provision for immediate payment. The future needs consist of things that can be postponed. They may involve additions either in structure or planting. They may include "do-it-yourself" possibilities. Maintenance is an item for future cost consideration.

Establishing a priority of development is a product of organized thinking for the future of the home and home grounds. *Immediate needs* should be itemized and head the list. In outdoor features they will include walks, drives, lawn, annual flowers, basic foundation plantings, and trees for earliest possible shade. Other *essentials that can wait*, for brief periods at least, will include many items or projects than can be done by the homeowner and the family if they so desire. Included in this group are shrub borders, hedges, gardens, fruit plantings, and recreation facilities. The home propagation of trees, shrubs, and perennial flowers may be part of this category of development. The future development plans may include a listing of *things that will be eliminated later*. Included

FIGURE 15-2   **This rock wall was not an immediate need nor was it developed immediately. Some things can wait.**

here are the sand box area after the children grow up, fences after shrubs are established, and vegetable gardens reduced in scale when family food needs are decreased. *Luxuries* for both the home and outdoors complete the priority listings. Specimen plantings and hobby materials fall into this category.

Avoiding Costly Errors   Mistakes can be extremely costly. An important purpose of any plan is to avoid mistakes. Clear thinking followed by appraisal and reappraisal of ideas will usually result in fewer errors of judgment both in home construction and planning the grounds. It is well to think in terms of how the occupants intend to live during the majority of the years ahead.

Mistakes in site can result in *drainage* problems both from the standpoint of growing lawns and other plantings and seepage into the basement. The *topography* is another source of error in selecting or utilizing a site. *Views and exposure* are influenced by topography, adjacent buildings, and plantings. A new subdivision on rolling or hilly land can present all these problems in certain areas. However, it is usually possible to arrange the shape and locate lots with respect to the entire new community. Leadership by the builder himself or community spirited individuals can realize an organized plan which respects each neigh-

FIGURE 15-3   Two building sites. *Top*: **The developer sheared all vegetation, including trees.** *Bottom*: **The builder left existing trees for immediate shade.**

bor's right to panoramic views, privacy, and drainage. Within the individual lot or farmstead, mistakes in site of the permanent features can also be costly or annoying. Too often a picture window is located at a certain position because the builder's plan called for it. It may present a view of a busy highway, a dusty road, or into a neighbor's picture window. The window is sometimes more obviously for the picture it presents from the outside viewing in. Houses built on hilltops should have adequate space for windbreak trees or shrubs especially in areas of severe winters or cold winds. Locations near streams or rivers that are capable of overflowing their banks during flash floods or spring thaws are to be avoided for home sites. Property values decrease in proportion to undesirable or hazardous surroundings. Time spent in thoroughly evaluating a site is usually time well spent.

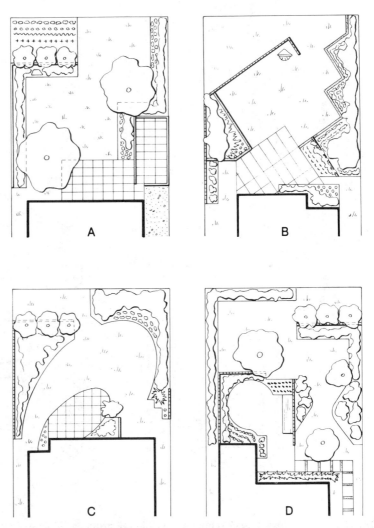

FIGURE 15-4   **Four different plans for private areas of a home. (Adapted from Wm. R. Nelson, Jr., *Landscaping Your Home*, Circular 858. University of Illinois, College of Agriculture, 1963.)**

Planning the Home Grounds   **321**

Mistakes in physical development are generally less serious, monetarily, than are mistakes in site. However they can be a source of minor, but continuing, irritations. Walks or drives that are rarely used because of incorrect placement are examples. It may not be their non-use which is the irritation but rather the paths which are formed where the walks should have been. A garage that is not large enough to accommodate the latest model of automobile may be quite annoying. Adequate foresight to consider additional storage space in the garage for bicycles, lawn mowers, and garden equipment can often eliminate much harassment later. A farmstead with inconvenient locations of farm buildings can be very costly both of time and labor. In planning a home

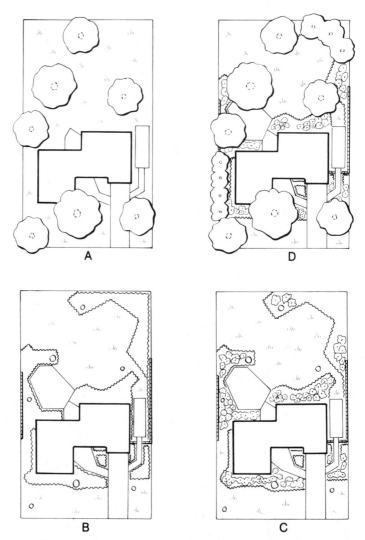

FIGURE 15-5   **Stages in development of a home grounds. (Adapted from *Landscape Design*, A Teacher's Manual, R. F. Stinson, D. R. McClay, G. Z. Stevens, Department of Agriculture Education, Pennsylvania State University Teacher Education Series, Vol. 9, No. 3t., 1968.)**

and home grounds, careful thought and projected vision will reveal many things. How convenient is the kitchen to the patio? Is the children's play area where it can be readily observed from work areas in the house? Do downspouts pour out on a walk, path, or terrace? Are there steps in unexpected places? Are all pits, depressions, basement window wells, and differences in grade adequately barricaded or marked? These and a host of other questions may be asked to insure future satisfaction and safety.

Mistakes in planting are costly of time, labor, and plant materials. A tree that has to be removed after 10 or 20 years of growth is a real loss to the home grounds. Greater care in planning the location at planting time will often avoid later disappointment. Trees planted too close to the house and under power lines are common errors. They may have to be removed later or be badly disfigured to prevent damage. In selecting an area for the vegetable garden, convenience and soil productivity are important factors. Shaded areas and close proximity to competing plants are to be avoided. For continued productivity trees should not be planted in or near the garden area. Trees and shrubs that are planted too close together may look all right when small but compete to the detriment of both as they grow larger. Large-growing shrubs placed beneath windows of the house will require frequent or drastic trimming if not removed entirely to prevent screening of outside views. Wrong cultivars or wrong kinds of plants are common errors in planting. Fruit trees grown solely for ornament and spring bloom are a poor choice. They will either require considerable care or result in large quantities of unusable pest-infested fruits that drop to the ground and create a disposal problem. Small fruited ornamental crabapples or a sterile cultivar would be a much better choice if bloom is the effect intended.

OBJECTIVES OF THE PLAN

The overall objectives of the plan for the home grounds can be stated in four items: (1) to be attractive, (2) to be interesting, (3) to be convenient, and (4) to be functional. These objectives overlap and interlock as they are accomplished.

To Be Attractive

Public attractiveness of the home and home grounds is an outward sign of character. People judge by appearance. When the residents of a community exhibit orderliness, neatness, and harmony of their home grounds, they are well accepted by the community. It reflects a spirit of community pride and molds neighborly relationships. From a material view, attractiveness of a home and its surroundings influence property values. People place priority on desirable settings and are willing to spend more to be a part of it. Attractiveness to the public also influences business and business relations. The successful farmer, whether he is a dairyman, hog breeder, poultryman, or general farmer, prefers having an attractive farmstead. It is especially important to his business if he has much contact with the public on his farm. Other home-business

FIGURE 15-6   **General appearance of a business establishment influences the surrounding properties. (Photograph by K. A. Denisen.)**

establishments owe part of their business success to the attractiveness of their home and surroundings to customers and potential customers.

Attractiveness to the family contributes to pride in their home, increases family unity, and creates an atmosphere for character development. Effort and capital expended for improvement will be enjoyed in the future as well as the present. The ties which unify the home can be found in family projects and cooperative endeavor. When all members work, plan, and use the home together, they act as a unit in developing, utilizing, preserving, and defending the home. Family projects that are well administered and executed can develop responsible attitudes in children. Work is performed with a willing spirit while play is not neglected. Consistent family unity and an abiding interest in the home provide the atmosphere for healthy character development. The interest of all ages in the home and home grounds provides an opportunity for release of energy for constructive use.

To Be Interesting   *Individual tastes* vary widely. In home grounds planning and dvelopment, they add interest to the setting. Stereotyped plans should be avoided. This is especially true in areas or subdivisions of similar architecture. It adds interest in such areas to note different ideas on tree and shrub selection, placement, and use, different plans and objectives in foundation plantings, and different allocations for service, recreation, and relaxation areas.

Accent areas created by color, texture, form or size in the landscape add interest to both urban homes and farmsteads. These can be obtained either with buildings or plant materials. A background of trees provides a peaceful setting to the house and grounds. Entryway plantings which include an accent at the entrance to a house, focus attention and add interest to the door. Avoiding stereotype in selection or maintenance of these shrubs or small trees is important, however.

To Be Convenient   The home grounds or farmstead should be designed to *conserve time and steps*. Efficiency in performing routine tasks in the home and in business are important to both productivity and leisure.

Convenient location of buildings on the farm can be a great aid to efficiency of production and conservation of time and labor. Easy access from feed storage to feed lot, use of portable feeders, constructing portable swine and poultry houses, and close proximity of milk storage to the dairy barn are important considerations in the development of livestock farms. Vegetable and fruit farms should have easy access to

FIGURE 15-7   The children's play area should be convenient.

325

water for spraying. Machinery storage and repair shops should be conveniently located on vegetable, fruit, grain, and cotton farms. Convenience of office to sales-yard is important for many home-business establishments.

Easy access to vital areas is important in planning the home grounds. Where required clothes lines or clothes poles when readily accessible avoid unneccessary exertion in laundering. To be most useful, the outdoor living area should be easily reached from the kitchen via rear or side entrance. The children's play area should be convenient for the use of children. A good view of the play area is especially important when children are small and need constant supervision.

Adequate provision for inclement weather can be extremely vital to convenience and comfort. A firm and substantial lawn turf is a valuable asset in wet weather. It decreases the frequency of muddy tracking into the house. It reduces compaction of soil and formation of paths when the soil is wet. A good lawn also reduces the amount of dust in the air during dry and windy weather. Trees and shrubs used for windbreaks greatly modify the severity of cold winds during winter months in temperate climates. Hard-surfaced drives and walks add greatly to the convenience of driving and walking during wet weather and for snow removal in cold climates. The use of eave troughs on buildings eliminates the pounding effect and erosion by water run-off from the roof. When eave troughs are not used, place shrubs a little out and away from the building and use a gravel ditch beneath the eaves. Convenient location of water taps is a good provision for drouth periods when lawn irrigation may be necessary. Placing shrubs in front of water taps often obstructs their use.

To Be Functional   The home grounds may be functional in many ways and for various purposes. Among them are recreation, relaxation, enjoyment, entertainment, outdoor living, and fruit and vegetable production.

Recreation areas for children will usually have added facilities that include sand box, wading pool, swings, and other equipment. Many play items can be improvised and some may be very temporary. The children's play area may be enclosed by either a temporary or permanent fence or by dense shrubbery or hedge plantings. It should provide shade in some part of the area. Children, like adults, are impressed by the attractiveness, interest, and convenience of their play area. In their eyes it is one of the most useful parts of the home. Mutually attractive play areas offer encouragement for children to allocate their play time between their own and their young friend's homes. The recreation area for teenagers and adults may develop from the children's play yard even while the children grow. It will usually include a sports area on the lawn for badminton, croquet, and other lawn sports. Terraces and paved drives provide for roller skating, basketball, and other games.

Relaxation is extremely important to a highly civiiized society. The private area of the home grounds is an excellent place for hobbies. It can provide a serene and restful setting free of distractions. For the person who works in an office, at a factory, or away from nature and the soil,

FIGURE 15-8   **Substantial turf and adequate shade contribute to the desirability of the play area.**

the quiet setting of his home grounds can give a new outlook and refreshed attitude toward life and his life's work. It can afford mental relaxation and a place for thought and meditation.

Enjoyment can include many things, but here refers to the beauties of Nature in the home grounds. Blooming flowers, a neat lawn, attractive shrubs, heavy shade on a hot day, and frosty twigs on a cold day all can add to the enjoyment of daily living.

Entertainment of the informal type in a rustic setting is increasingly popular. The extent to which a home area is developed for entertainment is entirely a matter of personal taste and anticipated expense. It may vary from a well-kept lawn with inexpensive lawn furniture to a luxurious patio or terrace. The area should be easily accessible. Surfaced areas of a patio or terrace may be constructed of brick, stone, or concrete. Trees, shrubs, flowers, or grass may be growing in the surfaced area, emerging from the soil between stones or bricks. However, they can be grown in pots, planters and urns. Further adaptations in development for entertainment would include outdoor lighting, erection of a screenhouse, and more extensive terrace furniture. For both daylight and evening entertainment, provision should be made for insect control. Easy access to the house and screened areas hold special significance under these circumstances.

Planning the Home Grounds   **327**

Outdoor living is a popular activity. Facilities for outdoor living can be included in a home grounds with very little added expense. A shaded area, a picnic table, and some lawn chairs are the essentials for outdoor living. It can be a summertime living room and is available for "spur of the moment" use. This living area is frequently used for picnics and outdoor cookery. It may have a charcoal grill, barbecue pit, or fireplace. Portable charcoal grills have several advantages over permanent installations because they can be removed after use, do not wear out a certain area of the lawn, and can be moved for wind protection and shade.

Fruit and vegetable production is often an important use of the home grounds. It can be a source of fresh and high quality produce especially for crops which deteriorate rapidly after harvest. A fruit and vegetable garden reduces the food budget. Many vegetables and berries are quite expensive to purchase even when in season. A garden provides nutritional advantages. Fruits and vegetables are good sources of vitamins, minerals, and carbohydrates. An abundant harvest means an abundant supply of these food nutrients. The garden is also a means of physical exercise and provides an opportunity for family participation.

FIGURE 15-9   **A rustic setting can be utilized both for entertainment and family outdoor living.**

DEVELOPING THE PLAN

A plan should be recorded on paper, revised, visualized, and changed again if necessary. Two perspectives, vertical and horizontal, should be considered for proper location of permanent features and plant materials.

The Plat or "Blueprint"   *Square-lined* paper is ideal for developing a plat of the area. First, a scale of distance should be established, e.g., $\frac{1}{4}$ in. ($\frac{1}{2}$ cm) on the plat equals 1 foot (30 cm) actual measurement will give ample space for indicating all pertinent objects and materials. The permanent features of the plan are then located. These include buildings, walks, drives, and other installations. It is important that these be indicated according to scale both in size and location. The plat is now ready for designation of the various areas and the functional location of plant materials.

The *areas* for specific purposes or uses are shaped, outlined, and designated according to individual and family desires. The *approach* area is in front of the house facing the street, road, or entrance to the grounds. As the term implies it is the first view the grounds present when a person approaches the home. It may be large and spacious as often found on a farmstead, short and wide as sometimes found with many one-story rambling floor-space houses on city streets, or long and relatively narrow where back yards or rear areas are reduced in size to give greater area in front. The size and depth of the approach area implies a degree of privacy acquired through space and distance. The *foundation* area includes the immediate surroundings of the house and garage. This area, though small, receives special attention since it is

FIGURE 15-10   **The foundation area receives early attention.**

located by the central feature of the plan, the house. It varies in size and shape because of locations of walks, width of eaves, attached garages, and the direction which the house faces. Walks which parallel the house isolate strips of ground that are allocated to the foundation area. Spreading eaves encompass more area around the house and increase the area which logically belongs to the foundation. Because attached garages appear as a continuation of the house, they too require a foundation area for plant materials. Houses that face the sun may have larger foundation areas, since it is easier to find variety in plants that thrive in sunlight than plants that thrive in shade. The foundation area encircles the house, but most plantings are located on the approach sides. The *outdoor living* area is planned for privacy. It is away from the approach area, often being separated by the house itself. It may be

FIGURE 15-11   **The tool shed, like other service areas, should be conveniently located.**

Horticulture for the Home

located to either side of the house, however, with screening accomplished by fences, shrubs, or trees. Since many activities of the family are centered in the outdoor living area, it usually occupies a large portion of the home grounds. It may be subdivided on the plat into play area, patio, picnic area, etc., for present or future development. From the family standpoint this is a major area of the home grounds. It is most useful when provided with a good lawn, adequate shade, and privacy. The *service* area in an urban plan includes such features as the garage, drive, garbage receptacle, and tool shed. Its location on the plat is determined primarily by location of the permanent features. Convenience, as previously discussed, should be the guiding factor in its location. On a farmstead plan, the service area is usually the largest area. It includes all the grounds between the farm buildings and around the well, drives, and lanes. It is especially important that it be usable and accessible during inclement weather. Some parts of it may be hard surfaced because of the large amount of traffic. Other home-business enterprises also require a larger service area than that needed for a home grounds. Adequate parking space is one important consideration for these establishments. The *garden* area's location on the plat should be influenced by shade, soil, and accessibility. Size of the garden is an important consideration. Too large a garden for proper care and attention leads to discouragement. A poorly tended garden detracts from the home grounds. The garden size may fluctuate as years progress, but it is

FIGURE 15-12   A fence gives privacy from intrusion. Shrubs give privacy from view.

important to locate it properly, since many other plantings are determined by its location. Fruit trees are not necessarily restricted to the garden area but some may be used in the outdoor living area for shade or ornamental purposes.

Functional locations of plant materials is the final step in developing the horizontal plat or blueprint. The functions or purposes of plant materials include shade, screening, privacy, background, accent, framing, windbreak, and specimen plants. *Shade* is obtained primarily from deciduous trees in temperate climates or from broadleafed evergreen trees in subtropical and tropical climates. The shape and ultimate size of trees for shade are important criteria of selection and placement. *Screening* is frequently attained by effective use of shrubs, although vines on trellises, lattices, and fences are also effective screening materials. Both clipped and unclipped hedges are widely used for screening. *Privacy* in the home grounds results from topography, build-

FIGURE 15-13   **This pine is a stately accent point for a new home. Note the island of soil retained when the grade was changed.**

ings, fences, and plants. Fences are not necessary to give privacy. Plants along the lot line, even intermittently, in urban areas are useful in designating boundaries. Shrubs, vines, trees, and flowers are adapted for such purposes. Privacy in home-businesses and farmsteads is obtained by the use of screening plants which separate the outdoor living area from the service yard. *Background* plantings generally consist of trees or shrubs along the back or distant end of the grounds. They give a feeling of depth and soften the harsh lines produced by man-made structures. In warm weather they add coolness to the surroundings. Some backgrounds are so distinctive that they become landmarks of the community. *Accent* points of the landscape are those which have outstanding features and catch the eye. High, towering trees against the sky are accents. They are breaks in the regular pattern and may be used to coincide in rhythm with the accents of the house. *Framing* plants are those which encompass the main point of interest in a framing view. The house is the main point of interest in a home plan. The trees and shrubs on either side, both in foreground and background, form the framing view of the house. *Windbreaks* in an urban area consist mostly of shrub plantings near walks and doorways. In rural and suburban areas, windbreaks may serve an extremely useful purpose to reduce the icy blast of winter winds. Evergreens are especially helpful in producing a haven of still air around the house and in the farmstead. Deciduous trees and shrubs are less effective but do reduce wind velocities considerably. *Specimen* plants are used for special attraction in the home grounds. Because they are interesting in themselves and may detract from other things, they are not as well adapted to planting in the approach area. The main point of interest in the approach area should be in the house alone.

The Vertical View

It is sometimes difficult to visualize the completed plan from the plat or horizontal view. A vertical view helps in relating the plant materials to the house and grounds. The starting point for such a view can be a *sketch or photograph* of the completed, unlandscaped house. This photograph or sketch should be at least 8 by 10 in. (20 by 25 cm) in size to give adequate space for inserting crayon sketches of plant materials. It should include boundary lines of the lot or farmstead.

An *overlay of transparent* material placed on the photograph is very useful for sketching in the plants with grease pencil or soft crayon. These marks can be removed easily for changing locations of shrubs, trees, and flowers. The ultimate or maintained size of plants should be sketched to scale. The approximate shape of the shrubs and trees should be adhered to for accuracy of the completed plan. Coloring may be used to show differences in foliage color, bloom, and general effect. The vertical view should be checked with the plat at frequent intervals for reference, proper placement, and modification. This is a critical phase of the planning; since, here the plat is visualized from a vertical position. Changes can be made more easily on the drawing board but become more difficult, more costly, and at a sacrifice of plant growth if made after planting has been done.

Planning the Home Grounds

FIGURE 15-14   *Top*: **Unlandscaped one-story house.** *Bottom*: **Plant materials sketched in to complete vertical plan. (Photographs courtesy of** *Better Homes & Gardens* **magazine.)**

FIGURE 15-15 *Top*: Unlandscaped two-story house. *Bottom*: Plant materials sketched in to complete vertical plan. (Photographs courtesy of *Better Homes & Gardens* magazine.)

Replacement of plants is an important consideration even when formulating plans. In the course of a decade or two, many things will happen to the plant materials which will tend to change the entire scene. The first of these is growth which can lead to overgrowth. Sometimes pruning is practical for overgrown plant materials but often replacement is the only alternative. Injury may malform plants. This may lend interest in itself or may demand replacement of the same kind or with some other choice. Senility comes to some trees and shrubs sooner than with others. For variety the owners may like an alternative selection of plant materials as replacement becomes necessary. This requires some planning, however, since these changes are not all made at once.

It is important that all plats or blueprints with the horizontal view and all vertical plans be retained. These can be consulted again as the materials grow, develop, and fill in their destined areas and functions. The overall plan can even be brought up to date at intervals and maintained as the current setting and key to planting.

THE FINISHED DESIGN

Balance   Any design should have balance of weight, color, height, texture, interest, and other factors that make up the home grounds. Balance may be obtained by adjusting these factors in a relationship of either formal balance or assymetric balance. They have the common feature of completion, or fulfillment, or repose. However, the manner of fulfilling the desire for balance differs greatly between the two types.

Formal balance is characterized by geometric precision with like objects equidistant from the center. The lines created by plant materials and architectural additions are sharper and more severe. Straight lines, circular groupings, concentric lines, squares, triangles, and radial lines are used frequently in formal plans. The effect of formality is intensified by sharp contrasts in color and texture of plant materials. In a formal garden, balance is also obtained by using the technique of mirror image. This implies equal treatment on both sides of an area equidistant from the central attraction. Balance in a formal garden is relatively easy to attain by placement of plant materials, archways, lily ponds, formal benches, and lattices. A pattern of repetition, duplication, and precision is established and mathematical exactness is employed in completion of the plan. Maintenance of formal balance is as specific and exacting as its design and establishment. This is a feature which limits its usefulness in many home plans. Loss of a plant requires removal of its counterpart to maintain balance. Two new plants are sometimes required instead of one to have equal treatment for exact balance. Complete uniformity in clipping and pruning is required for distinct, sharp lines. These are time consuming operations when applied to an entire home grounds planting. Because of this time consuming and costly maintenance, and because of the characteristically harsh, exact lines, most home plans are not of formal balance design. This does not preclude the use of certain formal features, which are easier adapted for home grounds, for example, a clipped hedge, a sheared and pointed ornamental evergreen, or color

FIGURE 15-16   **Formality is indicated by equal balance.**

FIGURE 15-17   **A formal garden. (Photograph by K. A. Denisen.)**

Planning the Home Grounds   **337**

contrasts. Formal balance is frequently used to dignify public parks, botanical gardens, and arboreta.

Asymmetric balance is balance without formality. In the home grounds, it is characterized by smoothness rather than harshness. The use of smooth, flowing irregular lines in shrub and flower borders and foundation plantings presents an attraction softened and mellowed by broken shadows and diffused light. Complementary plantings equal in interest weight are used instead of mirror images. This involves the use of plants that are not identical but placed in similar places and given

FIGURE 15-18   Asymmetric grassed walk. (Photograph by K. A. Denisen.)

similar uses. They are complementary, i.e., capable of substitution one for another, in such features as color and texture, size and form, and in seasonal changes. Another characteristic of asymmetric balance is its flexibility. The use of annuals interspersed with perennial flowers gives variety from year to year, thus avoiding the monotony of duplication. Shrubs, trees, and vines can be eliminated, relocated, rejuvenated, or replaced and asymmetric balance may still be retained. Its flexibility extends also to the various areas in the home grounds as the mode of life changes for its occupants. The children's play area can easily be removed and incorporated into lawn or other areas. Fruit trees and the garden area can be eliminated. These changes can be effected without loss of asymmetric balance or a major rearrangement of existing plantings. The asymmetrically balanced grounds are functional in design. Open lawn areas are functional, unrestricted, and adaptable for many purposes.

FIGURE 15-19   **There are many and varied types of plants available for planting. (Photograph by K. A. Denisen.)**

The absence of beds and dramatic features in central areas leaves room for the feeling of spaciousness. Care and maintenance can be less exacting in the asymmetric plan. Shrubs may also need pruning, but a delay in the operation is not as disrupting or serious to the overall effect as under the formal system.

Plants to Use There are many and varied types of plants available for planting the home grounds. Even though two home plans were identical in layout, it is probable that they would look quite different merely because of variability in selection of plant materials. The adaptability of plant materials to climatic conditions is an extremely important consideration. The woody plants native to an area are often a good choice. Ultimate size and shape are other factors to keep in mind. Tall houses call for the use of tall growing trees. Low, rambling houses require smaller trees and shrubs for background and foundation plantings. Some trees and shrubs are grown for conspicuous flowers and decorative fruits. Dense foliage on others can provide shade or screening. Selection of plants for the home grounds is dependent primarily on their intended functions and the qualifying characteristics which they possess. Succeeding chapters offer examples of plant materials which may be used.

STUDY QUESTIONS

1. Why is it important to have a plan for the home grounds? Is it important for a renter? Why?
2. Why should a home owner let his thoughts and plans "simmer" for a while before completing plans for development? Are changes at that time going to represent ultimate desires of what is wanted in the home plan?
3. What are some considerations in site than can prove to be costly in the long run?
4. Of what value is a photograph of existing property in planning its development?
5. What are the objectives of a plan for the home grounds? Have you suggestions for additional objectives? List them.
6. Should the outdoor living area be given greater emphasis in most home plans? Why?
7. Of what real value is an overlay of transparent material for the proposed plan?
8. Distinguish between formal and asymmetric balance in a landscape plan.
9. Some people like to propagate their own plants for landscaping. Suggest means for propagating trees, blooming-type shrubs, annual flowers, and perennial flowers.
10. What is the principal purpose of ground cover plants? Give examples.

SELECTED REFERENCES

1. Brooklyn Botanic Garden. 1976. *Creative Ideas in Garden Design* (No. 49). Brooklyn Botanic Garden, Brooklyn, New York.
2. Nelson, W. R. 1963. *Landscaping Your Home.* University of Illinois College of Agriculture Cooperative Extension Service Circular 858 (151 pp.).
3. U.S. Dept. of Agriculture Yearbook. 1972. *Landscape for Living.* Washington, D.C.
4. Your favorite nursery catalog (illustrated).

16 The Lawn

The lawn is a carpet for the landscape. It is a vital part of the home grounds and should have top priority among new plantings. The lawn is basic to home grounds development. It improves the appearance of a new home and makes the grounds usable for a living area, which increases its value both to the occupants or to those who are in the market for homes. Beauty, convenience, and usefulness add monetary value to real estate.

Good lawns don't just happen. They are the product of thorough and constant management and care. Efforts expended on a lawn usually show results. Since the approach area of the home grounds is the most visible, it is natural to express the owner's best achievements. It is an area of impression for friends, neighbors, and visitors. The lawn in the secluded areas of the grounds is also extremely important to the home and its occupants. It is here that outdoor living demands a floor of dense turf, since traffic in the recreation area is heavy. Entertainment and relaxation both are incentives for an attractive lawn.

ESTABLISHING A NEW LAWN

Grading   *Drainage* is one of the objectives of grading for the lawn. The slope of the lawn should be at least 1 percent and away from the house, especially for homes with basements. During rainy weather, water is prevented from collecting near the house, and there is less danger of water seeping into the basement. Both on the slope away from the house and in the level parts of the grounds, it is very important to avoid "pockets" or depressions which hold water unusually long. An extremely localized water-logged condition can result and make it difficult to establish grass. If there are any large depressions that are usually wet it is sometimes necessary to use tile for drainage, since there may be a hardpan causing the difficulty. The presence of springs will cause marshy conditions. In selecting a home site this is an important item, since it can be the cause of water seepage into the basement.

Hardpans in a lawn can actually be created during the leveling and grading process. Heavy earth-moving equipment, considerable clay near the surface, and a high content of soil moisture are conditions favoring formation of artificial hardpan. The clay becomes extremely compact, will be almost like brick as it dries, and seriously restricts aeration and root growth.

FIGURE 16-1   The lawn is a carpet for the landscape.

Accent of the house and grounds is a factor for consideration in grading. The main attraction is the house; however, series of retaining walls and terraces on a steep bank tend to distract from the house. Gradual slopes emphasize the house and blend the lawn with the surroundings. It is important to keep original trees on the grounds. They take many years to grow, are a great asset to a new house, and are the envy of those who are waiting for young trees to grow sufficiently to be useful. These trees should be carefully protected during the grading operations to avoid trunk and root injury. If the level of the lawn is lower than previously, an "island" of soil should be left in a circular area around the tree or trees to protect its roots (Fig. 15–13, page 332). If the level is higher a "well" constructed around the trunk of the tree will permit better aeration and water penetration.

Disposal of excavation subsoil that was removed for the basement is sometimes a problem. One of the serious problems in lawn establishment is created when the poor subsoil from the excavation is spread over the fertile topsoil in the interests of expediency. Such a procedure can cause many future problems and is eventually overcome only by hauling in more good topsoil or a large amount of organic matter. The most desirable procedure is to remove the topsoil from the area to be filled, then deposit subsoil from the basement, and finally cover this with the original topsoil. It can be accomplished with earth-moving equipment. While the original expense will be greater, some serious problems of lawn establishment will be eliminated.

Soil Fertility   *Fertilizers* are used to bring the nutrient level of the soil up to optimum conditions for growth. Soil tests of the lawn area will indicate the immediate nutrient needs for building up fertility prior to seeding. Nitrogen is usually needed in greatest quantity, and lawns generally show a pronounced response to this nutrient even when adequate amounts are available. Commercial fertilizers of both inorganic and organic materials are available for lawns, however, they differ considerably in rate of availability. The inorganic forms, in most instances are readily available and give quick response. The organic forms are more slowly available; therefore, their effect will last through most of the season. Since immediate fertility is important in getting a lawn started, the inorganic forms of fertilizer are generally used during seedbed preparation. Organic forms may also be applied at planting time or later for continued response. Inorganic fertilizers should be applied prior to seeding. A complete fertilizer is considered most desirable at this time, since it can be worked directly into the soil.

Liming is often desirable. Most lawn grasses thrive at near the neutral point, although a pH range of 6.0 to 6.5 is generally favorable. At the time a sample of soil is tested for available nutrient content a pH reading can be obtained. Continuous fertilization with ammonium sulfate and other acid type fertilizers may reduce the pH significantly and a definite response is obtained when lime is applied. Lime application prior to seeding should also be worked into the soil.

Kinds of   Among the thousands of species of grasses, only a relatively few are
Grasses desirable lawn grasses. These may be placed in three groups: basic grasses, nurse grasses, and special purpose grasses. *Basic* grasses are perennials which eventually predominate in the lawn. *Nurse* grasses are short-lived kinds which germinate readily, grow fast, and later disappear when the basic grasses dominate. They help basic grasses get established by holding the soil in place and providing some shade. *Special purpose* grasses, as the term implies, are for unusual or special needs. Grasses for shady areas, dry conditions, heavy traffic, or carpet-like effect are included in this group. A description of several important turf grasses follows:

Kentucky bluegrass (Poa pratensis): This is the most used lawn grass in North America. It grows along roadsides and non-crop areas without being seeded by man. Yet, it is not native to North America. Its wide range of adaptability is evidenced by its rapid spread through eastern United States within a generation of its introduction. Although grown and cultivated on the eastern coast, it beat the early settlers across the Appalachian Mountains to Kentucky. Daniel Boone and his fellow settlers found it flourishing in the meadows of their new territory.

FIGURE 16-2

Lawn grasses. *Top*: left, Kentucky 31 fescue; right, creeping red fescue.
Middle: left, lawn mixture consisting mostly of annual grasses (poor mix);
right, lawn mixture consisting mostly of perennial grasses (good mix).
Bottom: left, *Zoysia japonica*; right, Meyer zoysia.

Kentucky bluegrass is not tolerant of dense shade but will succeed in
partial shade if well established. It thrives from moderately short clipping, about $1\frac{1}{2}$ to 2 in. (4–5 cm), under good growing conditions. In dry
weather during periods of high temperature, it becomes dormant and
turns brown. Cool temperatures along with moisture are necessary to

end the dormant condition. Irrigation before dormancy occurs will maintain this grass in a green condition. It makes a very attractive and durable lawn when it predominates as the basic grass. Selected strains of Kentucky bluegrass are being used because of certain specific advantages. 'Merion' bluegrass will tolerate more drouth before losing its green color. 'Pennstar' has resistance to *Helminthosporium* leafspot. 'Arboretum' has an erect growth habit.

Fescues (*Festuca* spp.): These grasses vary considerably between cultivars. They have various uses based on their characteristics although any of them can be a basic grass under special circumstances. Creeping red fescue does well in shady areas and has a low fertility and water requirement. It will tolerate more traffic with shady conditions than will the bluegrasses. Chewing's fescue has a clump-type growth habit. Alta fescue is a very coarse-bladed type which is vigorous, stocky, and retains its dark green color throughout the summer even with a moisture deficit. It is deep rooted and makes a tough turf. It seems well adapted for athletic fields and playgrounds, however, many people object to its extreme coarseness for home lawns. Kentucky 31 fescue is a selected strain which is not as coarse and retains its color well during hot, dry weather.

Ryegrass (*Lolium perenne*): This grass is distinct from rye, the cereal crop. Perennial ryegrass is an early germinating, rapid growing grass. It gives quick response and encouragement to the person starting a new lawn. Although perennial, it will give way to Kentucky bluegrass and thus is not considered a permanent lawn grass. Italian ryegrass is a strain of Perennial ryegrass. Domestic ryegrass is fast growing but usually gives way to permanent grasses within two or three years. Annual ryegrass is sometimes used as a nurse grass or for a quick, though temporary lawn.

FIGURE 16-3   **Propagation of bent grass by plugging. Zoysia and Bermuda grass may also be planted in this manner.**

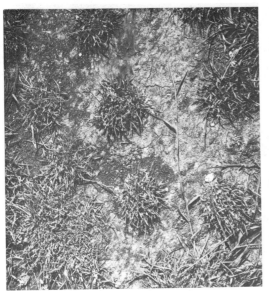

FIGURE 16-4   Bent grass spreads by stolons and thrives with close clipping.

Zoysia (Zoysia spp.): A grass which is native to Japan, zoysia is propagated by "plugs" of soil containing plants and rhizomes. It is extremely fast growing. In a favorable growing season, 2-in. (5 cm) diameter plugs spaced 1 foot by 1 foot ($\frac{1}{4}$ meter $\times \frac{1}{4}$ meter) will usually merge to fill out the lawn. It is quite drouth tolerant but in cold climates may not be sufficiently winter hardy. It has the added disadvantage in temperate climates of turning a whitish brown following a frost. This limits its period of attractiveness to the frost-free period of the growing season. It will remain green during hot, dry weather in the summer months. It is not well adapted for lawn use in cold climates.

Bent grass (Agrostis palustris): The very smooth carpet effect of the golf course putting green is due to the closely clipped bent grass, a special purpose grass. It is very specific in its maintenance requirements. It is propagated either by seed, plugs of soil with growing plants, or by planting dormant plants which soon give rise to new plants by stolons. Bent grass is shallow rooted, requires frequent watering, and thrives with close clipping, about $\frac{1}{4}$ to $\frac{1}{2}$ in. ($\frac{1}{2}$–1 cm) high thus finding little use in home lawns.

Bermuda grass (Cynodon dactylon): Propagation of Bermuda grass is by stolons. This grass is coarser in texture and deeper rooted than bent grass which makes it more adaptable for lawn use. While limited in adaptability to many areas by severity of winters, it does thrive under conditions of mild temperature and abundant rainfall. It spreads rapidly and, in some areas, can become a weed.

Bahia grass (Paspalum notatum): This is coarse grass used frequently for turfgrass on roadsides. It is seed-propagated, grows fast and must be mowed frequently (every 4 to 5 days) to control the growth of its tall seed heads. It should be mowed with a rotary mower. This grass rather

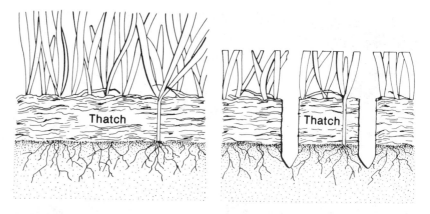

FIGURE 16-5   Thatch is an accumulation of grass clippings and other debris that has not decomposed. Its decomposition is hastened by use of a verticutter or aerifier so water will penetrate the thatch.

readily crowds out other grasses and weeds. Bahia is adapted along the Gulf of Mexico and southeastern and southwestern U.S. coastal areas.

St. Augustine grass (Stenotophrum secundatum): This is a vigorous growing turfgrass usually requiring some form of mechanical thatch removal. It is propagated by sprigs and is quite tolerant of shade. Adaptation is to the same areas as Bahia grass.

Centipede grass (Eremochloa ophiuroides): This grass grows best on acid soils and adapted to most southern states. It is propagated by seed and requires little care and infrequent mowing. Chief problems are iron-deficiency chlorosis with over-applications of nitrogen fertilizer.

Lawn Seed Mixtures   *Reasons* for mixtures of various kinds of grasses for seeding are: (1) to include a nurse grass, (2) to include special purpose grasses for adverse conditions, and (3) to favor a good stand of permanent grasses. Generally about 25 percent nurse grass seed in a mixture is desirable for adequate stand. Special purpose grasses are sometimes included as a matter of course in many mixtures to provide for the adverse conditions that may prevail in certain parts of the lawn. Shady areas, sandy areas, and dry areas may benefit by the presence of some of these grasses even if only in relatively small amounts. A good stand of permanent grasses is not only dependent on the presence of those grasses in the seed mixture but also on their ability to survive and prosper. The presence of faster germinating grasses gives encouragement to the person establishing a lawn. It focuses attention on the new lawn and may indicate a need for irrigation, erosion control, pest control, or increased fertility.

Requirements of a good lawn mixture include: (1) a high percent of permanent grasses, (2) freedom from weed seeds, and (3) high viability. The permanent grasses should comprise at least 50 percent of the mixture. Since they will ultimately be the predominant grass, they should be given an opportunity to develop a good stand as soon as possible. While most permanent grass turf will "fill in" by rhizomes in time, a dense turf can be accomplished sooner with adequate seed of the basic grasses. It is extremely important to read the label if a grass seed mixture is being purchased. Unfortunately a few seed companies have used misleading advertising about the contents of grass seed mixtures. The mere presence of a certain desirable grass in a mixture is not enough. It must be present in sufficient amount to be worthwhile. Most cheaper mixtures are low in cost because of the cheap, short-lived grasses which comprise them. It pays to read the label. Home mixing is easily done and frequently with considerable saving. Desired grasses of good quality are available from reliable seedsmen. They can be mixed by placing in a closed container and rotating the container. A mixture for general use in full light with occasional shade is as follows:

			'Arboretum'	25%
Kentucky bluegrass	75%	{	'Merion'	25%
			'Pennstar'	25%
Creeping red fescue	25%			

Freedom from weed seeds can often avert a serious weed problem. One per cent of weed seeds may sound like a very small amount. Yet if that much weed seed of dandelion or plantain were present, those weeds would practically smother the new seeding. Although these weeds can be killed with chemicals, the new seeding is also susceptible to chemical injury in the seedling stage. The content of weed seeds should be less than $\frac{1}{20}$th of 1 percent and preferably even lower. High viability is important not only for good stand, but also because it generally is accompanied by good vigor and strong seedlings.

Seeding   *Time of seeding* is often a critical factor in germination, stand, and durability of the new lawn. Most lawn grasses in temperate zone areas are cool season plants. Early spring and early fall are ideal periods for seeding. As soon as the soil can be prepared in the spring, grass seed can be planted. It germinates rapidly in the cool, moist soil and gets a good start before summer heat slows its growth. By early summer it also has a deeper, more extensive root system which, with its larger crown, enables the plant to withstand more severe conditions of moisture shortage. Early fall planting may actually be late summer planting in colder climates where the growing season is shorter. The proper time for early fall planting is determined largely by temperature. When the nights become cool, grass germination is favored. Fall rains are extremely helpful to a new seeding, and it may be necessary to irrigate at this time of year in some areas. Many people prefer fall seeding because the weed problem may be less serious for the new lawn. Since many of the troublesome weeds in a new seeding are annual weeds or warm season types, they do not winter over with the grass. The grass thus starts with an advantage the following spring. The grass is established and most weeds are not. Also, the grass will start growth as soon as the soil warms up and conditions are suitable. Weeds will not germinate until conditions favor it; consequently, they have severe competition from the grass with its established root system, food reserve, and large photosynthetic area. The lawn can be mowed by this time which is especially helpful in eliminating erect growing annual weeds. The stoloniferous grasses can be planted by plugging or sprigging during any part of the growing season when moisture conditions are satisfactory.

Seedbed preparation is important to both germination and stand. Since grass seeds are so small, a fine seedbed is needed for intimate contact of seed and soil. The soil should be moist but not wet. Avoiding the use of extremely heavy equipment in soil preparation will keep soil compaction at a minimum. The area to be planted is first plowed, spaded, or "roto-tilled" to incorporate organic matter and break up hard lumpy soil. Further preparation with a disk harrow, toothed harrow, roller, or garden rake will increase the fineness of the seedbed. Excess working of the soil will usually result in compaction and loss of granulation.

Rate of seeding will vary with kinds of grass. A grass mixture as suggested for lawns in sunny locations with intermittent shade can be sowed at the rate of 1 to 2 lb ($\frac{1}{2}$ to 1 kg) per 1000 square feet ($92\frac{1}{2}$ square meters) with good results. If soil and climatic conditions appear less favorable, the higher rate is preferred. On the basis of numbers of seeds per pound (kilogram) uniformly distributed, very low quantities are required. There are approximately 2,500,000 bluegrass seeds per pound (or 5,000,000 per kilogram). One pound ($\frac{1}{2}$ kg), distributed evenly over 1000 square feet ($92\frac{1}{2}$ square meters), means there are 17 seeds per square inch. This is a far greater number than is needed or desriable if each were to become a grass plant. It must be remembered that condi-

FIGURE 16-6   **Power seeders provide a more uniform rate of seeding. (Courtesy W. E. Knoop, Horticulture Department, Iowa State University, Ames, Iowa.)**

Horticulture for the Home

tions for germination are rarely ideal for near 100 percent emergence. Also other grasses in the mixture may have fewer seeds per pound (kilogram) and this reduces the number of seeds per square inch (square centimeter). Lack of uniformity in distribution will require an increased rate to give minimum coverage for all areas.

Seed sowing is done by machine or by hand. Equipment commonly used includes the "cyclone" seeder, and the seed and fertilizer distributor. The "cyclone" seeder throws the seed in a circular pattern as the operator walks back and forth across the lawn. Overlap of the distribution pattern is important for even coverage. Wind will hamper its usefulness giving a confused pattern and variability in seed density. The seed and fertilizer distributor may be used effectively if the calibrations are sufficiently accurate at low rates. With all machine seeders, it is important to keep the seed mixtures agitated. Light seeds tend to accumulate at the surface and heavy seeds tend to settle. Hand seeding should be done during calm weather if possible. Seeds are scattered in strips back and forth across the area applying at about one half the desired rate. The remainder is then scattered in the same manner but at right angles to the first direction.

Rolling or raking lightly after seeding will give more intimate contact with the soil. Seeds do not need to be buried. A light rain or light sprinkler irrigation will settle the seeds, reduce loss by blowing, and establish mosture contact.

Irrigation   It is important to keep the seedbed moist but not wet. It will probably need light, daily sprinkling. On cloudy days, water should be withheld to avoid a wet surface. Excessive watering may wash the seeds to other parts of the lawn and result in uneven stand. It may also cause soil erosion and increased leaching of nutrients. In areas which are dependent on irrigation for most of their water, it may be necessary to irrigate before seedbed preparation and seeding. The soil should be moist at least 24 in. (60 cm) in depth to facilitate settling of the soil. This soil moisture provides for root growth to adequate depths and serves as a moisture reserve for the developing lawn.

Mowing
the New Lawn   Fall sowed lawns should not be mowed before winter as it exposes them to adverse conditions. The spring seeding should be permitted to become established before mowing. Growth of grass roots and crowns is dependent on the food supply from the leaves. Since the planting is new, there is little stored food in the crown for root growth and new shoot development. If mowing starts immediately after mowing height is reached, the roots will be more shallow and the plants will be dwarfed. Mowing, like pruning, is a dwarfing process. Growth rate will determine the length of time before mowing starts. The nurse grasses should not be permitted to grow taller than 4 or 5 in. (10–12$\frac{1}{2}$ cm) before mowing. Excessive growth of nurse grasses can defeat their purpose. They may shade or crowd the basic grasses. The height of cut should be about 2 in. (5 cm) for the first cutting. This will cut principally nurse grasses which

FIGURE 16-7   **The new bluegrass lawn should be well established before mowing.**

have served their purpose. It is important that the clippings are raked off to prevent smothering of the new grass. Care should be exercised during the first mowing to avoid digging or gouging the soil surface which is not firm as yet. This is especially important with a power mower propelled by wheel traction.

It is advisable to avoid foot traffic on the new lawn until the second or third mowing and even then to avoid heavy use. The grass has a better start with the turf becoming more firm each successive week of good growing conditions during the first season. The height of cut may be reduced to $1\frac{1}{2}$ in. (4 cm) after the second or third mowing. This length of grass has less tendency to crumple and lie flat from foot traffic than a length of 2 or more inches (5 or more cm). Many golf fairways, which may have considerable traffic, are cut at 1 in. ($2\frac{1}{2}$ cm). This permits greater bounce and accuracy of bounce for the ball and avoids flattened areas of grass from the golfers' feet.

Special Techniques in Establishing New Lawns

Sodding is sometimes the most expedient means of getting a new lawn. It is a faster, more expensive method, but is often the most practical especially on slopes that are subject to erosion. It may also be the easiest method of establishing grass in shaded areas. Sod is sliced from the surface of a good turf by means of a sod cutter. This machine has a horizontal blade that undercuts the sod about $1\frac{1}{2}$ in. (4 cm) below the surface. The sod strips are rolled up for ease of handling and to reduce drying. They should be covered if they are to be stored for a few

FIGURE 16-8   **Riding rotary mower. (Courtesy W. E. Knoop, Horticulture Department, Iowa State University, Ames, Iowa.)**

days. Stacks of rolled sod should not be sprinkled with water unless absolutely necessary because it tends to wash the soil from the roots. If left in stacks and rolls for more than two or three days the grass turns yellow due to chlorophyll disintegration. It should be laid soon after cutting. Soil preparation for sodding is the same as for seeding. The roots of sod have been severely pruned in the cutting operation and should have a friable soil in which to replace their roots. It is often desirable to wet the surface of the soil before laying the sod. The lawn-bed should be level and the strips of sod evenly laid. Attempts to stretch strips of sod will make depressions in the lawn. Extra soil should be placed under thin parts of the sod where some of the soil may have been lost. Small sections of sod are used for patching where needed. The sodded lawn is rolled after laying; any depressions noted at this time can be filled by rolling back the strips involved and filling with soil. The sod is watered after laying and is generally watered at intervals to keep the soil and sod moist but not wet. Sod on sloping areas that are kept too wet may actually "creep" down the hillside before new roots adequately

The Lawn   **353**

FIGURE 16-9

Mulching and sodding are special techniques. *Top*: Cheesecloth is used to hold seeded bank in place. *Bottom*: Sod is fitted and rolled on a well-prepared sod bed.

anchor it to the soil. Heavy watering may soften the sod and soil so that it will not support even occasional foot traffic.

Mulching of the seedbed is sometimes done to avoid erosion on newly seeded slopes. It is important to use materials which decompose or allow grass to grow though the mulch. Commonly used and effective materials are cotton cheesecloth, burlap, and small grain straw. Single layers of cheesecloth or burlap are stretched over the area to be mulched and are anchored by stakes or guys. Cheesecloth disintegrates more rapidly than burlap, gives better aeration, and reduces likelihood of damping-off. If straw is used, it should be free of weed and grain seeds. The straw layer should be thin so light can penetrate to emerging seedlings. It is well to dampen it immediately after application to prevent blowing. On steep areas, it is very helpful to place strips of woven wire or chicken wire across the straw to hold it in place. This may be easily removed after the grass is established without injury to the seeding.

Tiling is sometimes necessary before grass can be grown successfully. Under inadequate drainage conditions, the lawn area may be in a nearly permanent water-logged state during wet seasons. For grass, which is a shallow rooted crop, the tile lines should be spaced fairly close, about 30 feet (9 meters) in a loam soil. A good outlet is important. This may involve consultations and agreements with neighbors of urban and suburban areas. Tiling is often of mutual concern and may develop into a group undertaking.

Using soil conditioning chemicals may hasten the development of a new lawn if an extremely compacted soil is encountered. It is a means of improving soil structure by aggregation of the fine particles. Soil aeration and water penetration are increased; this promotes root growth, an essential to good turf. Incorporation of considerable organic matter will also open up the heavily compacted soil.

CARING FOR THE ESTABLISHED LAWN

Fertilizing   One of the distinguishing characteristics between a well-cared-for lawn and a poorly maintained lawn is the intensity of the green color. The particular shade of green is associated with the nutrient content of the soil. A deep green color indicates an abundance of nitrogen with adequate amounts of other essential nutrients. Nitrogen is the nutrient required in greatest quantity and is most frequently the element in shortest supply. Consequently most fertilizers for lawn use should be abundant sources of nitrogen. It is doubtful if complete fertilizers are necessary at each time of fertilizing.

Inorganic fertilizers are generally less expensive than organic types on the basis of cost per unit of available nutrient. However, there are other important considerations. The nutrients being in available form will result in extremely lush growth during the spring months. Then occurs a period when these readily available nutrients are exhausted either by plant use, leaching, or conversion to unavailable forms. The result, a lawn which slows down considerably in growth or even indicates a nutrient shortage a few weeks later. A second application is usually needed to supply nutrients for the remainder of the season. The early spring application generally results in little if any burning of the grass blades; however, when a later application is required even distribution is important to prevent uneven growth or burning due to the high salts concentration from excessive application. Very few blades of grass are

FIGURE 16-10   Research is being conducted on grass plots. (Photograph by K. A. Denisen.)

showing in early spring and the rather frequent rains wash the fertilizer into the soil. Even excess deposits soon become dispersed under these conditions. However, the summer applications of inorganic fertilizer are considerably more critical. The soluble salts may be dissolved by dew or mist and cause burning. Large areas of the lawn may become brown and discolored and will remain in this condition until new growth again covers it. Summer applications of fertilizer may be followed by a sprinkler application of water. This washes the fertilizer off the grass blades into the soil. The need for reapplying fertilizer in midseason and the danger of burning or injuring the grass have prompted the development of slowly available forms for lawns.

Organic chemicals and organic matter are the prinipal forms of nutrients used in the slowly available fertilizers. Sewage sludge that is heat processed, dried, and pulverized is an organic matter form which is becoming widely adopted. It has a low analysis which increases the rate of application above that for many inorganic fertilizers. Spring applications usually supply nutrients well through the summer. There is no danger of burning from uneven distribution or when applied to grass that is wet. Some high analysis fertilizers which have slowly available nutrients are becoming more widely used. The nutrients are mostly organic in form, and the effect is an extension of nutrient availability over a longer period. Spring applications often give good growth and color throughout the season. Very few nutrients from this fertilizer are available immediately following spring application. There is actually less need for additional nutrients at this time since it is a period favorable for

FIGURE 16-11   **A fertilizer distributor aids in even distribution of nutrients.**

growth. Moisture is abundant, days and nights are cool, and stored food is being utilized in vegetative growth. The period when additional nutrients are needed most is after this period of lush growth. The slowly available fertilizers usually release nutrients in greater amounts as this period approaches. Synthetic forms of urea are used for slowly available lawn fertilizers. This material, through bacterial action, releases nitrogen as the only soil nutrient. Because of its high analysis, about 44 percent nitrogen, its rate of application is less than it is for most fertilizers. Spring applications will usually supply nitrogen through the entire growing season.

Raking   *Prevention of smothering* the grass is the principal reason for raking. Fallen leaves from trees, if allowed to accumulate, will form a tight mat over the lawn when they become wet. This cuts off oxygen from the grass, respiration cannot proceed normally, and the grass dies or is seriously weakened. Excess clippings following a delayed mowing should be removed. They are unsightly when long and also may be so heavy and thick that they prevent light from reaching the live grass. Under conditions of moist weather and disease prevalence, it is desirable to use a grass catcher on the mower to remove all clippings. Spring raking of the lawn is not generally practical except in certain areas where refuse has accumulated or leaves have fallen during the winter.

Some *types of rakes* may actually harm the turf. Tine rakes or rigid toothed garden rakes may dig up clumps of turf and usually remove

The Lawn

some decaying organic matter. Broom rakes made of flexible wire, plastic, or bamboo strips are most desirable for lawn use. The lawn sweeper which actually brushes the leaves or clippings into a hopper is a more expensive item of equipment for removing refuse. Properly adjusted, it does the job neatly and precisely.

Mowing   *Height of cut* is an important consideration. In general, grass can be cut shorter in the spring than in the summer. The growth rate, or rate of leaf replacement, is more rapid in the spring. However, many people prefer to adjust their mower once and leave it at that setting. This can be safely done with a basic grass of Kentucky bluegrass or the fescues if set at about $1\frac{1}{2}$ to 2 in. (4–5 cm). The mower is adjusted on a level floor and the distance from the cutting blade to the floor is measured with a ruler. Mowers with pneumatic tires must be maintained at a constant tire inflation to maintain a uniform height of cut.

Frequency of mowing is also important and is related to height of cut. If grass is allowed to grow to such height that nodes appear near the level of cut, it may be seriously set back by mowing. The principal reason for frequent or timely mowing is to prevent undue elongation of grass stems. A general rule to follow is to mow before the grass is twice the length of the height of cut. With a mower set at $1\frac{1}{2}$ in (4 cm) the grass should be cut before it is more than 3 in. ($7\frac{1}{2}$ cm) high. If mowing is unduly delayed, it may be well to raise the level of cut $\frac{1}{2}$ in (1 cm) to avoid cutting down to yellow portions of the stem. The level can be reset to the regular position at the succeeding mowing.

Types of mowers are reel and rotary. Reel mowers may be propelled by hand or power units. The sharp reel blades cut against the stationary basal blade and produce a smooth effect. Rotary types are electric or gasoline driven. They cut by means of a whirling rotary blade with a chopping rather than shearing action. The principal advantage over the reel type is the ability to cut tall grass or weeds readily. In general, they do not produce the smooth effect the reel type does.

Edging is a part of the mowing operation which is often neglected. It adds the finished touch to a smooth, well mowed, thrifty lawn. The required edging is kept to a minimum by a large open lawn. Fences and trees usually require edging since the mower cannot approach closely enough to cut all the grass. Borders and beds can be maintained with cultivated strips along the lawn edge, or a root barrier can be placed in the soil. Furrowed edges, removal of a triangular strip of sod, are used for reducing grass growth next to concrete walks and drives.

Irrigating   Water may need to be applied to grow grass in some areas. When using supplemental irrigation, a common error is to wait until it is too late to do maximum good. If the basic grass is bluegrass and the objective is to keep the lawn green, watering should be started when a moisture deficit begins to develop. If the grass becomes dormant and turns brown, it takes a combination of moisture and cool weather to regain its green color. Enough water should be applied for deep

FIGURE 16-12   **A barrier of used bricks prevents grass rhizomes from invading the flower border.**

penetration, because grass roots go as deep as 2 feet (60 cm) in most soils. The greatest absoptive area is in the upper 12 in. (30 cm). The soil is a reservoir for water; therefore, deep watering means less frequent irrigation, and the moisture needs of the grass plants are more nearly fulfilled. Sprinkling or surface watering in small amounts does very little for the grass; it doesn't reach sufficiently deep where the soil remains dry. Sprinkling lightly does encourage weed seed germination at the surface. Consequently, lawn weed problems can be increased with light water applications.

Warm season grasses like zoysia and buffalograss do not become dormant during dry weather. However, watering is important to them even though they tolerate brief periods of drouth. Survival of root hairs and feeder roots, and retaining their ability to withstand foot traffic are important reasons to water.

Pest Control   *Weeds* are serious pests in most lawns. After discovery of 2,4D as a weed killer, control of broadleafed weeds was less of a problem. Because of the selective nature of this compound, it will kill most broadleafed weeds without injury to established grass. Dandelion, plantain, sorrel, and knotweed can be controlled with 2,4D applications. If weeds are weakened but still persist after application of 2,4D, a second application ten days to two weeks later will usually finish the job. Precautions must be observed in using 2,4D and all weed killers. Seedling grass is susceptible to injury and may be killed by 2,4D. In the home lawn, spot spraying of individual weeds or groups of weeds is safest unless there is general infestation of the entire lawn. Spray drift to desirable sensitive plants, which include many flowers, trees, vines, shrubs, and vegetables, can be extremely detrimental. It is wise to use low volatile forms of 2,4D such as the amine and sodium salts or the low volatile esters. Much can be learned about the compound through reading the label. Spraying should be done on a relatively windless day and the sprayer nozzles should be relatively close to the ground to avoid drift to nearby susceptible plants. Plants, commonly grown in the home grounds, that are very susceptible to 2,4D include grapes, lilacs, roses, beans, and tomatoes. If the same sprayer is to be used later for insecticide or fungicide applications, it is extremely important to clean and neutralize the sprayer thoroughly. Household ammonia, $\frac{1}{2}$ cup per gallon of water (60 cc per liter), can decontaminate the sprayer. The tank and hose should be thoroughly rinsed with water before the ammonia water is poured into it. The cover is placed on the tank and it should soak for 12 to 24 hours. The mixture is run through the hose and nozzles before dumping. Many gardeners prefer two sprayers, one for herbicides and the other for foliage spraying on desirable plants.

Grassy weeds are usually more difficult to control since 2,4D is generally ineffective on them. Crab grass is one of the most serious of this group. It is an annual grass which germinates in late spring when the soil and air temperatures are about 68°F (20°C) or above. Seeds may continue to germinate all summer. Crab grass is favored by light showers or light irrigation; they make germination conditions ideal and do not benefit its competitors, the basic lawn grasses. Hot weather also favors crab grass and slows down growth of the cool season grasses. Two methods of chemical control, preemergence and post emergence, can be used effectively although they are both quite exacting in time of application. Pre-emergence herbicides include bandane, benefin, and bensulide. They kill germinating seeds. Second and third applications at 3- to 4-week intervals are needed for continued control of late germinating seeds.

Perennial grassy weeds like quackgrass (couchgrass) are not usually serious problems in a maintained lawn. The usual height and frequency of mowing is too severe for continued growth of most rhizomous perennial weeds.

A good fertility program is helpful in preventing the entrance of many weeds. It is more difficult for weeds to gain entrance in a lush growing, dense turf.

Moles tunnel in lawns. They may be controlled by trapping in the runways of their ridged tunnels or gassing by placing cyanide crystals in the tunnels. Poison bait can be used. It is important to give immediate attention to the problem, otherwise they can form a network of tunnels and pathways of dead grass throughout a lawn in a few days.

Earthworms are actually beneficial to soil structure and organic matter content. If the population is high, their castings on the surface make bumps on the lawn. These bumps may actually make walking difficult when they harden; they also can cause jolting of the lawn mower. If birds are abundant in the area, they may take care of the problem especially in moist weather when the earthworms come to the surface.

Diseases are most serious in damp humid weather. Overwatering, prolonged periods of moisture on the leaves, and even late afternoon or evening irrigation may favor fungus growth and spore production. Diseases spread easier and more rapidly in the fine leafed grasses such as bent grass than in the coarser bluegrasses and fescues. *Brown patch* is one of the more serious fungus diseases of turf. It spreads outward in all directions from the initial point of infection; however, it may by-pass certain areas leaving patches of green and patches of brown on the lawn. Control may be obtained by spraying with a turf fungicide, such as Tersan. *Dollar spot* is usually more serious on bent grass plantings of golfing greens. A regular fungicide spray program is needed for disease control on these highly specialized turf plantings. *Curvularia* is a hot weather disease that is favored by moist conditions. Grass turns light green, then almost white as the inoculum spreads over a lawn. Mowing the lawn during sporulation increases its spread especially if followed by irrigation and hot weather. Unless controlled it will kill large areas of lawn when conditions favor growth of the organism. Most turf diseases can be controlled with a fungicide application when trouble first appears. In the use of all chemicals, it is important to read the label, follow directions, and observe precautions.

SPECIAL PRACTICES FOR MAINTAINING OR RENOVATING LAWNS

Aerifying   Many lawns become extremely compacted from heavy foot traffic or from occasional movement of heavy equipment across the turf. These soils are poorly aerated, water penetration is greatly restricted, and the general thriftiness of the lawn is reduced. Under such conditions paths are easily formed because the soil is hard instead of firm and cushioned. Aerifiers have been very successful in improving compacted soils. They remove small plugs of soil at frequent intervals and leave openings through the almost impervious soil surface. These openings do more than provide aeration in the upper root zone of the grass; they improve water penetration and give more rapid response to surface applications of fertilizer. Both hand operated and motor propelled aerifiers are available.

Spot Seeding   Spreading of grass seed is sometimes an annual process if a lawn is used considerably and there is likelihood of thin turf in some areas. Early spring is a desirable time for this operation. The areas sowed may need to be marked off to prevent foot traffic until grass is again

FIGURE 16-13   **Aerifiers improve aeration of the upper root zone and facilitate water penetration in compacted turf.**

established. Care should be observed in using herbicides, mowing, and performing other operations in these spots to give the new grass every opportunity for a good start. Small bare areas will usually fill in with rhizomous grasses like bluegrass in a relatively short period if kept mowed and not subjected to heavy use.

Installing Walks When Paths Persist   Paths often decree the need for a walk. If a less conspicuous walk is desired, stepping-stones may be the answer. Limestone slabs of various shapes may be sunk into the turf at average step intervals. Grass is left between the stones to give a more informal line than either a gravel path or concrete walk.

STUDY QUESTIONS

1. Why does the lawn have highest priority among new plantings in the home grounds?
2. Is drainage an important factor on most lawns? Why?
3. How can soil fertility be improved in establishing a lawn with reference to organic matter, fertilizers, and pH?
4. Distinguish between nurse and basic grasses. What are special purpose grasses? Suggest kinds of each type.
5. How can soil plugs of certain species take over or change a lawn?
6. What is the purpose of a seed mixture versus one good basic grass? Elaborate.
7. In establishing a lawn on your home grounds, how would you proceed in the following: selecting the seed mixture? preparation of the seed bed? sowing the seed? irrigating? time of first mowing?

8. When are the special techniques of sodding, mulching, and tiling used?
9. In fertilizing an established lawn what are the advantages of organic chemicals versus inorganic chemicals?
10. Should all clippings be removed following each mowing? What are the arguments pro and con?
11. How frequently should a lawn be irrigated?
12. List several chemicals useful in controlling weed, disease, and insect pests in the lawn.
13. Is rolling a lawn a sound practice? When?
14. Is aerifying a lawn a sound practice? When?
15. What is thatch? How can it be controlled?

SELECTED REFERENCES

1. Beard, J. B. 1974. *Turfgrass: Science and Culture.* Prentice-Hall, Englewood Cliffs, N.J.
2. Brooklyn Botanic Garden. 1976. *Home Lawn Handbook* (No. 71). Brooklyn, New York.
3. Couch, H. B. 1970. *Diseases of Turfgrass.* Kreiger Publishing Co., Huntington, New York.
4. Hanson, A. A., and F. V. Juska (editors). 1969. *Turfgrass Science.* American Society of Agronomy, Madison, Wisconsin.
5. Knoop, W. E. 1976. *The Golf Professional's Guide to Turfgrass Maintenance* (51 page mimeo). Iowa State University, Ames, Iowa.
6. Knoop, W. E. 1975. *Pesticide Usage Reference Manual.* Golf Course Superintendents Association of America. Lawrence, Kansas.
7. Madison, J. H. 1972. *Practical Turfgrass Management.* Van Nostrand-Reinhold, New York.
8. Madison, J. H. 1973. *Principles of Turfgrass Culture.* Van Nostrand-Reinhold, New York.
9. Prasher, Paul. 1976. *Home Lawn Care in South Dakota* (5 page mimeo). South Dakota State University, Brookings, S.D.

17 Ornamental Woody Plants

The woody plants, trees, shrubs, and vines, form the permanent year-round plant features of the home grounds. They will show annual changes because of growth, from storm damage scars, and by absence through loss or removal. However these changes are generally not drastic; for years and generations they may be landmarks or identifying features of a home or farmstead. Avenues have become famous for their trees. Parks and playgrounds only reach their height of popularity when well populated with adequate-sized shade trees. Botanical gardens and formal public grounds rely on the woody plants for the basic arrangement of areas and their background of beauty. Spring bloom of woody plants, summer shade and coolness, fall colors of leaves and fruits, and winter serenity and havens of shelter allow year-round appreciation of woody perennials. It would, indeed, be bleak without them.

TREES

Selecting Kinds of Trees

Growth habits of trees are of major consideration in deciding on various types for various locations and uses. The principal direction of growth affects the shape of the tree. When terminal growth far exceeds lateral growth, an *excurrent* form results. Most conifers are in this group, in addition to Lombardy and Bolleana poplar. When lateral branches equal or surpass terminal growth, the form is called *deliquescent*. Most deciduous and broadleafed evergreen trees are in this group. A third form, *columnar*, is found in many tropical and sub-tropical trees. Consisting of a high, long trunk with no laterals, it is terminated by a whorl of large leaves and apical meristem. The palms are typical of this group.

The ultimate size and rapidity of growth are important to home grounds development. If no trees are present and shade is an immediate objective, a fast growing tree is often selected. Its ultimate size is sometimes unknown or ignored. Large trees may cause a one-story house to appear extremely low. Small growing or extremely spreading trees may accentuate the size of a two-story house. Harmony of scale is desirable. The silhouetted form of the tree based on its natural growth habits is a good guide to use in selecting and placing trees.

Some trees have characteristics which may be annoying or obnoxious. The sycamore is a fast growing tree which gives good shade and has good form, yet some people are annoyed by its annual shedding of bark tissue during the growing season and consider it an untidy tree. Sucker-

364

FIGURE 17-1   **Year-round interest is found in many woody plants. Dormant twigs of the staghorn sumac resemble deer antlers both in silhouette and texture. During the growing season this sumac has attractive foliage and seed heads.**

ing by certain poplars, wild cherry, and wild plum may make them undesirable for lawn trees or background near a garden. Excessive seed producing types are sometimes called "weed" trees because of the large number of seedlings produced annually. The elm, honey locust, box elder, and basswood may be included in this group. Their placement in a lawn is preferable to location in or near a tilled area, since the frequent mowing of the lawn prevents their growth beyond the succulent seedling stage. Tree seedlings can become a problem in protected areas like hedges, borders, and gardens.

Purpose or use of trees is also important in selecting kinds of trees. Their placement in the home plan should be determined by kind and use. Trees used for framing and accent points are not necessarily good shade trees. Very large growing types should be an adequate distance from the house to avoid rubbing by branches and damage in the event of severe wind storms. Trees with a tendency to weak crotches like some elms or those with somewhat brittle wood like willow and cottonwood are best located at least 50 feet (15 meters) from the house. Trees can have dual uses in the home plan. They may be used both for ornamental purposes and fruit production. The apple, cherry, pear, and tree nuts are adapted for dual purposes. Some may be used for both specimen plants and shade; ornamental flowering crabs, mountain ash, birch, and weeping willow are examples. On a farmstead, fast growing types like Chinese elm, catalpa, willow, and poplar may be used for temporary

FIGURE 17-2 The silhouettes of trees are a guide to selection. These silhouettes are based on the average height and spread at ten years of age at Morton Arboretum. Rate of growth is also an influencing factor in kinds of trees used. (Courtesy of E. L. Kammerer and The Morton Arboretum, Lisle, Ill.)

windbreak while the permanent windbreak trees are becoming established. When removed they may be cut into fenceposts.

Size of home grounds is influential in determining adaptable kinds of trees. Low branching types like the conifers ultimately reduce the amount of lawn area available. This may be very critical on a small lot. Consequently this shape tree in a small lot should be restricted for

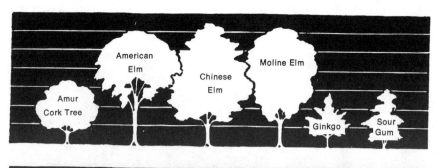

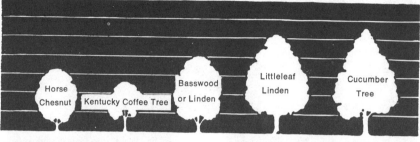

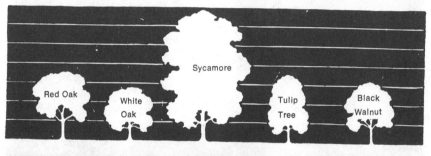

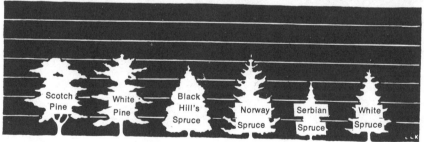

backgrounds, while dwarf forms are used in the foundation plantings, or their use avoided entirely. Low branching types are best adapted to greater expanse as may be found in larger city lots, suburban locations, or rural sites. Ultimate size of the trees should be in accord with the size of the home grounds. Large trees on a small lot will restrict the growing of other plants and tend to accentuate small size of the lot.

Hardiness and adaptability are extremely important considerations. Consulting the sources of known information on zones of hardiness (Chapter 5), minimum temperatures, and maximum temperatures combined with the tree's drouth tolerance, heat and cold tolerance, plus

FIGURE 17-3   The selection of trees is often determined by purpose or use. *Top*: *left*, Lombardy poplar for accent concealing light poles; *right*, American linden (basswood) for avenue shade tree. *Bottom*: *left*, ornamental crab for specimen; *right*, plum for bloom and fruit production.

rate of growth, puts less guessing and speculation into tree selection. Trial and error can waste time in acquiring useful adapted trees. Trees are investments whose returns are measured in satisfaction.

Spacing   Desired effect is the principal factor to consider in spacing ornamental trees. For large spreading trees, broad spacing is used. For a compact

FIGURE 17-4 **Fruit of the mountain ash is a delight to birds.**

row of trees, closer spacing will give the desired effect. For a dense background use close spacing. For an open lawn with occasional shade, broad spacing of trees along the edges of the home grounds gives the desired effect.

Competition is the influential factor that determines spacing effects. Adjacent trees with interlocking branches compete for light and tend to grow upward resulting in die-back of lower branches and self-pruning. This occurs in the forests. Roots will also become interlaced and compete for moisture and nutrients. Competition for light, moisture, and nutrients is reflected in growth of other plants located beneath or near closely spaced trees.

Fertilizing   *Inherent fertility* of the soil plays an important role in development, survival, and growth of native trees. Because of the far-reaching roots of

FIGURE 17-5   Attractiveness during the dormant season as seen in the deciduous honey locust and evergreen blue spruce.

trees, both in radial directions and depth, they are capable of utilizing great quantities of available nutrients. Compared to other plants, trees are large, but so are their root systems and absorptive capacities. Where moisture and climate are favorable, most trees will survive under conditions of low nutrient supplies. Growth may be greatly reduced but the trees adjust themselves to this condition by modification of their living standards. They absorb less and grow less, but still survive. Given more favorable nutrition, these same trees may grow into great giants of plant life.

Organic matter of the forest floor together with inherent fertility provided native trees with their only source of soil nutrients. Decomposing leaves and other debris still form an important source of nutrients to trees. However, those trees growing in lawns may not be so favored. In lawn care leaves are raked off the lawn to avoid smothering the grass, consequently, a good source of organic matter is lost to the trees. This must be compensated by other operations of good lawn maintenance which aid in tree maintenance, i.e., fertilizing and watering.

Commercial fertilizers are especially useful to ornamental trees. They contribute to increased growth rate of young trees making them useful at an earlier age. They can be used to maintain good leaf color in older trees. Direct applications of a complete fertilizer to young trees can be done by spreading it lightly around the base of the tree within a circle formed by the outer tips of the branches. This is best done in early spring and is usually followed by rapid shoot growth with large leaf area. This provides the machinery for a high rate of photosynthesis and an abundant reserve of carbohydrates, good insurance against winter

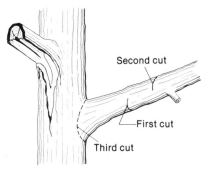

FIGURE 17-6   **Method for removal of large limbs. The limb on the left was not properly cut and skinning and tearing resulted.**

injury. On the other hand, late fertilization may stimulate fall growth, result in decreased carbohydrates, and increase winter injury. Since large trees are not expected to continue growth at a rapid rate, they are not fertilized heavily. Lawn fertilization and the consequent leaching of some of these nutrients are usually adequate for good nutrition of mature trees.

Irrigating   The first year is often the most critical from the moisture standpoint. Much of the absorptive capacity of the root has been lost in transplanting, and moisture should be readily available to the plant to aid in overcoming root loss. It is advisable to water periodically during the first year even though drouth conditions may not prevail. Surface application is satisfactory. Occasional probing may be done to observe the level of moisture and avoid a waterlogged condition. Watering of mature trees is done by surface application, or pipes can be inserted into the soil at 12- to 18-in. (30–45 cm) depths for subsurface application. Mature trees need be irrigated only during periods of drouth or in areas which depend on irrigation for plant growth.

Pruning   Pruning begins at planting time with the primary purpose of equalizing tops and roots. Reducing the number of buds reduces number of growing points, and growth is restricted to those areas. An important point in training avenue and shade trees is to develop a sufficiently high trunk enabling people to walk and to see beneath the branches. This takes time to accomplish. Some lower branches of the young tree should be retained for leaf area. As the terminal leader grows in height and in turn branches, the lower branches can be removed. This gives a straight, smooth trunk with more rapid growth than if all but the terminal bud or shoot were removed earlier. Pruning is also used to increase denseness of ornamental trees. Deciduous trees are headed back and evergreens may be sheared to stimulate lateral bud and branch development.

Corrective pruning may be needed when age and succeeding growth present problems. With increased height of the tree and increased weight of wood, low hanging limbs may need to be removed. Conifers and other single leader trees sometimes lose their leader through injury or disease. A branch trained upright and tied in this position will gain leadership and correct the shape of the tree. This should be done as soon as possible following loss of the single leader.

SHRUBS

Selecting Kinds of Shrubs

Bloom during part of the growing season is a frequent objective in growing shrubs. Some are grown for spring bloom, others for summer bloom. Spring blooming types include the very early forsythia and

FIGURE 17-7　　**Bloom in woody plants.** *Top* (trees): *left*, Dolgo crab (*Malus*); *right*, red-bud (*Cercis*). *Bottom* (shrubs): *left*, lilac (*Syringa*); *right*, elderberry (*Sambucus*).

FIGURE 17-8   Foundation shrubs. *Top*: *left*, Sabin juniper in evergreen hedge; *right*, the new shoot "candles" of Mugho pine. *Bottom*: *left*, tamarix, a deciduous shrub; *right*, Pfitzer juniper for foundation planting.

flowering quince which bloom before their leaves are conspicuous and the broadleafed evergreen azaleas and rhododendrons. Those blooming later in the spring include lilac, spirea, honeysuckle, and mock orange. All spring blooming types produce their flowers from flower buds or mixed buds differentiated the previous season. Summer blooming shrubs produce their flowers from new shoot growth; included in this group are the hydrangea, rose of Sharon, snowberry, and Anthony Waterer spirea. These blooming habits influence the time and techniques of pruning shrubs.

Fruiting is another objective of shrubs in the home grounds. Edible fruits from shrubs commonly used as ornamentals include blueberries, flowering quinces, highbush cranberries, sand cherries, gooseberries, and currants. Fruits that remain on the shrubs into the winter often are desirable as bird food. Snowberry, privet, highbush cranberry, and barberry are useful in this way.

Hedges and borders do not require conspicuous bloom or fruitful characteristics. Selection for hedges may be based on foliage color and texture, growth habits, denseness, response to pruning or clipping, and desired ultimate size when not regularly pruned. Japanese barberry, privet, and honeysuckle are readily adaptable for hedges. Most shrubs are satisfactory for borders, but final height, spread, and longevity should be considered. Nursery catalogs, recommended lists of state and federal experiment stations, arboreta, and local plantings offer suggestions for plant materials to use in shrub borders.

Foundation plantings of shrubs include both deciduous and evergreen types. Spreading and procumbent junipers (Pfitzer, Sabin, Andorra), spreading and upright yews, shrub pines (Mugho), and broadleafed shrubs (boxwood, azalea, rhododendron, gardenia) are evergreen shrubs that are used considerably, where climatically adapted, for foundation plantings. The wide range of deciduous shrubs for one-story houses includes mostly the less rangy, smaller growing types including Anthony Waterer spirea, flowering quince, Japanese barberry, and snowhill hydrangea. Those used for taller houses may include taller and more robust types like honeysuckle, rose of Sharon, beauty bush, and highbush cranberry. Ultimate size is extremely important in selecting shrubs for foundation plantings. Continuous need for pruning shrubs under windows, near walks, and by doorways may become an annoying task in time.

Hardiness of shrubs, like trees, is important to their survival and adaptability. Native shrubs may be very useful in many geographical areas and generally present no hardiness problem. The exception to the hardiness factor may be found in some shrubs which have natural winter protection in their native habitat. There are many adapted types and varieties of plants introduced from other areas or developed by plant breeders that are sufficiently hardy. Many plants that are known to be less hardy can be grown by taking special precautions in location and protection. Shrubs along the foundation of the house are subjected to milder conditions than are individual exposed specimens in open areas.

Spacing   *Ultimate size and form* is important in shrub spacing. Tall and upright types can be closer without competition than short and spreading forms. The manner of growth, whether by suckers or branching, influences spacing. The ultimate size and shape of shrubs should be known in order to do an intelligent and effective job of both location and spacing.

Desired effect or ultimate objective will influence spacing. A specimen plant should be without competition from other shrubs since its purpose is to grow and develop into the best possible, yet characteristic, shape and form. Shrub borders must be considered from the standpoint of adjacent plants. Tall plants are placed to the rear of low ones. Spreading plants should be given more space than upright types. Either more or larger plants can be placed in wide areas of the border or, conversely, fewer or smaller plants in narrow parts. Shrubs for hedges are planted quite close for compactness. If the hedge is to be clipped frequently,

FIGURE 17-9   Desired effect of blooming shrubs influences location and spacing. *Top*, Snowhill hydrangea; *Bottom*, mock orange (***Philadelphus.***)

plants are placed 12 to 18 in. (30–45 cm) apart depending on the height at which it will be maintained. Informal, unclipped, hedges require greater spacing between plants, or light will not penetrate interlocking branches adequately, and defoliation and dieback will result. Spacing in an unclipped hedge is about the equivalent of spacing in a border. Some competition will occur as the shrubs reach mature size but consistent annual pruning will keep them at a nearly constant size.

Care of Shrubs

Protection from various types of injury is one of the principal needs of shrubs in the period immediately following transplanting. Wire guards are helpful in preventing trampling by foot traffic or damage from the lawn mower. The guards aid in drawing attention to newly located shrubs so they can be seen and avoided. Animal repellents for small domestic animals and rabbits are sometimes helpful. Nicotine sulfate and sulfur are commonly used materials for other purposes which may serve as repellents. Their odor is offensive to these animals although none are applied in such quantities as to be toxic to man, animals, or plants. Mulches are a protection for shrubs either by stabilizing temperatures, conserving moisture, or avoiding erosion.

Irrigation may be as important for shrubs as it is for any other plants. Shrubs vary widely in the depth, spread, and extent of their root systems. They may compete with other shrubs, with trees, and with grass for the moisture supply. Surface watering is satisfactory. If sprinklers are used, it is important that it be early enough in the day that the foliage will be free of surface moisture by darkness. This is especially critical in humid areas where moisture favors fungus growth. Irrigation can be the deciding factor of winter survival in some areas. Death by desiccation due to drying winds is an effect of winter drouth. In areas dependent on irrigation, failure to irrigate before soil freezes in the fall is usually disastrous to survival.

Fertilizing shrubs increases twig renewal which increases the bloom on flowering shrubs, the berries or fruit on fruiting shrubs, and the foliage on all shrubs. Fertilizing can be overdone to the extent that such rapid and excessive growth may get out of bounds. In most cases, lawn fertilization is adequate for shrubs in the lawn. In borders and foundation plantings, fertilizer applied at the same rate as the lawn requires is adequate.

Pruning Shrubs

Training for shape and purpose is a principal reason for early pruning of shrubs. Selection of several stems or branches for equal size and vigor is desirable for shrubs in borders and informal plantings. This tends to produce an open center or vase type of growth. With some shrubs a central stem or trunk can be forced to produce nearly horizontal laterals and obtain a compact effect. This is in effect an espalier form or cordon system of training. Tying and pruning are both utilized in attaining this type of growth. These are mostly used along walls and fences and in formal gardens. Training for hedges involves severe pruning at planting time and the following year or until the base of the hedge plants are

FIGURE 17-10   Shrubs for hedges are closely spaced and severely pruned at planting time. (Photographs courtesy of *Better Homes & Gardens* magazine.)

dense with foliage. A hedge that is dense above but sparse below is less attractive as a hedge. Early training by severe heading back develops strong basal branches which are more easily maintained with dense foliage.

Clipping and shearing several times during the growing season is essential to good appearance of a clipped hedge or formally trained evergreen shrubs. Frequent clipping is used to maintain constant size and prevent excesses in growth. Shrubs allowed to go too long without clipping or shearing will develop long shoots which shade other shoots and leaves depriving them of food and nutrients. When they are eventually clipped, the shrubs have acquired a more open, straggly appearance.

Renewal is the process most frequently employed for pruning informally trained shrubs. It is a thinning out procedure in which the old wood is removed near the base and young new basal growth

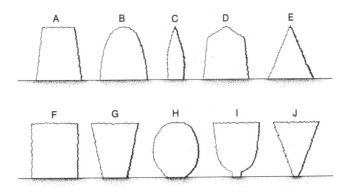

FIGURE 17-11   Hedge forms. A through E are desirable forms. They admit light to the lower branches, and leaf growth is dense to the ground. F is satisfactory in full sun. G through J are shaded at the base, and bare stems and sparse foliage result.

FIGURE 17-12   A trial area of evergreen and deciduous hedge plants. (Photograph by K. A. Denisen.)

encouraged. It is possible through severe heading back to rejuvenate large shrubs that have overgrown their boundaries.

Removing Unwanted Shrubs   If for some reason shrubs are no longer desired or weedy shrubs persist, there are several methods by which they may be removed. Cutting off at the base does not kill most shrubs. This is usually only an extremely severe pruning process, which results in reactivation of latent buds or stimulation of adventitious buds to produce new tops.

Grubbing will remove not only the top but the crown, subterranean stems, and the major portion of the root system. It is an effective, but laborious and expensive method. It is generally impractical for removal of poison ivy and other allergy-producing plants because of skin contact.

Chemical killing is often effective but must be done with great care to avoid injury to desired plants. Ammate when applied to foliage is effective in killing most woody plants. The hormone herbicide, 2,4D is also effective when sprayed on the foliage and repeated if kill is not complete. This herbicide can also be applied to subterranean parts to perform killing action. Soil sterilants such as neburon and sodium chlorate are effective but deny the use of the treated area for nearly all vegetation for periods varying from several months to two or more years depending on rate of application. It is only in extreme situations that these materials are used in the home grounds. In using any chemical killers of shrubs, it is extremely important to read the directions on the label, follow the directions, and observe all precautions. The information on the label specifies its use and the precautions for man, animals, and susceptible, desirable plants.

ROSES

Roses are under separate discussion because of their unique treatment, popular appeal, and special growth problems. They are grown both as a commercial florist crop and an ornamental in the home grounds. They are a leading item in nursery production.

Selecting Types of Roses

Roses may be placed in groups based on growth habit as follows: (1) shrub roses, (2) ramblers or climbers, and (3) tree roses. The *shrub roses* include floribunda, rugosa, hybrid tea, and hybrid perpetual types. Floribunda roses produce abundant bloom of moderate to small size on long stems which arise several in a group from individual branches or canes. They are used in beds and borders and their miniature blooms are excellent for corsages and flower arrangements. Examples are the cultivars Fashion and Vogue. Rugosa roses have shorter pedicels (flower stems), smaller blooms, and more flowers per cluster than do floribundas. Petals are serrated giving a soft texture. These rose plants generally grow taller and more erect than floribundas. They are especially well adapted to foundation planting and shrub borders. They are attractive as specimen plants or in unclipped hedges. This group includes some of the hardiest of roses. Red and Pink Grootendorst are popular cultivars. Hybrid tea roses are the most widely grown type. They have an ever-blooming habit which means they have blooms, buds, or flower primordia most of the time from a few days after growth begins in the spring until growth ceases in the fall. Their blooms are large, particularly if only one flower develops per stem. Pinching out, or disbudding, secondary buds as they appear increases the size of the "king," or dominant,

FIGURE 17-13   **Typical selections of roses.** *Left*: **The hybrid tea, McGredy Yellow.** *Right*: **The climber, Paul's Scarlet.**

bloom. Hybrid tea roses give gratifying response to good care and management and are the pride of many gardeners. There are literally dozens of good cultivars of hybrid tea roses available. Outstanding introductions include Peace, Chrysler Imperial, McGredy Yellow, J. H. McFarland, and scores of others. Hybrid perpetuals are hardier than hybrid teas and produce prolifically of large blooms usually just once during the growing season. They are extremely attractive while in bloom and serve as foliage shrubs the remainder of the year.

Rambler roses or climbers are sometimes classed with vines. Actually they do not vine but support their long stems, trunks, or branches with vigorous spines. Tying or fastening is usually necessary to hold them in place on fences, trellises, and archways. Ramblers with annual blooming habit include Paul's Scarlet, a quite hardy variety, and Christine Wright (pink): everblooming ramblers include Blaze (red) and Dream Girl (white with pink tinge).

Tree roses are actually floribunda or hybrid tea roses grafted or budded on an upright stem. The stock is of a sturdy, vigorous tree-type rose. They require special care and treatment both in pruning and winter protection (Fig. 5-8, page 122). Tree roses are used principally as specimen plants.

The Rose Bed

Because of the special care and attention required by hybrid tea and floribunda roses, they are generally grown in a rose bed rather than in a shrub border or flower garden. Rugosa and hybrid perpetual types are more readily adapted to random placement.

Spacing and arrangement will vary to a certain extent with vigor, fertility, site, and length of growing season. In general most hybrid tea and floribunda roses can be spaced about 18 in. (45 cm) apart in beds without serious competitive effects. The taller growing types should be placed to the rear of a bed which is adjacent to a fence or shrub border. They should form the center row of a larger bed which is viewed from both sides. Either the square or triangle arrangement of plants produces effective results. Color of blooms is another factor for consideration in placement of plants. Clashing colors for adjacent plants should be avoided.

Summer mulching is a common practice in rose beds. Chopped corn cobs, peat, leaves, and polyethylene film are all used effectively for various reasons. Mulch will conserve moisture, may reduce weed growth, make a clean place to work, create a near and tidy effect, and aid in moisture absorption. Organic mulches increase the humus content of the soil. Some rose gardeners find summer mulches very helpful in conserving moisture during extended absences or vacation periods. However, organic mulches, being a haven for moisture, may encourage fungus diseases. A continuous spraying or dusting program with fungicides will alleviate this problem. Good air circulation also helps.

Fertilizing is definitely related to good rose production. Roses respond to fairly heavy applications of commercial fertilizer. Complete fertilizers of 4–12–4, 5–10–5, or similar analysis can be applied at the rate of $\frac{1}{2}$ cup (100 cc) per plant per month during spring and summer.

FIGURE 17-14 **A rose bed with a Russian olive hedge as background. The roses have a summer mulch of chopped corn cobs.**

This system of feeding roses utilizes the everblooming tendency to give great satisfaction in flower production. Such high rates of fertilization must also be accompanied by proper pruning and irrigation practices. High analysis fertilizers can also be used but the rate should be decreased accordingly. Chemical fertilizers may be applied in a ring around each plant or broadcast in a rose bed. It is frequently applied before irrigating to hasten the downward movement of nutrients. Fall fertilization is not generally advocated because of the increased shoot growth prior to winter's low temperatures.

Irrigating during some periods of the growing season is necessary in most climates for maximum bloom and good growth. Both soakers and sprinklers are used. Soakers give a slow steady flow of water that does not accumulate on the surface or wet the leaves. Sprinklers can apply water faster, but wet the foliage which favors fungus diseases and washes off dust or spray residues. When sprinkling is used, it is best to apply water sufficiently early in the day to allow moisture evaporation from the foliage before sunset. The principles of irrigation apply to roses; that is, they should be watered before there is a water deficit and should receive enough water to penetrate to the entire root system. Roses require about an inch ($2\frac{1}{2}$ cm) of water per week with more needed in hot or dry weather.

Pruning
Roses   Both dormant and summer pruning are necessary for maximum production of everblooming roses. *Dormant pruning* is usually done in early spring before growth starts. Some dormant pruning may be done in the fall to facilitate applying the winter mulch in areas where winter

protection is needed. Since roses are planted quite closely in beds, and since long stemmed roses are desired, the dormant pruning is quite severe. Thinning out to three or four sturdy canes and heading back each cane to 12 to 18 in. (30–45 cm) are means of encouraging vigorous shoot growth.

Summer pruning is done to avoid accumulation of manufactured food in the seeds and fruit (hips) and to encourage development of strong shoots. When the petals begin to drop from rose blooms, they are removed. The hybrid tea stem is cut back to strong buds. The tendency of many home gardeners is to leave too many buds resulting in short stem growth. Along with removal of the bloom it may be necessary to remove considerable stem to have both vigorous and fewer buds. Long-stemmed roses are produced with more severe summer pruning. More blooms but shorter stems and smaller flowers result from pruning which only removes old flower heads.

Special pruning techniques are required for ramblers and tree roses. Since ramblers are used for climbing and covering, their long canes should be retained for support and the secondary canes utilized for flower production. These laterals or secondary canes are headed back in much the same way as the primary canes of hybrid teas. Annual blooming types will only require one summer pruning after bloom, whereas everblooming ramblers are treated very much like floribundas and hybrid teas. Pruning is not so severe since long stems are not essential. Rambler roses are principally for outdoor effect. Tree roses are primarily for specimen effect so long stems are not desired. The top is thinned as needed for compactness of head which is one of the features of tree roses. Pruning is primarily to remove old blooms and maintain desired size.

Pest Control

Principal insect pests of roses include the rose slug, rose chafer, aphids, spider mite, and general insects like leafhopper and grasshopper. Principal diseases are black spot, leaf spot, and mildew. All-purpose rose sprays or rose dusts are popular. They should contain at least two insecticides, two fungicides, and a miticide for control of most pests; however, the ingredients should each have a wide range of effectiveness. A spraying or dusting program is a precaution against loss or injury from these pests. Preventive rather than remedial measures best suit the interests of the rose grower. Rodents, especially mice, may be a problem under the winter mulch. A liberal application of sulfur dust around the base of the plants and over the mulch is one of the most effective repellents.

Winter Protection

Mounding with soil prior to mulching is a common practice in cold climates. Soil is a good insulator and even in very cold winters, most roses rarely die back further than the soil line. Mounds are most effective if 8 to 10 in. (20–25 cm) high. The plants are covered with 8 to 12 in. (20–30 cm) of organic mulch like small grain straw, hay, or soybean straw. Other methods of mulching include putting a covered box or shelter over the bed and placing a dry mulch of chopped corncobs, fine hay, or straw around the plants. The dry mulch is a good insulator, keeps the plant stems dry, and eliminates the need for mounding. Tree roses and ramblers need special protection for thir stems in cold climates. Ramblers may be cut down from their supports and the canes covered with soil. Tree roses are loosened around their roots, and the entire plant is tipped over and laid on the ground (Fig. 5-8, page 122). Soil is placed over the stem and top and the entire mounded area is mulched. Great care is required in restoring the plant to an upright position in the spring.

VINES

Selecting Types of Vines   Function, location, and means of support are determining factors in selecting vines. *For walls*, clinging types of vines are used. Their means of support may be by small adhesive discs as found on the Engleman creeper and Virginia creeper or by aerial roots like those of English ivy and wintercreeper (euonymous). Adhesive discs function on the principle

FIGURE 17-15   **Virginia creeper on a brick wall.**

Ornamental Woody Plants   **383**

FIGURE 17-16   **Clematis vines are adapted to trellises and have an abundance of large showy blooms.**

of a suction cup on smooth surfaces, and aerial roots actually penetrate porous walls of brick, stone, cement and stucco. They climb readily on screens.

Vines used *for fences and wires* actually form narrow borders for background or for separating areas. Tendrils, modified branches, of the grape support the vine by twining around wires and other supports. The twining stems of bittersweet and the trumpet vine offer support on fences and wires. For bittersweet they are the means of attaining great heights by twining around the trunks of trees.

Vines *for trellises and lattices* are supported by twining stems, e.g., the well known clematis and honeysuckle vine. The thorns of rambler roses aid in their support on these structures.

Vines may be used *for ground cover.* Many kinds are adaptable, but in general, they should have heavy foliage and vigorous growth. Most ground covers are needed in shady areas where grass does not prosper, so shade tolerance is also an attribute. Baltic English ivy, Virginia creeper, and honeysuckle vine make effective ground cover.

Several of the vines previously mentioned may be used *for flowers and berries.* Clematis, honeysuckle, and trumpet vine produce an abundance of attractive flowers. Bittersweet produces decorative berries and grape produces edible fruit.

Culture of Vines   *Spacing* between plants is widely variable depending on desired effect. For complete cover of a wall, uniform spacing of plants will maintain an even, gradual spread. If the wall is two or more stories high, plants should be spaced 4 to 6 feet ($1\frac{1}{2}$–2 meters) apart and the vines trained upward. If the wall is less than two stories, they may be spaced 6 to 10 feet (2–3 meters) apart and the vines trained in fans. Grapes on fences should be 8 to 12 feet ($2\frac{1}{2}$–$3\frac{1}{2}$ meters) apart for most training systems. Spacing from the supports or areas of spread is also important. For vines on buildings, it may be desirable to plant the vines 2 to 4 feet ($\frac{1}{2}$–$1\frac{1}{4}$ meters) from the building for better moisture conditions than within the 2-foot ($\frac{1}{2}$-meter) area. In general, vines used on trellises, lattices, and fences are planted directly beneath their supports.

Fertilizing at rates equivalent to those of shrub borders and lawns is usually adequate for good growth, attractiveness, and production. Excess fertilization can readily lead to overgrowth.

Irrigating may be done especially if the vines are located in foundation plantings, near shrub borders, or beside vegetable gardens. The root systems of most vines are quite extensive.

Pruning vines is often a matter of keeping the plant and foliage within bounds. On the walls of a building it may be necessary to prune both stems and leaves from around windows several times during the season. Vines on fences likewise must be cut back severely when they reach out to other areas, structures, or plants. Training is important to guide the main arteries of the vine in the decreed directions. It may be necessary to root prune if vines persist in overgrowth. Pruning operations, as in the vineyard, influence flowering and fruiting. Vines tolerate considerably more severe pruning than do trees and shrubs and yet will remain productive of flowers and fruit.

PLANTING ORNAMENTAL WOODY PLANTS

Bare-Root Nursery Stock *Size of hole* is often inadequate for many planting operations. This is especially true if the soil is quite compacted or the drainage is poor. The excavation for the root should be wider than the root span by 2 to 4 in.

FIGURE 17-17 Bare-root nursery stock is planted in a roomy hole, with roots spread out, and slightly deeper than it grew formerly. Watering insures a quicker start. (Photographs courtesy of *Better Homes & Gardens* magazine.)

(5–10 cm) so the roots can be arranged without constrictions. It should be deep enough that the plant can be set slightly deeper than its original depth. The soil at the bottom of the hole should be loosened to permit easier root penetration.

Root placement involves spreading out the roots, as nearly as possible, in several directions to guide the development of main roots and provide

FIGURE 17-18   **Large calipre balled-and-burlapped birch trees in nursery sales. (Photograph by K. A. Denisen.)**

better anchorage. It is important not to force or bend the roots. Broken and damaged roots should be pruned off to reduce the likelihood of infection.

Graft and bud unions should be placed below the soil surface except for dwarf fruit with an intermediate dwarfing stock. This encourages rooting from the variety stem and also reduces the possibility of shoots arising from the root stock. The latter reason is important with roses, since roses are sometimes grafted on suckering root stocks.

Filling the hole should begin with a light layer of top soil over the roots. This should be followed by slight pressure of the hands for firming. Additional soil is then added and firmed so that the hole is about two thirds filled. Water and commercial fertilizer may be applied at this point. Fertilizer should be applied in small amounts, $\frac{1}{2}$ cup (100 cc) is the maximum at planting for most trees and shrubs. An important precaution is to avoid direct contact of roots with commercial fertilizers. Many people prefer to delay fertilizer application until the hole is nearly filled or after planting. A watering basin may be constructed around the base of the plant for future watering. It is a good practice to leave the area around the stem free of grass or other vegetation for one or two seasons to reduce competition and give better water penetration.

Pruning should follow immediately after planting.

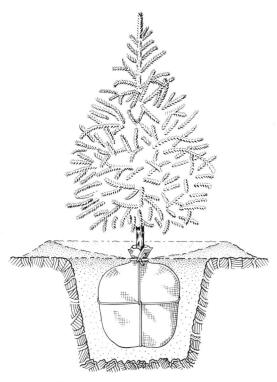

FIGURE 17-19   **Balled-and-burlapped nursery stock is planted at the same depth it grew previously. Burlap should be loosened at the top after planting for better water penetration.**

Ornamental Woody Plants   ***387***

Balled-and-burlapped nursery stock is used principally for evergreens but is also used for leafed-out deciduous woody plants. These kinds of nursery stock are carrying on transpiration and have considerable demand for moisture about the root system. A balled-and-burlapped conifer will retain moisture in the soil ball for several days as its rate of tranpiration is relatively low. Another advantage is the relatively small amount of root disturbance to the balled-and-burlapped plant.

The plant should be set at the same level it has been grown previously. This is important in digging the hole. The soil should be loosened below

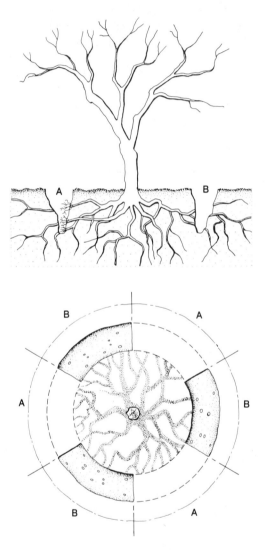

FIGURE 17-20   **Preparation for moving a large tree consists in digging a trench around the tree during a two-year period. A, three sections dug the first year followed by filling with straw or other fibrous material. B, sections dug second year preparatory to moving. (Adapted from *Landscaping Your Home*, Wm. R. Nelson, University of Illinois, Circ. 858, 1963.)**

and to the sides of the ball of earth. The burlap is left around the ball to decompose in the soil. The top part of the burlap should be cut open and rolled back slightly before soil is completely filled in. This increases water penetration from the top. A watering basin should be constructed around the newly planted tree or shrub and excess soil removed.

"Canned" nursery stock refers to that grown in tin cans. The purpose is much the same as for balled-and-burlapped stock, however, it is less expensive to produce. The digging and burlapping operation is eliminated; the plants are transported in cans with soil intact. Transplanting is a bit more difficult since the can must be cut away from the soil ball.

Moving Large Trees

It is possible to move quite large trees, up to 6 in. (15 cm) or more trunk diameter, by exercising extreme care and with adequate equipment. It is an expensive operation. Deciduous trees are moved when dormant. The larger the soil ball that can be taken, the greater is the possibility of tree survival. A general rule to follow, when possible, is to provide a foot of soil ball diameter for each inch of trunk diameter. Many large trees have been successfully moved with much less soil; however, their chances of survival are decreased. Also more severe pruning is necessary since more roots are lost. Large trees should be braced or guyed in position until developing new roots hold them firmly in position.

STUDY QUESTIONS

1. What are important considerations in selecting trees for the home grounds?
2. Is spacing of trees as important as spacing of shrubs or vines? Why?
3. How can trees benefit from the usual operation of the lawn, herbaceous ornamentals, and the vegetable garden?
4. Is insect or disease control on trees a serious problem?
5. List the uses for shrubs in a home plan. Suggest kinds of shrubs for each use.
6. Do shrubs need to be irrigated or fertilized?
7. What is the ideal shape for training and pruning hedges? Why?
8. Is root pruning ever practiced in shrubs? Why?
9. What are the types of roses that can be grown in your home grounds? From a nursery catalog make cultivar selections.
10. List several rules for pruning roses including dormant pruning, summer pruning, and training climbers.
11. Devise a winter protection plan for a rose bed in your locality.
12. When are vines used in the home grounds?
13. What types of support do vines require?
14. Distinguish between bare-root, B and B, and "canned" nursery stock. When is each type used? Do they vary in price? Why?

SELECTED REFERENCES

1. L. H. Bailey Hortorium Staff. 1977. *Hortus Third*. Macmillan, New York.
2. Bailey, L. H. 1939. *Cyclopedia of Horticulture* (3 vols.). Macmillan, New York.
3. Buck, G. J., and E. C. Volz. 1955. A handbook for rose growers. *Iowa Agr. Exp. Station Bull.*, *P117*, Ames, Iowa.
4. Christopher, E. P. 1954. *The Pruning Manual*. Macmillan, New York.
5. Denisen, E. L., and H. E. Nichols. 1962. *Laboratory Manual in Horticulture*. Iowa State University Press, Ames, Iowa.
6. Thompson, A. R. 1961. *Shade Tree Pruning*. Tree Preservation Bull. No. 4, National Park Service, Washington, D.C.
7. U.S. Department of Agriculture. 1977. *Shrubs, Vines and Trees for Summer Color*. Home and Garden Bull. 181. Agr. Res. Serv., Washington, D.C.
8. U.S. Department of Agriculture. 1972. *Insects and Diseases of Trees in the South*. Forest Service Southeastern Area Forest Pest Management Group. Atlanta, Georgia.
9. Wyman, D. H. 1965. *Trees for American Gardens*. Macmillan, New York.
10. Yearbook of Agriculture. 1949. *Trees*. U.S. Department of Agriculture, Washington, D.C.

18 Ornamental Herbaceous Plants

Herbaceous ornamentals in the home and home grounds contribute color and interest to the scene. They consist of flowers and foliage plants which, combined with the trees, shrubs, vines, and lawn, complete the ornamental plantings. Both garden flowers and house plants play important roles in beauty, attractiveness, and happiness of a home. They offer unlimited opportunity for the family and its individual members for a diversion or to acquire knowledge of plants. House plants, in themselves, cover a wide range of kinds and species giving adequate experience in nearly every type of propagation by the selection of several different species. Differences in soil preferences, light requirements, temperature limitations, and drouth tolerances emphasize the all important role of environment in growing plants. Ornamental plants provide an opportunity for utilizing the beauties of nature in outdoor plantings, in bouquets and arrangements, and for flower production or a touch of green when grown in pots, a modified environment. Flowers and house plants have decorative value as well as being a source of active enjoyment to the household. The extent of their use is determined principally by the satisfaction and pleasures which they provide.

GARDEN FLOWERS

Garden flowers include all flowering herbaceous plants which are grown out of doors strictly for ornamental purposes. Many shrubs have annual periods of bloom, however, garden flowers provide additional color and extend the distribution of bloom throughout the season. They are usually more adaptable for cutting than are flowers borne on woody plants.

Adaptability to the Home Grounds and Family

Flexibility of the plan is a desirable characteristic of home grounds development. Flowers contribute flexibility to the pattern of planting because of their herbaceous nature. They can be easily moved, removed, or allotted increased space. The maximum and minimum areas in flowers can be determined by: (1) amount of area in shrubs, (2) size of the play area and proximity to flowers, (3) amount of grounds devoted to lawn, and (4) available time for care and maintenance. Shrubs and flowers may actually compete for space and attention resulting in a jarring, "overbusy," unrestful effect. Shrubs are more permanent, but while they are becoming established, flowers are very

391

FIGURE 18-1   Herbaceous flowering plants are flexible in distribution of bloom. *Left*: They can be forced for out-of-season bloom, hyacinth. *Right*: Crocus can provide early outdoor spring bloom.

useful for filling otherwise empty areas. Flowers are used to soften the lines between shrubs and lawn. The two types of ornamentals can be used together in a restful, harmonious manner. The size of the play area and its proximity to flowers influence the amount of flowers both from the space standpoint and the danger of injury by trampling. The amount or proportion of the grounds in lawn will certainly affect the amount of area available for flowers. Those who want a large, open, spacious lawn must sacrifice flower, vegetable, and shrub areas. Those who desire considerable flower area are likely to have less area in lawn and other plantings. A very important reason for flexibility as offered by flower growing is the variable allocation of time for their planting, care, and management. If time on the home grounds is at a premium, flower plantings can be decreased. When children are small, there is usually little time for gardening by the homemaker. When the children grow older, the flexible nature of flower growing permits an easy shift to increased flower production.

Yields in pleasure should prevail in adapting flowers to the home grounds. The tastes of individuals should be a guiding light and determining factor for size of plantings and selections. For those who crave activity in their leisure hours, flower growing is an excellent means of occupying time and thoughts. It provides excitement for those who can enjoy flower plantings and anticipate their ever unfolding mysteries of growth. With proper selection and adaptation, flowers can provide beauty and pleasure with only occasional, but timely attention.

The priority in home grounds development expresses the original plan for flower adaptability to the landscape. However, top priority is nearly always given to the lawn. If no trees exist in the new grounds they are planted, since they may require a long time before their usefulness is realized. Shrubs usually follow since they, too, require time for growth to appreciable size. Annual flowers are usually planted the first year for they give quick results while adding color to a usually rather drab

Horticulture for the Home

FIGURE 18-2   **Geraniums transplanted to a bed give quick results in a home plan. (Photograph by K. A. Denisen.)**

incomplete surroundings. Perennials follow in the planting sequence. They may actually play an important role, lending interest and color through the seasons, before flowering shrubs are well established.

Maintenance and replanting are extremely important considerations in adapting flowers to the home and family. In our busy society, time is often at a premium and many people find they prefer those plants which take care of themselves to a considerable degree. This is true of many of the perennials which may be effective competitors against weeds, which bloom in spite of infrequent attention, and which continue to survive under seemingly adverse conditions. Day lilies, daffodils, and phlox are typical of this group. This may also be an advantage of some annuals that seed themselves. Thinning and only occasional weeding may be all that is required in beds or borders which have these types of annuals. Morning glories, zinnias, and marigolds are typical of this group. Petunias would qualify but give such a range of segregated colors that they usually lose their appeal when they seed themselves. In general, most perennials benefit from being divided and moved at two- to four-year intervals. Most annuals are either started anew from seed scattered in their location each season, or they are started in plant growing structures and then transplanted to give an earlier and longer season of bloom.

Lack of attention and its consequences is an important projection of planning in selecting plants for adaptability to the home and home grounds. Since some flower gardens are established with enthusiasm but maintained with disdain, it is wise to give serious consideration to what will happen with neglect. If weeding is the problem, sturdy and competitive plants may be the answer. If encroachment of lawn grasses into the flower plantings is a continual occurrence and the flowers are severely overcome, it may be better to extend the lawn and mow the area for expediency. If removing dead blooms of flowering plants is not done, consider how it will look. It may still give more enjoyment and beauty than if the blooming plants were not present, or it may be so unattractive as to suggest elimination of that particular flower. When perennials need winter protection and there is not time to prepare them for winter, they should not be grown. There are usually some kinds of perennials that are hardy. Many perennials are sufficiently protected by leaving their dead herbaceous tops over the crowns during the winter.

FIGURE 18-3   **Window boxes add color to a home.**

Flowers enhance the beauty and attractiveness of the home grounds through proper placement and performance of specific functions. They are used in (1) borders, (2) beds, (3) gardens, and (4) with architectural features that require plants.

Borders of flowers may be along property lines, next to buildings, adjacent to walks and drives, or in the foreground of shrub borders and

FIGURE 18-4   **Flower borders reflect the tastes of their planners. *Upper*, informal mixed border in the private area of a home grounds. *Lower*, formal border in a public park. (Lower photograph by K. A. Denisen.)**

Ornamental Herbaceous Plants   **395**

hedges. Flowers may even be intermingled with shrubs in the shrub and flower border or foundation planting. The border may be either formal or asymmetric in design. Informal mixed flower borders consist of annual, biennial, and perennial plants. They can be planned for continuous bloom through the season by judicious placement of kinds through the length and breadth of the border. The rules of placement are few and simple, nevertheless important. Tall plants should be at the rear or more distant side of the border. Low growing and edging plants should be to the front. Plants of a kind are best located in distinct clumps or flowing strips. Symmetrical borders are frequently located in rectangularly shaped areas, those spaces between the driveway and hedge or between the walk and garage foundation. Straight rows of flowers with increasing height from front to rear give a progression effect. Contrasting colors for each row emphasize parallel lines. Intermittent, regular placement of contrasting plants gives a rhythmic effect of islands. Borders can also contain a single row of large flowers of the peony and iris scale. During their non-blooming period their foliage can be a desirable part of the landscape. Peony plants in rows have an informal hedge appearance. Iris foliage with its erect and pointed leaves can be effective in avoiding monotony of texture at various places on the grounds.

FIGURE 18-5   **Tulips are more effective in mass bloom and more easily given proper care when planted in beds.**

FIGURE 18-6   **Gladiolus corms planted in rows for production of flowers for cutting.**

Beds are used with certain flowers that require specific care and management. Just as hybrid tea roses are placed in beds rather than in shrub borders, so certain herbaceous flowers are placed in beds. Tuberous begonias and gloxinias are exacting in their culture. They are more apt to receive proper care and treatment in a bed that is adapted specifically for tuberous begonias and gloxinias. Tulip beds are easier to plant, manage, and replant than are tulips scattered, even in clumps, through a border. Annual flowers which follow the tulips to give summer bloom to that area are more easily cared for in tulip beds than they are in scattered areas of an informal border. Beds for shade-loving plants are often located next to foundations, under trees, or on the shady side of buildings.

Gardens may be specifically to grow flowers for cutting or propagation. They can be conveniently made part of the vegetable garden area. Their placement in rows for cultivation, pest control, and adequate spacing for full development makes the garden an excellent source of cut flowers. Since borders are primarily for beautifying the grounds, cutting flowers from borders may decrease their utility. Those who desire large quantities of cut flowers should seriously consider a garden for flower production. Gladiolus is well adapted for flower production in the straight rows of a tilled garden.

Architectural features that may require plants, and particularly flowers, are low fences, gateways, archways, steps, window boxes, and patios. These structures can be made to more intimately connect with the surroundings by use of the blending and softening effects of floral

Ornamental Herbaceous Plants   **397**

FIGURE 18-7

Patio arrangement of potted plants lends itself to change during the season at the owner's discretion. (Photograph by K. A. Denisen.)

life and foliage. A patio of concrete, brick, or stone construction is bare and austere without plant life. If soil in beds is not provided, boxes, urns, planters, and pots can be utilized to good advantage. These containers will even add an element of flexibility to the patio, terrace, or steps.

Kinds of Flowers and Their Characteristics

Perennials and biennials are commonly grouped together. Perennials predominate in this group, but there are some important and desirable biennials. Several kinds are described briefly, and many others are listed in tabular form.

IRIS

This widely grown perennial may be obtained in many colors and shades. It blooms in late spring from flower primordia differentiated the previous fall. Fleshy rhizomes, which often protrude above the soil, are storage organs and the means of vegetative propagation. Healthy foliage and good enlargement of rhizomes the previous season are forerunners of large, showy blooms in the spring. The amount and size of bloom is determined by the amount of stored food. Iris plants with small rhizomes and small leaves produce small flowers. Probably these plants need to be reset for improved production. Continued production of rhizomes eventually results in crowded conditions, which is the principal reason for dividing and resetting iris. When planting, the rhizome should be near the surface with roots extending downward. A ridge for the rhizome with a trench to each side for the roots is ideal for planting. If perennial weeds become established, it is usually expedient for time and effort to reset the iris planting.

PEONY

This is an old, yet favorite, perennial. Its huge bloom in late spring may be single or double, i.e., a single concentric line of petals or multiple rows of more compacted petals. It has a wide range of color. Propagation is by means of crown division. Plants will remain productive without resetting for as much as 10 or 15 years but can be divided for increasing the planting in 2 or 3 years. Its foliage is attractive and dense, giving it a bushy appearance during the nonblooming period.

SPRING-FLOWERING BULBS AND CORMS

The spring-flowering bulbs are planted in the fall for early spring bloom. Tulips can be planted deeper than other bulbs, and they actually produce better flowers with deep planting. Most recommended planting depths are from 4 to 7 in. (10–18 cm). If planted 8 to 12 in. (20–30 cm) deep they are less likely to produce small bulblets beside the mother bulbs. These small bulblets produce small flowers; consequently, tulips need resetting frequently when shallow planted to reestablish larger flower size. At greater depths, the mother bulbs are less susceptible to injury by summer heat. Annuals are planted in the tulip bed following the blooming period of tulips utilizing the space and making it more attractive. Tulips enter a rest period in early summer after translocating food to the bulb for the next year's production of flowers.

Narcissus species include daffodils and jonquils. They are early spring blooming, but unlike the tulip, retain their leaves and continue growth after the blooming period and during the summer months. They are excellent for cut flowers and their vivid shades of yellow and stark white are striking features of the soft green landscape in early spring. Daffodil and jonquil bulbs divide yet may not require thinning or resetting for

FIGURE 18-8   Peony and iris are well-known perennial flowers.

FIGURE 18-9   Flowering bulbs. *Top*: *Left*, tiger lily; *right*, narcissus. *Bottom*: *left*, amaryllis; *right*, daffodil.

several years. They "take care of themselves" with better results than do tulips.

Hyacinths are generally shorter lived than tulips and narcissus. This is because of their more exacting requirements of drainage, mulching, and noncompeting plants. They are excellent for forcing, i.e., growing out of season in pots and flats. The other bulb crops can be adapted for forcing.

Spring blooming corms include crocus and scilla. The crocus is one of the first blooms in the spring, and is often found blooming in the last snow of the season in the colder climates. It is extremely hardy. Scilla blossoms are much smaller than crocuses and they bloom later. They are used to give a splash of color under trees, shrubs, and out of the way places.

SUMMER-BLOOMING BULBS AND CORMS

The vast group of lilies include many colors, sizes, and degrees of hardiness. They are a popular group of bulbous plants which give dramatic satisfaction to the gardener. Fortunately many hardy types are now available so that lily growing is within the realm of practically all

Horticulture for the Home

FIGURE 18-10   Olympic hybrid lilies are hardy, vigorous, and productive.

climates. Regal, Olympic hybrid, and tiger lilies can be grown in extremely cold climates with only a light mulch during the winter. In many cases they flourish with no added mulch. Auratum, Madonna, and orangecup lilies thrive best in milder climates; but with special precautions, they can be grown successfully in areas of quite severe winters.

The gladiolus is the most widely cultivated of the flowering corms. It is non-hardy and the corm divides during the summer, which are the reasons why it is dug each fall, stored for the winter, and replanted the following spring. The gladiolus is susceptible to injury from thrips, a tiny sucking insect, and rust, a fungus disease. A pest control program utilizing a dust or spray containing a fungicide and contact insecticide, each with a broad range of effectiveness, is essential to continued success with this flower. "Glads" produce beautiful long spikes of brilliant bloom under good management. It is a flower crop which will not succeed when neglected. Large corms ($1\frac{1}{2}$ to 2 in. [3–5 cm] in diameter) produce the largest blooms.

Day Lily

The Hemerocallis or day lily is not a bulb, but is tuberous rooted. It is extremely hardy and vigorous. It competes exceedingly well with weeds and grass. It may present a problem of overgrowth and tend to take over borders and beds. This flower responds to good treatment, but also blooms under quite adverse conditions. Its blooming period is characterized by the emergence of tall spikes terminated by an inflorescence of numerous buds. These buds usually open successively, one each day; thus the term "day lily." Continued progression of each bloom is facilitated and size of bloom is increased if old blooms are removed before they form seeds and fruit.

MINT

There are many kinds of mint, some of which have very attractive blooms. The firecracker plant (Oswego tea) has a brilliant red bloom with narrow, nearly erect petals which give an "exploding" appearance. The plant is vigorous, square-stemmed, and blooms in early summer. Spearmint and peppermint are often grown for their attractive waxy foliage and for seasoning and garnishing. Their bloom is not especially conspicuous or attractive.

CHRYSANTHEMUM AND DAISY

These flowers occur in many colors, sizes, and forms. They respond to lifting and resetting annually or in alternate years. Most types are responsive to photoperiod, some requiring quite short days for bloom and others long days. They are extremely responsive to pruning. The large ball and Fiji "mums" are enlarged with disbudding of all the laterals on the stem. The garden mum and cushion mum produce a greater abundance of bloom from laterals following pinching-back of the terminal growth early in the season. Chrysanthemum and daisy are readily propagated from stem cuttings in the crown region of the plant.

HOLLYHOCK

Most hollyhocks are biennial. They produce a bushy type of growth the first year and send up a long spike with lateral blossom buds the second year. They occur in a wide variety of colors in both single and double blooms. Hollyhocks are very effective as background or accent plantings or specimen flowers along walls, fences, or the vegetable garden. Once established, they will seed themselves and provide an interesting array of colors every year.

PERENNIALS FOR THE SHADE

Lily-of-the-valley provides excellent ground cover in shady areas due in part to its heavy production of rhizomes by which it propagates itself. Its exquisite and scented white bloom is a favorite among many gardeners. Ferns are well adapted to shade, even dense shade, as witnessed by their lush growth in dense forests. Violets do well in shade.

OTHER IMPORTANT PERENNIALS

Many popular, attractive, and extremely useful perennials and biennials have not been mentioned. This should not be misconstrued, for they are not less important or inferior to those discussed. There are hundreds of species from which the flower grower can select his plant-

FIGURE 18-11   **Delphiniums are tall growing perennials and generally require staking. They are frequently used as accents. (Photograph by K. A. Denisen.)**

ings. Those discussed are quite representative of perennial and biennial flowers. Numerous perennial flowers are listed in Table 6.

Annuals give a tremendous riot of color to borders and gardens. They have the advantage over perennials of producing this bloom the same season they are planted. They provide variety from year to year by changing kinds, locations, and colors. Their principal disadvantage is the

FIGURE 18-12 **Violets as ground cover in the shadow of an apple tree.**

need for replacement, starting anew, each year. For this reason, annuals generally require more time and care during the season. Seeds of annuals may be sowed directly into their location or may be started indoors or in a hotbed for earlier bloom. Weed control in the annual flowers is extremely important. It is necessary that most weeding be done by hand especially for direct seeded flowers. Furthermore, it is often difficult to distinguish between weeds and flower seedlings until some time after emergence. Brief descriptions and characteristics of some annual flowers follow.

AGERATUM

This low-growing, usually blue, annual is used for edging and borders. It is adapted to a wide range of soil and climatic conditions. For early bloom, it is started indoors. The compact bushy growth of the plants is terminated by fluffy blossoms from their first appearance in the spring until fall frosts.

CHINA-ASTER

Although the China-aster is very satisfactory for borders and for cut flowers, it may be difficult to grow in some areas. It is subject to a serious virus disease called aster yellows. Some areas are relatively free of this disease, whereas in other sections it is nearly impossible to grow this flower. Another disease pest is aster wilt; however, there are certain varieties resistant to this disease. Asters usually bloom in late summer when sowed out-of-doors but will bloom much earlier if started indoors. They are well adapted to transplanting.

Common name	Botanical name	Height (feet)*	Spread (feet)*	Color	Use
Spring Flowering					
American columbine	*Aquilegia canadensis*	$1\frac{1}{2}$–$2\frac{1}{2}$	1	red, yel.	border, shade
Bleeding heart	*Dicentra spectabilis*	2	3	pink	border, shade
Crocus	*Crocus* spp.	$\frac{1}{2}$	$\frac{1}{3}$	various	border, scattered
Dwarf iris	*Iris pumila*	$\frac{2}{3}$	1	blue, yel., wh.	edging
Iris (bearded)	*Iris* spp.	2–3	1	various	border, bed, accent
Lily-of-the-valley	*Convallaria majalis*	1	$\frac{1}{2}$	white	bed, shade
Narcissus (daffodil & jonquil)	*Narcissus* spp.	1	$\frac{1}{2}$	wh., yel.	border
Oriental poppy	*Papaver orientale*	2	3	wh., red, yel.	border, accent
Peony	*Paeonia officinalis*	2–3	3	wh., pk., red	border
Periwinkle	*Vinca minor*	$\frac{1}{2}$	$\frac{1}{2}$	blue, vi.	edging, shade
Scilla (squill)	*Scilla* spp.	$\frac{1}{2}$	$\frac{1}{3}$	wh., pk., blue	border, shade
Violet	*Viola* spp.	$\frac{1}{2}$	$\frac{2}{3}$	vi., blue, wh.	gr. cover, shade
Summer Flowering					
Adams needle	*Yucca filamentosa*	3–6	2	white	border, accent
Balloonflower	*Platycodon grandiflorum*	2–3	$1\frac{1}{2}$	wh., blue	border, shade
Bluebell	*Campanula rotundifolia*	$\frac{1}{2}$–1	$\frac{1}{2}$	blue	border, shade
Canna	*Canna* spp.	3–6	$1\frac{1}{2}$	red, yel.	border, bed, accent
Canterbury-bells (biennial)	*Campanula medium*	3–4	$1\frac{1}{2}$	blue, pk., wh.	border
Dahlia	*Dahlia* spp.	2–6	2	various	border, accent
Day lily	*Hemerocallis* spp.	2–4	$1\frac{1}{2}$	yel., red, or.	border
Delphinium (larkspur)	*Delphinium* spp.	4–6	2	wh., blue, vi.	border
Dianthus (pinks)	*Dianthus* spp.	1	1	red, wh., pk.	border, edging
Foxglove (biennial)	*Digitalis* spp.	2–3	1	various	border
Gaillardia	*Gaillardia aristata*	$1\frac{1}{2}$–2	$1\frac{1}{2}$	yel., red, or.	border
Gladiolus	*Gladiolus* spp.	3–4	$\frac{1}{2}$	various	border
Hollyhock (biennial)	*Althaea rosea*	5–6	2	various	border, accent
Lily (many kinds)	*Lilium* spp.	2–6	1–2	wh., yel., red, or.	border, accent
Monarda (Oswego tea)	*Monarda*	2–3	2–3	red, pk.	border
Phlox	*Phlox paniculata*	2–3	$1\frac{1}{2}$	wh., pk., red	border, shade
Plantain lily	*Niobe* spp.	$1\frac{1}{2}$–2	2–3	wh., blue	border, shade
Prickly pear cactus	*Opuntia missouriensis*	1	2–3	yellow	poor soil, dry
Purple coneflower	*Echinacea purpurea*	3	$1\frac{1}{2}$	pk., pur.	border
Rosemallow	*Hibiscus* spp.	3–5	2–3	various	border
Shasta daisy	*Chrysanthemum maximum*	$1\frac{1}{2}$–2	$1\frac{1}{2}$	white	border
Tuberous begonia	*Begonia* (tuber hybrida)	1	$1\frac{1}{2}$	various	border, shade
Autumn Flowering					
Chinese lantern	*Physalis alkekengi*	2	3	orange	border
Chrysanthemum	*Chrysanthemum* spp.	2–4	$1\frac{1}{2}$	various	border
Japanese anemone	*Anemone japonica*	2	$1\frac{1}{2}$	wh., pk.	border, shade
Showy stonecrop	*Sedum spectabile*	$1\frac{1}{2}$–2	2	wh., pk.	border
Sunflower	*Helianthus decapetalus*	4	2	yellow	border

TABLE 6 (*cont.*)

Common name	Botanical name	Height (feet)*	Spread (feet)*	Color	Use
Nonflowering					
American shield fern	*Dryopteris spinuloca*	2	2		arrangements, shade
Cinnamon fern	*Osmunda cinnamonia*	4–6	2–3		background, shade
Maidenhair fern	*Adiantum pedatum*	1	2		ground cover, shade

* For *Height* and *Spread* in meters multiply feet by 0.3.

CORNFLOWER

The cornflower, or bachelor's button, is an old favorite among flower gardeners. It is quite hardy, a cool season flower. Seeds may be sown out-of-doors in the spring but preferably the previous fall. Because it will reseed itself, cornflower is one of the group of annuals which will remain in an area for many years. While useful both in borders and for cut flowers, it is unique that their flowers may actually increase in size when placed in water.

LARKSPUR

This annual form of the delphinium will also reseed itself and does best when sown in the fall for early spring emergence. It is a cool season

FIGURE 18-13 *Left, Nicotiana* **has a long tubular corolla which is quite showy.** *Right,* *Verbena* **is a low growing annual used for edging. (Photographs by K. A. Denisen.)**

type and likes a cool, moist soil. Long spikes of bloom at medium to tall heights make it useful for background, borders, and cut flowers. The foliage is generally of finer texture than the perennial delphinium.

MARIGOLD

Marigolds vary greatly in size of plant and bloom (Fig. 18-4, page 395). Dwarf cultivars are usually 6 to 8 in. (15–20 cm) in height, are very compact in shape, and are extremely effective for edging. The French marigold cultivars have medium growth and texture; they tend to more upright growth than the swarfs and are commonly found in borders. The tall African marigold has a more spreading habit and is useful among shrubs or as background in borders. This is the most highly scented type and if used for cut flowers will soon fill the room with the marigold aroma. Unfortunately this aroma is unpleasant to many people. Marigolds have a long blooming season and usually survive light frosts. They add brilliant coloration to the landscape after many other flowers have ceased blooming.

PETUNIA

New cultivars and hybrids of this common annual make it continuously popular. The trumpet shaped corollas of the numerous blooms make it very conspicuous in a border. It produces dense foliage. Plant

FIGURE 18-14   A white petunia clump makes an effective planting. (Photograph by K. A. Denisen.)

breeders have made petunias available in both tall and dwarf forms which start blooming in early summer and continue to frost. They are tolerant of high temperatures. Seeds are extremely small so indoor planting is usually the most practical, resulting in an earlier start and easier cultivation. Petunias will reseed themselves but since they are

FIGURE 18-15   The giant size tetraploid cultivars of snapdragon are available for those who want larger blooms, heavier stems and greater height. (Photograph by K. A. Denisen.)

cross-pollinated, they will segregate for color and type. It is most feasible to plant true-to-cultivar seed each year. Several "double" flowered cultivars are available.

PORTULACA

This annual has been called moss rose. It has thick, cylindrical shaped leaves which are adapted to hot, dry conditions and other adversities of nature. Portulaca forms a ground cover with its carpet of foliage and is a prolific producer of various colored blooms. The low-growing habit, 6 to 9 in. (15–24 cm) makes it excellent for edging, cover beneath shrubs and larger flowers, and for filling in vacant areas of the border. It readily survives transplanting even when in flower.

FIGURE 18-16 **Zinnias are extremely heat tolerant. They are available in dwarf (above), medium, and tall sizes. (Photograph by K. A. Denisen.)**

Ornamental Herbaceous Plants **409**

SNAPDRAGON

In areas of mild winters, this plant is perennial. However, it grows from seed to a flower producing stage easily within a season to produce continuous bloom when old blooms are removed. Snapdragons make excellent cut flowers. The long spike, with its brilliantly colored "dragon jaw" florets, is attractive and interesting. In areas of mild winters, "snaps" can be mulched, or otherwise winter protected, and will result in much earlier bloom the next spring. Giant-size tetraploid cultivars are available for those who want larger blooms, heavier stems, and greater height. Snapdragons are good border plants especially for background and accent.

STRAWFLOWER

This annual has gained prominence because of its use as a dried flower in winter bouquets and arrangements. The lively colors of their petals are retained in the dried condition. Strawflowers are easy to grow, but seem to benefit from transplanting. They are quite tall (24 to 40 in. [60–100 cm]), fast growing, and very useful for borders.

SWEET ALYSSUM

The low growth, profuse and continuous bloom, and dainty flowers of sweet alyssum have contributed to its popularity as an edging plant. When planted thickly in clumps, they form compact masses of bloom

FIGURE 18-17   Pinks (*Dianthus*) are available in numerous shades of color and foliage types. (Photograph by K. A. Denisen.)

and foliage which are reminiscent of overstuffed pincushions. The flowers are either white or shades of blue to violet.

ZINNIA

Zinnias have been a challenge to both gardeners and plant breeders. As a result they are now available in dwarf, medium, and tall sizes; in numerous shades and hues; and in various flower forms. The plant is extremely heat tolerant and thrives under intense sunlight. Zinnias are very attractive in masses in a border, and also are frequently grown in rows for cutting.

A list of annual flowering plants, including those discussed, is given in Table 7.

HOUSE PLANTS

Growing plants indoors is the principal type of horticultural production in many homes. Apartment dwellers and families with little or no outdoor gardening area may derive pleasure from growing potted plants in windows or artificially lighted areas. For all families, regardless of the extent of outdoor grounds, plants in the home provide interior decoration either by their attractive foliage, flower production, or unique and grotesque habits of growth. Since growing plants have specific environmental requirements, it is of utmost importance that factors of the environment be given due consideration.

Environmental Considerations

Light is extremely important and may easily be the limiting factor for house plants during the winter months in cold climates. At this period of the year the sun is low in the heavens, days are short, and there are sometimes rather long periods of cloudy days. Some avid indoor gardeners take care of these situations with supplemental artificial lighting. This requires considerable candlepower (see Chapter 5), and as a result most plants are raised and grown by sunlight which filters through windows. Plants like geranium, cyclamen, and coleus, which require high light intensity, should be placed on the window sill or on a stand or table adjacent to the window. East and south windows in the northern hemisphere provide best light conditions for these plants during the winter months. West windows have a tendency to build up high temperatures in the direct rays of afternoon sun during the warmest outside temperature of the day. Light intensity can be reduced by drawing shades or placing in windows of the opposite exposure. Peperomia, sansevieria, philodendron, and English ivy do not require direct sunlight.

Temperature and temperature control play vital roles in growing house plants. High temperature is more frequently a problem than is low temperature. In general most house plants thrive at daytime temperatures below 70°F (21°C). Some plants have tolerance to high temperatures and these types should be used if rooms are consistently maintained at higher temperatures. Night temperatures are often lower

TABLE 7　　Some Annual Flowers and Their Botanical Names, Height, Spread, Color, and Use

Common name	Botanical name	Height, inches*	Spread, inches*	Color	Principal use
Ageratum	*Ageratum houstonianum*	6–20	10–12	blue	edging, border
Babysbreath	*Gypsophila elegans*	12–18	10–12	wh., red	border
Balsam	*Impatiens balsamina*	20–28	12–14	red, pk., wh.	border
Calendula	*Calendula officinalis*	14–18	8–10	yel., or.	border
Calliopsis	*Coreopsis* spp.	18–24	10–14	yel., red, br.	border
Candytuft	*Iberis affinis*	9–12	8–12	wh., pur.	edging
China-aster	*Callistephus chinensis*	12–24	10–12	various	border
Clarkia	*Eucharidium grandiflorum*	18–24	10–12	various	border
Cockscomb	*Celosia argentea*	16–40	10–12	red, yel.	border
Cornflower	*Centauria cyanus*	16–30	12–14	various	border
Cosmos	*Cosmos* spp.	30–48	10–12	wh., pk., yel.	border
Dahlia	*Dahlia* spp.	18–40	12–14	various	border
Forget-me-not	*Myosotis* spp.	8–12	10–12	blue, wh.	edging
Four-o'clock	*Mirabilis* spp.	20–24	12–14	various	border
Gaillardia	*Gaillardia* spp.	12–18	10–12	red, yel., or.	border
Globe-amaranth	*Gomphrena globosa*	18–24	10–12	red, wh., pur.	border
Godetia	*Godetia* spp.	18–24	12–14	various	border
Larkspur	*Delphinium* spp.	18–24	6–8	various	border
Lobelia	*Lobelia erinus*	6–8	6–8	wh., blue	edging
Lupine	*Lupinus* spp.	18–24	6–8	blue, red, wh.	border
Marigold	*Tagetes* spp.	6–36	8–16	yel., red, or.	edging, border
Nasturtium	*Tropoeoleum* spp.	10–12	8–12	yel., red, or.	edging
Pansy	*Viola tricolor*	6–10	6–8	various	edging
Petunia	*Petunia hybrida*	8–24	12–14	various	edging, beds
Phlox	*Phlox drummondi*	6–12	6–8	various	border
Pink	*Dianthus chinensis*	6–16	8–12	red, pk., wh.	edging
Poppy	*Papaver rhoeas*	12–16	6–10	red, yel., wh.	border
Portulaca	*Portulaca*	6–9	10–12	various	edging
Rudbeckia	*Echinacea* (*Rudbeckia*) spp.	20–24	10–14	yel., pur.	border
Salpiglossis	*Salpiglossis sinualta*	24–30	10–12	red, yel., wh.	border
Scabiosa	*Scabiosa atropurpurea*	18–36	12–14	various	border
Scarlet sage	*Salvia splendens*	14–36	8–12	scarlet	border
Snapdragon	*Antirrhinum* spp.	12–36	6–10	various	border
Spiderflower	*Cleome* spp.	30–48	12–14	wh., pk.	border
Stock	*Mathiola incana*	24–30	6–10	various	border
Strawflower	*Helichrysum bracteatum*	24–40	8–12	various	border
Sweet alyssum	*Alyssum*	6–10	10–12	wh., vi.	edging
Verbena	*Verbena*	9–12	10–14	various	edging
Zinnia	*Zinnia* spp.	12–36	8–12	various	border
Vining Flowers					
Morning glory	*Convolvus*	8–12 ft	20–36	red, blue, wh.	screening
Sweet pea	*Lathyrus odoratus*	4–8 ft	6–8	various	screening

* For *Height* and *Spread* in centimeters multiply inches by 2.5.

FIGURE 18-18 **Gardening can be done on a table top. *Left*, gloxinia; *right*, African violet.**

which is generally not harmful to most plants. If fluctuations are extreme, as encountered at windows during cold winter nights, it is advisable to move the plants to a warmer place, or protect with a paper shield.

Moisture can be a very critical factor in growing house plants both from the standpoint of humidity and soil moisture. The relative humidity of the atmosphere in many homes during winter months is generally low, perhaps down to 20 or 25 percent, while most greenhouses are in the 70 to 80 percent humidity range. Using a humidifier with the heating system is one method of making the atmosphere more favorable for plants. The immediate atmosphere of plants can be made more humid through the use of moistened sphagnum moss or wet gravel or rock placed among the pots on trays. Sprinkling the foliage with water at frequent intervals will reduce desiccation of leaves in a dry atmosphere. Jade plant, cactus, aloe, and screw pine are adapted to dry atmospheric conditions.

Air pollution is a result of escaped fuel gas, flue gas, and smog. Some house plants are susceptible to injury from these air pollutants. In the home, an extinguished pilot light or leak in the fuel pipes may cause only a slight escape of gas but can still cause plant injury. Flue gas is not generally as harmful as fuel gas since it is mostly oxidized; however, certain products of incomplete oxidation may escape due to faulty venting. Smog is a problem in many large cities.

Soils and nutrients may easily be the restricting environmental factors for house plants. Often it is desirable to restrict growth, since many plants tend to become too large. However, pruning either of roots or tops must accompany efforts at dwarfness. The plants must be kept in good health and appearance or restricting their size is not practical. Soils should have adequate organic matter for release of nutrients and to maintain good soil structure. Peat and compost are especially good sources of organic matter. Loam is a good basic soil for house plants, while sand is useful in opening heavy soils which compact readily. Fresh soil which has been sterilized is desirable for potting plants. Nutrients for potted plants are applied in very small amounts. There is danger in overfertilization. If the concentration of the soil solution becomes greater than that of the plant cells the moisture will actually go from plant to soil. Wilting and a water-soaked appearance of the leaves are indications of overfertilization. In general, one-half teaspoon of 10–10–10 fertilizer every two months per 6-in. (15-cm) pot is adequate. If it is observed that too many nutrients have been supplied, the plants should be repotted. Some root pruning may be necessary because of feeder root desiccation.

Planting and Care

Plant containers should have a means of drainage, such as a hole in the bottom of pots, otherwise extreme care must be taken to avoid waterlogging. Plastic or ceramic saucers and other shallow receptacles or jardinieres are employed for catching the gravitational water that seeps through the drainage opening. Painted or glazed pots prevent evaporation through the sides; as occurs with porous clay pots. Porous pots, conversely, provide better soil aeration.

Potting and repotting are important steps for growing successful, thrifty house plants. The hole at the base of the pot should be partially covered with a piece of broken pot, stone, or similar object to prevent loss of soil, yet providing drainage. Small plants should be placed in small pots. If greater size is desired, they can be repotted to larger containers after they grow. A small plant in a large pot has poorer root aeration. It is not capable of removing large amounts of water by absorption with its limited root system and low volume of transpiration. When potting a plant, the root system is evenly distributed in the pot, and the soil is applied around the roots and firmed gently. It is very important to leave sufficient space between the soil line and the top of the pot for watering; for a 4-in. (10-cm) pot, $\frac{1}{2}$ in. (1 cm) is adequate. After potting, the plant should be watered sufficiently to wet all the soil. Plants that become pot-bound should be repotted if a larger plant is desired or root-pruned if dwarfing or maintenance of size is preferred.

Watering is frequently overdone or quite often forgotten. Enough water should be applied at each watering so that moisture penetrates to

FIGURE 18-19   **Literally hundreds of plant species are satisfactory house plants. (Photograph by K. A. Denisen.)**

the bottom of the pot. The plant should not be watered again until the surface of the soil is dry to the touch. Adding water to a soil that is still moist and muddy from the previous watering will create a waterlogged condition. The soil should not be allowed to get crumbly dry except for those plants accustomed to near arid conditions. Foliage of pubescent plants should not get wet as it does not dry readily, thus favoring disease organisms. If the water contains large quantities of mineral salts, it may be desirable to use soft water. Plants susceptible to injury from wide fluctuations in temperature will do best if water applied is at near room temperature.

Pest control in house plants can usually be obtained by dipping the plants in a large container of appropriate spray material. Common insect problems include aphids, mealy bug, spider-mites, white fly, and scale insects. Fungus diseases can be avoided largely by humidity control, even watering, and pruning infected leaves. Insect control will usually prevent spread of virus diseases by insect vectors.

Selection of House Plants   Literally hundreds of plant species are satisfactory house plants because of their flowers, their foliage, succulent nature, or peculiar shape or habit of growth. Culture for some of these species is quite specific. For others, cultural conditions are less exacting and some

FIGURE 18-20   The amaryllis is a challenge to grow from a bulb both indoors and out-of-doors. (Photograph by K. A. Denisen.)

seemingly thrive on no attention. There seem to be about as many degrees of exactness or apparent neglect under which house plants thrive as there are types of growers. With this in mind, it is possible for everyone to grow some kind of house plant.

Flowering house plants include the bedding geranium, African violet, and everblooming begonia which bloom almost continuously; cyclamen, Martha Washington geranium, gloxinia, and poinsettia which bloom profusely and then require a rest period; amaryllis, narcissus, hyacinth, and tulip which are forced each year from bulbs; primrose, browallia, cineraria, and Jerusalem cherry which are grown from seed. Many of these types have attractive foliage in addition to their flowering habit. The African violet, gloxinia, and primrose require more organic matter in the potting mixture than do the other plant examples cited.

Foliage plants are of two general types, large or tall-growing and small or low-growing. The large or tall-growing types are often used in planters, jardinieres, or placed on the floor in pots. Plants typical of this group are snakeplant sansevieria, English ivy, rubber plant, diffenbachia, monstera, croton, fiddle-leaf fig, and Norfolk Island pine. The small or low-growing types include the foliage begonias, dwarf sansevierias, philodendron, peperomia, and jade plant. The foliage plants are generally easy to grow in soil of average fertility, aeration, and organic matter content. These are very popular and attractive foliage plants for the home and are able to tolerate occasional drouth conditions in the home. Ferns of various types are also used quite extensively. Their organic matter requirements are higher than for most other foliage plants.

Succulents are actually foliage types with modified leaves and stems. Many of them also flower. The cacti and many other cactus-like plants

FIGURE 18-21   **The Haageana begonia is a glossy foliage plant.**

FIGURE 18-22   **The aluminum plant resembles its namesake because of its shiny metallic appearance. (Photograph by K. A. Denisen.)**

FIGURE 18-23   **Succulents. *Left*, old man cactus and barrel cactus; *right*, Sedum.**

FIGURE 18-24 The mimosa, or sensitive plant. *Left,* before center leaf was pinched. *Right,* a few seconds after pinching. Leaves recover turgidity in ten or fifteen minutes if unmolested.

are included in this group. Echeveria, the stonecrops (*Sedum* spp.), bryophyllum, and aloe are typical succulents. Most succulents prefer a sandy loam soil and infrequent watering. They are generally slow growing, have a low rate of transpiration, and will grow for long periods before becoming pot-bound. The cacti thrive under warm conditions and may be placed out-of-doors in full sun during the summer months. Some cacti have very attractive flowers.

Exotic plants or those with peculiar or interesting habits are often adapted as house plants. The mimosa, or sensitive plant, reacts to touch

FIGURE 18-25 Terrariums (*left*) and hanging plants (*right*) add interesting dimensions to growing plants in the home. (Photographs by K. A. Denisen.)

Ornamental Herbaceous Plants

and physical shock by folding its leaflets and collapsing the base of its petiole so that in a few seconds it has the appearance of a wilted or dying plant. It recovers turgidity several minutes after cessation of the stimulus. The pitcher plant and venus fly-trap are insectivorous plants that are capable of trapping insects and digesting and assimilating them. The airplane plant propagates itself by long aerial stolons terminated by new plants. they are easily imagined as jet planes with vapor trails. Screw pine or pandanus tree has large aerial brace roots that seem to "lift" the plant out of the soil. The shrimp plant has flowers which have the appearance of cooked shrimp ready for dipping into sauce. Crown of thorns is woody and extremely thorny with small red blooms. It is said to be the plant from which the corwn of thorns was made for Christ on Calvary. Lady's slipper and cigar plant have flowers reminiscent of their namesakes. Orchids attract attention not only because of their beauty but also because of their extremely interesting culture. They live solely on organic matter media and moisture.

STUDY QUESTIONS

1. Why are seed-propagated plants usually started indoors in flats or pots?
2. Contrast the uses of borders and beds of flowers. If you plan to use flowers for cutting where should they be located?
3. How deep are tulip bulbs planted compared to crocus corms? Why?
4. Why do petunias that have seeded themselves from last year's bed so often prove a disappointment to the gardener? What is the genetic basis for this?
5. Why are annual flowers so often planted over a tulip bed?
6. If you were to start a hobby of growing flowers, which one would you select? Why? What are your second and third choices? Why?
7. How much of growing perennials is due to current care and maintenance and how much is due to their previous attention? Explain.
8. The hollyhock is usually a biennial. Why do gardeners continue to have them year after year without replanting? Do the colors and types change during that period?
9. What are the specific advantages of tall plants like delphinium and shorter plants like ageratum? Can they be placed in the same planting? Why?
10. The ferns are classed as "non-flowering." How do they reproduce?
11. The petunia and snapdragon are very tiny-seeded. How do they survive in nature?
12. What are the advantages of having house plants? Disadvantages?
13. What type should be grown where there is a dry atmosphere in a house? What modifications can be made for plants to be grown?
14. Give examples of exotic plants for the home. What special care might be needed?

SELECTED REFERENCES

1. Bailey Hortorium Staff. 1977. *Hortus Third.* Macmillan, New York.
2. Extension bulletins: Most state extension services have these bulletins which are frequently found locally in offices of county agents, farm advisers, or extension directors. Most of them are quite current.
3. Laurie, Alex, D. C. Kiplinger, and K. S. Nelson. 1968. *Commercial Flower Forcing*, 7th ed. McGraw-Hill, New York.
4. Seed and Nursery Catalogs of local or nationally known nurseries are an excellent source of information on kinds, types, and cultivars of herbaceous ornamentals. Some may have local Garden Centers which can supply both plants and information.
5. Taber, Henry G. 1977. *Small Plot Gardening.* Pm 720. Cooperative Extension Service, Iowa State University, Ames, Iowa.
6. Wayside Gardens Catalog. Wayside Gardens, Paynesville, Ohio. This is for sale only. In addition to being a catalog, it is an excellent source of information on culture, care, recommended areas, etc. of herbaceous ornamentals.

19 The Vegetable Garden

A characteristic feature of many homes is the family vegetable garden. It receives varying degrees of emphasis in response to local, national, and international crises. Wars and economic depressions tend to increase food production in the home, an expression or mobilization of the forces at hand. Inflation and loss of buying power tend to increase the trend toward vegetable production in home gardens. The most stable group in home vegetable production is the family which gardens both during and between crises of local, national, and international scope. This is the group which gardens not only to save money or aid national effort but because they aspire to high quality vegetables and take pride in the products of their toil. To them it becomes an annual family enterprise, an institution, an integral part of their family life.

The development and management of a vegetable garden provides a wealth of individual experience, learning, and appreciation. A small plot can be as effective as a large farm in unfurling the mysteries of soil and Nature. Gardening is principally an art developed through experience or based on scientific findings and governed by economic forces of production and consumption. A true gardener becomes a philosopher. He plants his seed, places his trust in the soil and the weather, works for his plants by weeding and spraying, and logically expects to reap the harvest. In the face of environmental adversity he will adjust his cultural techniques and benefit from the experience.

SELECTING THE SITE

Site and location of the home vegetable garden should receive careful consideration. It is most desirable to have the vegetable garden on the home grounds or farmstead. However, this is not always possible, and in urban areas various factors may decree its impracticability on the home lot. It may be necessary to rent a vacant lot or a plot in a larger area devoted to community gardens.

Good Soil   The best available soil in the home grounds or a rented plot is a basic requirement for the most productive garden. Areas which have been used to deposit the subsoil from basement excavation are not good sites for gardens. Vegetables thrive in soils which are not only high in available nutrients but high in organic matter. Structure, water holding capacity, and aeration play important roles in vegetable root develop-

FIGURE 19-1   **Gardening is fascinating.**

ment and ultimate productivity. The inherent fertility, soil pH, and depth of topsoil are factors to consider in locating the vegetable garden. In a farm garden, rotation with field crops and soil building crops contributes to continued productivity from year to year.

Topography   The slope of the land plays an important role in site selection because it influences erosion, irrigation practices, and surface drainage. Waterlogged areas, frost pockets, and steep slopes are not good areas for a vegetable garden. Direction of slope influences earliness of crops in temperate climates. Soil temperatures will rise fastest on slopes offering greatest exposure to the sun in the spring.

Leveling or filling with topsoil may be practical in modifying the topography of gardens on the home grounds. Terraces or retaining walls can be used in sloping areas to develop a permanent site for vegetable gardens.

Competitive Factors Trees, shrubs, and buildings located too close to the garden may reduce the amount of light sufficiently to make it a limiting factor in growth and production. Not only is light restricted by the proximity of trees, shrubs, and other plants, but competition for moisture and nutrients becomes a problem. The garden should be located so it receives at least six hours of full sun each day and is not undermined with the extensive roots of large trees, adjacent vines, or shrubs.

Perennial weeds can be extremely competitive. When renting a plot of ground for a garden, it is important to avoid areas infested with perennial weeds such as bindweed, leafy spurge, Canada thistle, and other perennial broadleafed and grassy weeds. It often takes a year or more of intensive effort to rid the garden of one of these weed pests. For a garden in the home grounds, every effort should be made to eliminate the weed problem as soon as possible. It may be desirable to sacrifice one year's crop to facilitate more rapid eradication of persistent weeds. Herbicides with soil "sterilant" properties are helpful in ridding the soil of these pests, but they render the soil toxic for prolonged periods.

Accessibility A garden is useful if easily available to the gardener and his family. When the principle purpose of the garden is to supply vegetables for the home, its usefulness is determined in large part by frequency of visits. It must be accessible if vegetables are to be harvested at their prime and when they are needed. Accessibility is often the greatest disadvantage of a garden away from the home grounds, whether it be in a remote part of a farm or "across town" in an urban area. If a principal purpose is for outdoor exercise, activity, or avocation, it is also vital that the family have easy access to the garden. When machine cultivation is contemplated, even occasionally, accessibility of entrance for the machines is a vital consideration. The annual plowing operation usually necessitates bringing in fairly heavy equipment. If lawn, borders, trees, or fences are in the path of such equipment, severe damage can result.

Some areas are entirely dependent on irrigation for vegetable production and many areas will benefit from supplemental irrigation in most years. Consequently, accessibility to water may be an extremely important consideration. The cost of water is likewise of concern since vegetable gardens are expected to give a return in yields which is in excess of production costs.

PLANNING THE GARDEN

Time and thought expended in planning the garden are usually well invested. Maximum utilization of space, time, materials, and talents results from planning.

Size It is important to be realistic in the planning stage. Too extensive a garden area for available help and facilities represents wasted time, effort, and talents and may dull enthusiasm for gardening. Some gardens are planted with good intentions but never reach the harvest because of neglect. A garden with size based on needs, anticipated production, and

FIGURE 19-2 **Flowers in the vegetable garden give a splash of color and provide a harvest of cut flowers. (Photograph by K. A. Denisen.)**

time available will usually result in a well planned, efficient producing home enterprise.

Family requirements have a direct relationship to garden size. A large family can produce sufficient vegetables to greatly reduce their food budget with an adequate sized garden. This reduction in food costs is not limited to the producing months or harvest period, because with canning, freezing, and adequate storage, the period of vegetable consump-

FIGURE 19-3   *Upper*, a community garden. *Lower*, watermelons require considerable space which makes them more adaptable to larger gardens and commercial plantings. (Photographs by K. A. Denisen.)

FIGURE 19-4   Young members of the family can acquire a wealth of experience and pleasure from participation in gardening. (Photograph by K. A. Denisen.)

tion can be spread over a considerable portion of the year. Smaller families require smaller quantities of vegetables, and thus, size of family is shown to be very instrumental in influencing garden size. Where children are large enough to participate in the planning and care of the garden, it can become a larger enterprise. It may even be expanded to an income project with marketing of excess production. This can offer greater incentive for youthful participation.

Vegetable consumption within the family and between families varies considerably. The amount of physical exertion by working members of the family, their nutritional requirements for a good diet, and variety in foods are considerations for determining size of garden for a given size family. A 20 by 40 foot (approx. 6.5×13.0 meters) garden will supply many of the vegetable needs for a family of 4 or 5 during the summer months. A larger garden will be needed if more vegetables are desired for canning or freezing. Some people prefer a 50 by 100 foot (15×30 meters) garden and others may prefer up to 100 by 200 feet (30×60 meters) with the idea of doing considerable canning and freezing and producing an abundant variety of vegetables. A storage cellar for vegetables has a tremendous influence on garden size. Potatoes, root crops, onions, pumpkins, squash, and cabbage can be stored in the fresh state for several months to prolong their period of availability (see Table 3, Chapter 12).

Kinds of Vegetables   *Cool season* vegetables include those planted earliest in the spring. Onions, lettuce, peas, and radishes are among the first seeded and earliest harvested vegetables in the out-of-doors. However, cabbage,

FIGURE 19-5   Cool season perennial vegetables. Seedstalks arising from rhubarb at left should be removed to direct manufactured food to crown storage and petiole growth. Asparagus is harvested only during the spring months, then spends the remainder of the season building up food reserves for next spring's production of spears.

cauliflower, broccoli, and head lettuce may be started in forcing structures two to six weeks in advance of outdoor planting dates. With proper hardening, cabbage and related vegetables (cole crops) will withstand temperatures below freezing and often as low as 25°F (−4°C). Transplants of these crops will produce an early harvest of good quality vegetables which are favored by the cool weather of early spring. In areas of hot summers, setting of transplants in early spring is about the only means of producing high quality head lettuce, cauliflower, and broccoli.

Other vegetables which flourish under cool conditions, but are less frost tolerant, include spinach, beet, carrot, potato and turnip. Some cool season types like Chinese cabbage, celery, winter radish, rutabaga (Swede turnip), and parsnip are started in the summer for maximum growth during the cooler fall months. The perennial cool season vegetables, rhubarb and asparagus, are productive of succulent, crisp stalks during cool months of spring. The later growth during the warm temperatures of summer is less edible. The leaves are permitted to produce food for storage in the crown for the next season's production. In the family garden, perennial vegetables are usually located along one side to facilitate plowing operations without injury to the crowns.

Warm season vegetables may be started indoors to get earlier production and to lengthen the growing season. This is a common practice with tomatoes, peppers, and eggplants. The cucurbit family which includes cucumbers, pumpkins, squash, watermelons, and cantaloup may be transplanted but require more exactness and care. Since their root systems become suberized or hardened at an early age they do not readily form branch roots when injured in transplanting. For this reason, cucurbits to be transplanted are sowed directly into pots or bands which may be removed with a minimum of injury to the root systems.

Other warm season vegetables are generally seeded directly into the garden soil at their permanent location. Sweet corn, snap beans, and tomatoes are often planted before danger of frost is past. A frost before emergence will not harm them. Should they be killed by a late frost, they can still be replanted and produce a good crop for the season. Many gardeners feel it is worthwhile to plant these warm-season crops before all likelihood of frost is past, since, in more cases than not, the frost will not occur and they will have several days advantage in earliness. When the weather is cold and generally unfavorable for warm season crops, it is better to await more favorable conditions. Warm season vegetables planted after danger of frost is passed include the cucurbits, lima beans, okra, and sweet potatoes. Not only are they harmed by light frost, but they grow slowly and may become stunted under cool weather conditions.

New vegetables which the family may include are both cool and warm season kinds. Since there are so many kinds of vegetables, it is rare for a family or its individuals to be acquainted with them all. New introductions from foreign lands, old vegetables with new uses and methods of preparation, and nutritional appeals for wider use of certain kinds all suggest acquaintance with unfamiliar vegetables. Some families adopt a plan of trying at least one new or unfamiliar vegetable each year to extend their knowledge and experience in vegetable production and consumption. Some less common vegetables which are familiar to some people but new to others include husk tomato (ground cherry), kohlrabi, kale, chard, eggplant, okra, Brussels sprouts, salsify, vegetable papaya,

FIGURE 19-6   *Left*, kohlrabi, a cool season vegetable with a thick, fleshy, foreshortened stem. *Right*, eggplant, a warm season vegetable with a large fruit. (Photographs by K. A. Denisen.)

endive, fennel, leek, and globe artichoke. It is important to follow proper or accepted methods of preparation or a fair trial may not be given new vegetables.

Successive, Succession and Companion Crops

Successive plantings are an attempt to extend and prolong the harvest season. They may be effective for many vegetables but are especially adapted to sweet corn, snap beans, radishes, peas, cabbage, and lettuce. These crops are harvested within a relatively brief period of time. A single planting of each may fall far short of the desired period of fresh vegetable consumption for the family, but may produce more than can be handled at one time. This situation is adjusted by either using cultivars which differ in days to maturity or by planting on successive dates or with combinations of both. Sweet corn hybrids and cultivars range from 65 to 110 days from planting to edible stage. In a sufficiently long growing season, an array of several carefully selected cultivars and hybrids could supply roasting ears over a long harvest period. The period of harvest could be further extended by making successive plantings at weekly to 10-day intervals of one of the more productive midseason to late cultivars. There is generally less interval between harvests than between planting dates. The higher summer temperatures hasten growth and maturity for the later plantings of the successive crops. Climatic conditions have a tremendous influence on the extent to which successive crops can be continued. In a relatively cool climate, suc-

FIGURE 19-7   **Successive plantings of sweet corn distribute the harvest season over a longer period. Four sizes are found in this garden.**

Horticulture for the Home

cessive crops of lettuce, peas, and radishes can be harvested over a long period while they are high in quality. Areas with hot summers are limited to spring and fall harvests for successive crops of cool season vegetables. Peas are not a desirable fall crop because of their response to photoperiod; long days favor pea production.

Succession plantings enable greater utilization of space available to the gardener. They consist of following one vegetable with one or more vegetables in the same row or rows during the same season. It can be called double or triple cropping. Short-term vegetables are best adapted to succession cropping although an area with an extremely long growing season can often utilize two moderately long-season crops during the same year. Combinations commonly used consist of cool season vegetables like radishes, lettuce, green onions, or spinach followed by snap beans, tomatoes, cucurbits, or late planted sweet corn. In late summer or fall, cool season crops may follow warm season crops. Turnips, Chinese cabbage, radishes, and lettuce can be planted after snap beans, early sweet corn, and cucumbers. The gardener with a very limited area for production can make very effective use of succession crops for maximum vegetable production.

Companion cropping is a means of getting more production from a given area. Two crops are grown at the same time in an area designated for one of the crops. It is essential to avoid competition of either crop with the other. One of the vegetables is usually a short-season crop. Radishes, lettuce, and green onions may be planted between the widely spaced rows of small tomato plants. These short-season crops will be harvested and removed before the tomatoes have grown sufficiently to compete for space, moisture, or nutrients. During the first year of an asparagus or rhubarb planting, companion crops, even of long duration,

FIGURE 19-8   **Companion cropping can be utilized successfully if neither vegetable competes with the other. (Photograph by K. A. Denisen.)**

can be grown successfully. Snap beans, lima beans, onions, and carrots will utilize the space without injury to the developing asparagus or rhubarb. It should be emphasized that crops grown in competition with each other are not considered desirable companion crops.

Arrangement of Vegetables in the Garden
A plat or sketch of the garden is very useful in planning the location of vegetables. It can indicate cultivars, planting dates, and successive crops. This sketch if started well in advance of the growing season will give adequate time for study and development. Planning is especially important when space is at a premium.

Direction of rows is not of great concern from the standpoint of vegetable growth. However, it is more economical and efficient for wheel and machine equipment when rows are parallel to the long axis of the garden. It means less turning space is required. Gardens on slopes should utilize good soil management practices by placing rows on the contour to avoid or reduce erosion.

Spacing between rows and within rows varies greatly because of the diverse nature of the many kinds of vegetables. Some crops, e.g., radishes, onions, carrots, and leaf lettuce, produce most abundantly at row spacings of 12 to 18 in. (30–40 cm), while melons, pumpkins, and squash may require from 4 to 10 feet ($1\frac{1}{4}$–3 meters) between rows. The gardener who needs to utilize his area efficiently must pay particular attention to spacing between rows both with the same kind of vegetable and with different kinds of vegetables. If machine cultivation is used, the row spacings will need to be more uniform to allow for machine widths, wheel spread, and cultivator shovels.

Type of vegetable and nature of growth influence location and arrangement in the garden. Perennials are best located along one side of the garden. Extremely tall plants like some sweet corn varieties and hybrids should be located so they do not shade smaller plants. The usual recommended spacing distances do not entirely compensate for this type of competition. Squash, pumpkins, and other vine crops will often trail considerably. They may offer severe competition to low growing adjacent crops but do not generally interfere with sweet corn production, since most trailing occurs late in the season.

PLANTING THE GARDEN

Soil Preparation and Fertilization
Plowing or spading in the fall incorporates organic matter before winter and generally improves soil structure. Alternate freezing and thawing tend to form soil aggregates which improve aeration and increase the rate of water penetration. In many cases, spring plowing is necessary because of late harvested crops the previous fall. Plowing and spading should be at a depth of 8 to 10 in. (20–25 cm).

A *fine seedbed* is required for vegetable gardens. This does not mean that soil aggregation is not desired. It does mean that there should be no large clods of soil and should provide for intimate contact of soil particles, moisture, air, and seeds. Many vegetable seeds are quite small. The

FIGURE 19-9   Onions are spaced carefully when seeded and thinned if necessary later. (Photograph by K. A. Denisen.)

smaller the seed, the finer the seedbed should be. Rotted manure and compost worked into the soil helps in seedbed preparation. Chemical fertilizers are applied, at least in part, during soil preparation. Most gardeners prefer a general broadcast application prior to final soil preparation. Another method is to place fertilizer in a band or trench alongside each row. It is often economical to have the soil tested to

TABLE 8　　Vegetable Planting Guide. Ranges of Figures are Due to Differences in Cultivar Characteristics. Amounts of Row for Family of Four are Based on Average Consumption Throughout the Harvest Season.

Vegetable	Warm or cool season	Days to edible stage (days)	Amt. of row for family of 4 (ft.)	(m)	Requirement 100 ft. row — Seeds (oz.)	(g)	Plants (no.)	In row (in.)	(cm)	Between rows (in.)	(cm)	Planting depth (in.)	(cm)
Asparagus	C	2–3 yrs.	50	15			50–65	18–24	45	36–48	100	8	20
Beans, lima	W	45–65	50	15	16	450		4–6	10	18–24	50	$1\frac{1}{2}$	3
Beans, snap (bush)	W	45–65	80	24	16	450		3–4	8	18–24	50	$1\frac{1}{2}$	3
Beans, snap (pole)	W	50–70	40	12	8	225		6–8	12	30–36	70	$1\frac{1}{2}$	3
Beets	C	60–100	40	12	2	57		2–4	6	12–18	40	$\frac{3}{4}$	2
Broccoli	C	60–90	20	6	$\frac{1}{4}$	7	50–65	18–24	45	30–36	70	$\frac{1}{4}$	$\frac{1}{2}$
Brussels sprouts	C	140–150	20	6	$\frac{1}{4}$	7	50–65	18–24	45	24–30	60	$\frac{1}{4}$	$\frac{1}{2}$
Cabbage	C	65–120	30	9	$\frac{1}{4}$	7	40–65	18–30	50	24–36	65	$\frac{1}{4}$	$\frac{1}{2}$
Cabbage, Chinese	C	80–100	20	6	$\frac{1}{4}$	7	70–100	12–18	40	20–30	60	$\frac{1}{4}$	$\frac{1}{2}$
Carrots	C	60–100	50	15	$\frac{1}{2}$	14		1–3	5	12–18	40	$\frac{1}{2}$	1
Cauliflower	C	80–100	20	6	$\frac{1}{4}$	7	50–65	18–24	45	24–30	60	$\frac{1}{4}$	$\frac{1}{2}$
Celery	C	100–130	30	9	$\frac{1}{4}$	7	200	5–6	10	20–30	60	$\frac{1}{8}$	$\frac{1}{4}$
Chard	C	60–75	25	8	1	28		6–8	12	15–18	40	$\frac{1}{2}$	1
Corn, sweet	W	65–110	300	90	3	85		12–15	30	30–36	70	$1\frac{1}{2}$	3
Cucumbers	W	60–90	30	9	$\frac{1}{2}$	14		18–24	45	48–60	140	1	$2\frac{1}{2}$
Eggplant	W	120–130	15	5	$\frac{1}{4}$	7	50–65	18–24	45	24–36	65	$\frac{1}{2}$	1
Endive	C	80–100	15	5	$\frac{1}{2}$	14		6–12	15	18–24	50	$\frac{1}{2}$	1
Kale	C	60–70	15	5	$\frac{1}{4}$	7		12–15	30	18–24	50	$\frac{1}{2}$	1

Kohlrabi	C	60–70	25	8	¼	7		6–8	12	18–24	50	½	1
Lettuce	C	40–60	40	12	¼	7		4–6	10	18–24	50	¼	½
Muskmelon	W	90–120	25	8	1	28		24–36	65	48–60	140	1	2½
Okra	W	55–110	20	6	1	28		18–24	45	24–36	65	1	2½
Onion, seed	C	100–150	80	24	1	28		2–3	5	12–18	40	½	1
Onion, sets	C	40–60	30	9			2 lb. sets	2–3	5	12–18	40	1	2½
Parsley	C	80–100	5	2	¼	7		3–5	8	12–18	40	⅛	¼
Parsnips	C	120–150	40	12	½	14		3–4	8	12–18	40	½	1
Peas	C	60–120	200	60	16	450		2–3	5	18–24	50	1½	3
Peppers	W	90–140	20	6	¼		50–60	18–24	45	24–30	60	¼	½
Potatoes, white	C	100–140	300	90	8 lb.	4 kg		12–15	30	30–36	70	4	10
Potatoes, sweet	W	140–160	80	24			60–80	12–15	30	30–36	70	4	10
Pumpkins	W	90–120	15	5	½	14		5–7 ft	200	8–12 ft	300	1	2½
Radish	C	30–60	40	12	1	28		1–1½	3	12–15	40	½	1
Rhubarb	C	1 yr	15	5			25–30	36–48	110	36–48	100	3	8
Rutabagas	C	80–100	20	6	½	14		4–8	10	18–24	50	½	1
Salsify	C	140–160	20	6	1	28		1–2	4	12–15	40	½	1
Spinach	C	40–70	30	9	1	28		3–4	8	12–18	40	½	1
Squash, summer	W	60–75	30	9	½	14		36–48	100	36–48	100	1½	3
Squash, winter	W	100–140	50	15	½	14		5–7 ft	200	6–10 ft	250	1½	3
Tomatoes	W	100–120	100	30	⅛	4	25–30	36–48	100	36–48	100	¼	½
Turnips	C	50–70	30	9	½	14		3–4	8	18–24	50	½	1
Watermelon	W	90–130	50	15	1	28		5–7 ft	200	6–10 ft	250	1	2½

determine specific nutrient needs. Gardens are sometimes over-fertilized. This means wasted nutrients by leaching and may even cause excess vegetative growth at the expense of fruitfulness.

Avoiding compaction of the garden soil is important to best root development and moisture penetration. Too frequent or excessive working of the soil prior to planting, excessive foot traffic, working while wet, and heavy machines are the chief causes of soil compaction. Incorporation of organic matter is one of the effective methods of renovating a severely compacted soil.

Seeding

Date of seeding out-of-doors varies widely from one location to another and for the different kinds of vegetables. The hardiest cool season crops can be planted early in the spring when the soil can be properly worked and prepared. This means the soil should be sufficiently dry after spring thaws or rains so that it does not become sticky or lumpy during soil preparation. Light frosts after emergence do not generally kill cool season vegetables and they germinate under cool soil conditions. As temperatures rise during spring, the other vegetables are planted in sequence with successive plantings of each vegetable determined by time interval. Calendar dates are only a general guide for planting vegetable seeds, since seasons vary from one area to another and from year to year. For many of the cool season crops, delayed planting may mean drastic reductions in yield when warm weather begins. It may be virtually impossible to grow good crops of some cool season vegetables in tropical areas.

Sowing the seed is done by hand or by simple machine. Suggestions are given for depth of planting in Table 8. Most seed packets list depth of planting, row spacing, and other pertinent information. When sowing by hand, a trench is made with a hoe for large-seeded vegetables or the hoe handle for fine-seeded kinds. The seeds are distributed evenly in the trench at rates that approximate the desired stand of plants. It is better to plant too many than too few seeds, since they can be thinned later. The seeds are then covered with soil from each side of the trench. It should be firmed slightly to promote intimate contact of seed and soil. If the family garden is large, it may be practical to use a mechanical seed drill. This drill is pushed by hand and seed is placed in a hopper for uniform distribution and depth of planting. The drill is adjustable for seed size, rate, and depth of planting. In a large garden, an investment in such a machine will amortize itself in savings of seed, reduced time required for planting, and almost entire elimination of the thinning process. If there is danger of the soil baking or forming a hard crust after seeding, a light application of compost, peat, or decomposed manure will keep the soil in good tilth.

Transplanting

Potted or band-grown plants are transplanted with little disturbance of their root system. The trench or hole should be deep enough for the transplants to be at the same or a slightly deeper level than when growing in the pot. It is a good practice to sprinkle some soil over the

FIGURE 19-10
Vegetables that require careful transplanting. *Left*: Cucumber plants seeded directly in pots are set out with soil ball intact. *Right*: Asparagus crowns are planted about eight inches deep on mounds of soil. They are covered gradually during the first season as growth from crown buds develops. (Right photograph courtesy of *Better Homes & Gardens* magazine.)

top of the soil ball to reduce drying. Soil is gently firmed around the soil ball for early establishment of roots in their new location. With all pot-grown plants, the method of inverting the pot and plant to free the soil ball results in little if any injury to the transplants (Fig. 6-13, page 151). The use of starter solutions is appropriate following transplanting.

Flat or bed grown plants receive a greater shock at transplanting. With no barrier for isolating each root system, the roots will intertwine. Separation of plants prior to transplanting results in loss of roots to a considerable degree. For this reason it may be desirable to remove some of the lower leaves of tomato, eggplant, and pepper plants to reduce wilting. Since flat or bed grown plants are more crowded prior to transplanting, they are apt to be leggy and weak stemmed. When this is the case, they can be set considerably deeper than their previous level. Tomato plants may be placed horizontally in a trench with the tip upright. Adventitious roots arise from the buried portion of the stem, transpiration is reduced, the plant grows more erect, and it recovers from the shock of transplanting more quickly. Water or starter solution is especially helpful in establishing these plants which have lost a large portion of their root system.

CARE AND MANAGEMENT

Unless seeding and transplanting are followed by good care and intelligent management, the earlier efforts are of little avail. The garden will require attention several times each week both for the cultural operations, harvests, and observations. The garden visits are not only for work but should allow time for enjoying the developing wonders of Nature under man's guidance. It is during these periods of observing and showing when many gardeners reap their greatest satisfactions growing plants, producing food, and mastering the art of gardening.

Cultivation   *Weed control* is a principal reason for cultivation. The competitive effect of weeds can have a very retarding influence on growth of vegetables. The first emerging weeds often have a head start on the seedling vegetables. This is logical, since they were seeded earlier. Removal of early weeds by hoeing, cultivation, or herbicides (pre-emergence to the crop) gives the garden vegetables a size advantage over future weeds. However, the advantage may be temporary and additional weed control is usually needed. Weeding within the row is essential to thrifty growth of vegetables. Once the row has been rid of weeds, the growth of vegetable foliage and competitive characteristics of the plants prevent serious reinfestations of weeds in the row. One advantage of close spacing the rows within recommended distances is the smothering effect of vegetable foliage upon late germinating weeds.

Other effects of cultivation are applicable to garden management. Improved water penetration will result following cultivation of a sun-baked soil. A cool soil when cultivated will warm faster and increase the rate of nitrification. Organic matter is covered or partially covered during cultivation which hastens its decomposition and nutrient availability. It is always important to avoid compaction, a soil evil which means loss of good structure. Compaction accompanies heavy traffic, cultivation of wet soil, and intense pulverizing.

Side-Dressing   Nutrients applied at seeding time may leach before the absorptive capacity of the plants is at an effective level. Therefore many gardeners make more efficient use of fertilizers by split-applications. Approximately one half the nutrient needs are supplied at planting time and the remainder about midway through the growing season of the vegetable concerned. This is most feasible for long season crops, those requiring more than 60 days from planting to harvest.

Fertilizers applied by hand in bands on either side of the row are usually placed in a shallow trench. In large gardens where garden tractors or other machines are used, a special side-dressing attachment may be used to apply fertilizer at "lay-by," the last cultivation before foliage restricts machine traffic. Special wheel fertilizer distributors pushed by hand are available either for surface or soil application. Tomatoes, sweet corn, cucumbers, and melons usually respond to side-dressing treatments.

Pest Control   No garden is immune to the attacks of pests. Weeds have been discussed, but insects, diseases, and nematodes take their annual toll from the family garden. There may be a wide range of pests and ills which attack vegetable plants in the small garden as in the large fields of a commercial grower. In fact, it may be easier for the large grower to control some of these pests, since he has the facilities for meeting both usual and emergency attacks. The commercial grower usually specializes in a few specific vegetables, while the family gardener may have 20 or more different vegetables. There has been a definite trend supported both by gardeners and pesticide formulators to combine various insecticides and fungicides in "general purpose," "all-purpose," or "one-pack" mixtures. They occur as either sprays or dusts and have a broad range of effectiveness. In making selections of this type of pesticide, one should note on the label that there are at least two insecticides and two fungicides. These should each have a broad range of pest control without considerable overlap. They should be low in toxicity to humans. Mixtures of this type are more expensive for the control of a specific pest; although specific control of each pest entails having numerous infrequently used materials on hand. Some gardeners prefer to make their own dust and spray mixtures. The gardener should be aware that there are *other means* of controlling pests in addition to chemicals. Good sanitation practices, roguing of infected plants, rotation within the garden, burning of pest ridden tissues, traps and barriers, and good growing conditions all play an important role in maintaining a pest-free garden.

Insect problems include many of the foliage- or tissue-eating types. Cabbage worm, corn ear worm, tomato hornworm, stalk and vine borers, cutworms, and grasshoppers are voracious eaters. Stomach poisons are effective on certain of these, but contact insecticides which are of lower human toxicity are generally replacing them for use on vegetables. Lindane and methoxychlor, either in an all-purpose spray or dust, or used individually where effective, are commonly used in the vegetable garden. Sucking insects like aphids, thrips, squash bug, and leafhopper are also susceptible to one or more of these insecticides. Malathion may be included for sucking insects because of its wide range of effectiveness and, in addition, its control of red spider and other mites. State or federal recommendations should be consulted and observed for most recent specific control methods.

Diseases of vegetables which attack foliage and fruits can generally be controlled by the fungicides captan, elemental sulfur, fixed copper compounds, manzate, and parzate. The most recent recommendations should be consulted and precautions regarding residues on edible parts should be closely observed. Plants having virus diseases should be rogued and burned to prevent further spread. Controlling the insect vector helps in preventing spread and may exclude viruses from their usual host crops. Rotted vegetables should be destroyed to reduce bacterial infection of other plants. Plant remains used for composting should not be used indiscriminately. Those known to harbor disease should be burned. Other stem and leaf tissues should be kept for a year or more in the compost pile or pit for more thorough decomposition and destruction of plant pathogens. Even so, seed treatment chemicals should be used when planting seeds in a soil which is fortified with compost. Resistant cultivars are effective for avoiding many plant diseases in the garden.

Nematodes can become a serious problem in gardens that are kept in the same location year after year. They are generally more serious in areas of mild winters where there is little or no freezing of the soil. This does not mean nematodes are not a problem in regions of severely cold winters. Some types are exceptionally hardy and may winter over in roots of perennial plants or in cysts in the soil. Crop rotation is helpful for combating nematodes, but the greatest weapon seems to be soil disinfectants applied prior to planting. Ethylene dibromide, methyl bromide, and vapam are effectively used in this manner. Very special precautions should be taken when working with these chemicals because they are gaseous and quite toxic to humans.

Special Techniques in Care and Management

Pruning and staking of tomatoes for increased earliness, cleaner fruits, and adaptability to small areas are commonly practiced. The stake is driven in the soil near the base of the plant and only the main stem is retained. All axillary shoots are removed as they appear. The top may be pruned off when the maximum desirable height is reached. There is less likelihood of diseased fruits since they are off the ground. However, cultivars with sparse foliage may have increased sunburn of fruits.

FIGURE 19-11   *Left,* pole beans supported on laths. *Right,* okra pods ready for harvesting. (Photographs by K. A. Denisen.)

FIGURE 19-12   **Two cultivars of peas given an opportunity to vine on string trellis. Cultivar shown at the top has larger pods not yet filled, whereas the one shown at the bottom has plump pods ready for harvest. (Photographs by K. A. Denisen.)**

Thinning operations in some of the root crops can be simplified. Carrot and radish seeds may be mixed together when planted. The radishes mature much faster and when pulled leave adequate space for carrot root development with very little additional thinning. Beet seeds contain multiple embryos. Thinning can be delayed until young beets can be harvested for greens. Judicious pulling of alternate plants or groups of plants for greens can be the thinning process for those left to produce large beets.

Forcing rhubarb for succulent, crisp stalks during winter months can be accomplished in a cool basement or cellar. Crowns are dug in the fall, placed in a pile with mulch covering to prevent drying, and subjected to two to four weeks of winter cold. They are then placed on the cellar floor with organic materials for maintaining moisture content and high humidity. A semidarkened room is desirable for it prevents large leaf development at the tips of the thick petioles. Forced rhubarb is a delicacy in the winter months. The crowns may be replanted outdoors in the spring but should be given a season to recover before the next harvest.

Beds of perennial or biennial vegetables may be practical for producing some of the novel garnishes like water cress, parsley, and chive,

FIGURE 19-13   *Upper*, two cruciferous vegetables. *Left*, cauliflower. *Right*, savoy cabbage. *Lower left*, pickling cucumbers; *lower right*, bell pepper. (Photographs by K. A. Denisen.)

or for seasoning herbs like horseradish, mint, sage, and others. Parsley is biennial but if winter mulched will seed itself and after two years can be a continuous source of parsley leaves if weeds are controlled. Chive and the mints as vegetable ornamentals are sometimes grown in flower borders. Water cress is adapted to shady and moist places. Horseradish is dug at harvest but small sections of rootstock should be left for regeneration of the bed. Leaves of sage, mint, and thyme may be harvested with little deleterious effect on these perennials.

Blanching of celery and asparagus is usually accomplished by ridging with soil. Asparagus is covered before emergence to produce white spears. Cerery is ridged, or if growing in a trench, the trench is filled. However, this operation is done late in the season, a few weeks before harvest. Most people have become accustomed to green asparagus spears and green celery stalks; consequently, blanching is less common; and when blanching is done, it is less intense.

Broadcasting of carrot seed can be used to increase the yield from a given area. Weed control is accomplished by spraying with a petroleum herbicide like Stoddard solvent or fuel oil. Carrots should not be sprayed after roots are greater than $\frac{1}{4}$ in. ($\frac{1}{2}$ cm) in diameter or off-flavors may result. The close proximity of carrot plants usually eliminates the weed problem after one application of the herbicide.

Mulching may be used for moisture retention, weed control, and clean fruits. Sawdust, chopped corncobs, and straw may be used. Their rate of decomposition will vary and nitrogen will need to be supplied accordingly. Polyethylene plastic and aluminium foil have also been used successfully; however, they must be punctured to permit water penetration.

STUDY QUESTIONS

1. Considering the overhead costs of a home garden, does it pay to grow your own vegetables? Make a balance sheet of costs of gardening versus cost of purchasing all vegetables you would grow.
2. Why is vegetable gardening so readily adaptable as a hobby?
3. How does it contribute to family pride and family unity?
4. Many communities have gardens for residents of the area. What are some of the problems that might be encountered? How can they be handled?
5. Can cool-season and warm-season vegetables be grown in the same garden? Explain.
6. How can an urban family with a small garden make use of the techniques of companion cropping and succession cropping effectively?
7. How does a home gardener avoid soil compaction in the garden? How can it be corrected?
8. What is "shock of transplanting"? How can it be reduced?
9. How are starter solutions used effectively?
10. Is cultivation of a vegetable garden essential if there are no weeds?
11. Why are tomatoes sometimes staked or caged?

12. When is thinning a recommended practice for root crops? Explain.
13. How is rhubarb forced for winter production?
14. Is blanching always needed for crisp celery? What is the modern alternative?

SELECTED REFERENCES

1. Asgrow Seed Company. 1977. *Seed for Today. Catalog of Vegetable Cultivars.* Asgrow Seed Co., Kalamazoo, Mich.
2. Askew, R. G., and S. R. Mills. 1977. *Everybody's Garden Guide.* Cooperative Extension Service, North Dakota State University, Fargo, N.D.
3. Edmond, J. B., and A. R. Ammerman. 1971. *Sweet Potatoes: Production, Processing and Marketing.* Avi Publishing Co., Westport, Conn.
4. Kling, G. J. 1977. *Harvesting and Storing Vegetables.* Cooperative Extension Service, Iowa State University, Ames, Iowa.
5. Knott, J. E. 1971. *Vegetable Production in Southeast Asia.* University of the Philippines, Laguna.
6. Prasher, Paul, and Dean Martin. 1976. *Vegetable Gardening.* EC 668, Cooperative Extension Service, South Dakota State University, Brookings, S.D.
7. Thompson, H. C., and W. C. Kelly. 1957. *Vegetable Crops*, 5th ed. McGraw-Hill, New York.
8. Ware, G. W., and J. P. McCollum. 1959. *Raising Vegetables.* Interstate, Inc., Danville, Ill.
9. Yearbook of Agriculture. 1977. *Gardening for Food and Fun.* U.S. Department of Agriculture, Washington, D.C.

20 Fruit Plantings

Fruits rank among the favorite foods of the world. They are eaten as ripe, immature, dried, canned, frozen, preserved, and pickled products. Their juices, flavors, or whole fruits are used extensively in cooking, baking, ice cream manufacture, and other culinary arts. Great technological advances in transportation, refrigeration, and controlled storage have brought high-quality fresh fruits within the realm of most homes in our modern economy. Their availability extends over all months of the year. The overlap of harvest, storage, and later harvests of other areas supply fresh fruits at all seasons. These factors have contributed greatly to changes in home fruit production. Home orchards have nearly disappeared from the scene in some areas. High yields and efficient production from ideally located fruit areas combined with factors of transportation and storage have led to this decline. Yet, there remains demand for fruit production on the home grounds. Several factors favor home fruit production but in a modified form. The trend is away from large fruit trees. This is caused principally by inadequacy of equipment for pest control and the high overhead cost of adequate equipment. Because of people's desires for many kinds of fruit, large numbers of trees of each kind are not practical for home fruit consumption. The home gardener's interest in dwarf trees of apple and pear has given a new approach to home production of tree fruits. Greater diet variety appeal has stimulated planting of other tree fruits like cherries, peaches, and plums for home fruit production. These fruits bear earlier than standard size apple and pear trees and, yet, do not generally grow as large. The small fruits contribute greatly to home fruit growing; they offer a quick return following their establishment. Improved materials for pest control, the use of dwarfed trees, earlier bearing trees, and more of the small fruits give greater assurance that it is as satisfactory to have fruit plantings as it is to have quality vegetables from a home vegetable garden.

ADAPTING FRUITS TO THE HOME

Dual Purpose Plantings   Fruit plants can serve dual purposes. Trees may be beneficial for shade, specimen, or background in the home grounds. Grape vines and perennial caned brambles are adaptable for screening and separating areas. They may be trained on fences, trellises, walls, and arbors. Raspberries form desirable backgrounds for the home grounds and may be effective barriers against trespassers and roaming pets. Strawberries are

FIGURE 20-1   Dwarf trees have given a new approach to home fruit production.

often included in the vegetable garden with perennial vegetables because of similarities in growth habit and culture. Runnerless types of strawberries make interesting and effective edging plants. Blueberries, gooseberries, and currants can be used as specimen shrubs. Nut trees require many years of growth before they come into bearing but can be utilitarian in the interval as ornamentals.

Orchards   The home orchard is better adapted to rural and suburban homes than in most urban areas. If considerable fruit production is desired, fruit trees can be given better care and more uniform treatment in an area devoted solely to their growth and production. It cannot be over-emphasized at this point that adequate pest control measures must be included for successful production of tree fruits. The ideal site, best and hardiest cultivars, and the most complete program of soil management will be of little avail should pest control be neglected or taken lightly. More home orchards have failed to give satisfaction from lack of pest control than for any other factors of neglect.

Good site is important from the standpoint of air drainage, exposure, soil drainage, and fertility. It should provide ready access from the home or home grounds. If supplemental water is used, either routinely or occasionally, provision for irrigation must be included. An adequate water source for spraying is a necessity.

Several kinds of tree fruits should be included if the home orchard is to supply the principal fruit requirements of the family. Early and late cultivars of each kind will extend and stagger the harvest period, important factors both for fresh use and for distribution of time needed for processing. It is not uncommon for home orchards to produce fruit in much greater quantity than family demand requires. This poses a pro-

FIGURE 20-2 The home orchard is better adapted to rural and suburban homes than in most urban areas. (Photograph by K. A. Denisen.)

blem of disposal. Marketing may entail further time and costs, discarding the crop or neglecting to harvest it results in waste, and giving the fruit away actually increases production costs of that retained. Adequate planning with respect to size of orchard, amount of each kind of fruit, and relative proportions of each cultivar are the factors for consideration in planning a home orchard.

Vineyards   Within three years after planting, a small vineyard of 10 to 20 vines can produce abundant quantities of grapes for fresh fruit and processing. There either seem to be few serious pests of grapes or home fruit gardeners do an effective job in controlling pests. In either event, grapes are usually better adapted to home fruit production than are tree fruits. The critical features of vineyard management besides pest control are pruning, fertility, weed control, and drainage. Trellis construction should be based on permanence of location, for a well managed vineyard can be productive for 25 years or more. Grape vines are readily adapted to contour planting. They are excellent perennial plants to grow on terraces. A fine network of feeder roots is found in abundance in and above the terrace where moisture is caught and retained for plant use. An exposure facing the morning, noon, or afternoon sun is more desirable than slopes away from the sun. Rows of trellises on a hillside sloping away from the sun may actually cast shadows on the lower leaves of adjacent rows and reduce the photosynthetic ability of the vineyard.

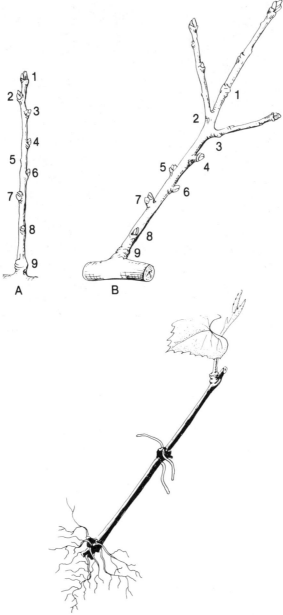

FIGURE 20-3   Upper, A is a 1-year-old twig of apple; B is the same twig one year later. Note where the growth has occurred. In B, buds 4, 5 and 7 are fruit buds. Buds 1, 2 and 3 are shoot buds. The lower portion of the illustration shows rooted cutting of grape. It is feasible for a home owner to propagate his own grapes especially if the cultivar is a vigorous one.

Berry Plantings   The bramble fruits and blueberries will produce significant quantities of berries within two years of planting. Gooseberries and currants will fruit in two to three years. All are quite readily established. With the good flavors and the quick returns, they are popular items in the home fruit garden. They are readily adaptable to most home grounds, even on small lots, for they do not require the large spacings between rows and between plants needed for fruit trees. Bush fruits are also adaptable to the home garden plans of renters who have a lease of more than one or two years' duration. Some type of trellis or support is usually recommended for the bramble fruits as a precaution against wind damage and heavily laden fruiting canes.

The home strawberry planting is a favorite of home gardeners and the entire family. Many people consider strawberries their favorite dessert. Besides their excellent flavor they are relatively easy to grow and give a quick return on a small investment of money and time. It is even possible to have big yields of everbearing strawberries within four to six months after planting if given special treatment, a summer mulch and runner removal. All plants, whether spring bearers or everbearers, will produce berries in the spring of the second year. Strawberry plantings in the home grounds fit well into a rotation with vegetables in the over-all garden scheme. Production of this fruit is within the realm of any home gardener, for the planting can be either extremely small or enlarged to become a profitable home enterprise. Disposal of an excess is not usually a problem because of its popularity as a fresh fruit and its adaptability to freezing, canning, and preserving. Home production is an important source of berries. The large amount of hand labor involved in commercial production keeps prices at a relatively high level. This and quality advantages offer further incentive for growing berries on the home grounds.

TREE FRUITS AND NUTS

Selecting Kinds and Cultivars   Climatic conditions of an area will have a determining influence on the kinds of fruits which can be grown. Of these various kinds, selections should be made on the basis of the individual family's desires for fresh and processed fruits. It is desirable for the trees to begin fruiting at an early age; consequently, this is an important factor for consideration when selecting kinds of fruits. Dwarf trees will bear earlier and do not need the wide spacing distance of standard trees. It should be noted, however, that some pruning is required both for maximum production and maintenance of the desired dwarf size. The stone fruits, also called "drupe" fruits, begin bearing at an earlier age than the "pome" fruits, those with a core (apple, pear, and quince). Pecans, almonds, and filberts are earlier bearing and less hardy nuts than black and English walnuts and butternuts.

Cultivar considerations include hardiness and general adaptability, fruit characteristics, and pollination requirements. For some fruits, there are great differences of cold hardiness between cultivars. The Wealthy and Haralson apple cultivars are hardy to extremely low winter temperatures, whereas, Delicious and Jonathan might winterkill under the same conditions. In areas of mild winters, the chilling requirement for breaking the rest period is less for adapted cultivars than for those adapted to the more severe winter climates. Sour cherries are hardier to

FIGURE 20-4   Peaches are a rewarding fruit crop for the home if winters are not too severe. Tree ripened peaches are superior to those shipped from distant sources. (Photograph by K. A. Denisen.)

cold than are sweet cherries. American type plums are hardier than European and Japanese types and are more resistant to drouth. However, European types are considered to be of better dessert quality. Some cultivars of peaches and apricots are quite hardy. Where hardiness is not an important factor, the home orchardist will make his choices on the basis of earliness of bearing, fruit quality, and pollination requirements. No single cultivar of any fruit is universally adaptable or its fruit universally acceptable. Those cultivars which are self-fruitful, that is, set fruit with pollen of their own cultivar, are readily adaptable to home fruit production. When they require pollen from other cultivars, it is important to make the proper selection of pollinizer cultivars. If possible the pollinizer cultivar should produce fruit of acceptable quality. Reliable nurseries will inform the purchaser of trees about the cultivars they list or carry which are self-pollinizing and which need pollinizer cultivars. It may not be necessary to include pollinizing cultivars in

FIGURE 20-5

Apple and plum trees in bloom. Self-pollinizing cultivars are best adapted to home fruit production. The long "spikes" of bloom on the plum at right are why this fruit has very heavy fruit set.

urban fruit plantings if there are other cultivars of the same kind of fruits in neighbor's gardens or orchards. Bees, the agency of fruit tree pollination, do not respect boundary lines. Lists of recommended fruit cultivars, their characteristics, and pollination requirements can be obtained from state, provincial, and federal experiment stations and educational services. Nursery catalogs often contain illustrations of cultivars which are the purchaser's guide in selection. Trees are a long-term investment. Due evaluation and consideration before their purchase is allied to the careful planning of the home grounds.

Establishing Fruit and Nut Trees

Good quality nursery stock is vital to the establishment of thrifty, productive fruit trees. Two-year-old trees are the most practical size to purchase for the home orchard unless the home owner prefers to make his own piece-root grafts or topwork on hardy stocks. Unbranched whips, one-year-old trees, may be planted but will usually take an additional year before fruiting. Trees that are over two years old will lose more of their roots when dug. They then require more severe top pruning at planting time to reduce water loss in proportion to reduced absorptive capacity. Dormant nursery stock that arrives before it can be planted should be "heeled in." A trench is dug in the soil, the trees and other woody plants are placed in the trench at an angle, and soil is thrown back over the roots. It is, in essence, a temporary planting operation which prevents drying of the roots. Root growth may actually begin during heeling in, especially if the period is prolonged. When mild temperatures are expected, it is better to store the trees in a cool place and keep the roots moist.

Location of the trees, whether it is in the lawn of the home grounds or in a home·orchard, should give consideration to placement of other trees, shrubs, and herbaceous plantings. Competitive effects can be felt by each of the adjacent plants, the strong as well as the weak. Spacing of fruit trees is extremely important. Many trees are so closely spaced that they begin to compete seriously at about the time they should be coming into full production. At this point the only recourse is to remove approximately half of the competing trees to make way for efficient production of the remaining trees. More careful planning at planting time can avoid these situations. The spacing chart in Table 9 gives approximate distances between fruit plants for efficient handling and production.

Planting operations must include provision for keeping the roots from drying out. They arrive from the nursery either with moist packing material around the roots or "bare-root" in a polyethylene plastic wrap. After unpacking they may be planted immediately in moist soil or placed in a container of water briefly and then planted. Straight rows are marked out for planting, and stakes are placed at the designated tree locations. A planting board makes it possible to mark the location, remove the stake, dig the hole, and then place the tree in the exact location of the stake. Deciduous trees are placed slightly deeper than they stood in the nursery row, but balled-and-burlapped evergreen stock is planted at the same depth it grew previously.

The most practical system of soil management for home orchards is sod. However, it is well to precede the establishment of sod with open cultivation at least in the immediate vicinity of the trees. This eliminates the competing effects of sod for moisture and nutrients. Sod may be established between the rows, or intercrops of vegetables and small fruits can be grown. Intercrops provide some return while the trees are growing and until they require the entire area for growth and production. When the trees begin fruiting, a sod-mulch system helps in conserving moisture and cushions the impact of falling fruit.

Fertilizers may be applied in late fall or early spring for spring availability. The fruit buds for most fruits are formed in late spring or early summer for the succeeding year's crop. Nitrogen plays an important role in fruit bud differentiation. It promotes shoot growth, a large leaf area, and a high rate of carbohydrate production. Other nutrients are also needed but in much smaller quantities than nitrogen unless they are especially deficient in the soil. Fertilizers can be applied to the surface of the soil in rings around the base of the trees. The outer periphery of the branches is the maximum distance from the trunk that fertilizer need be applied. Animal manure, green manure, and organic mulches of leaves, straw, and hay contribute to the organic content of the soil. Organic matter is especially valuable to the orchard in that it supplies nitrogen at adequate rates but spreads its availability over a long period.

There are various accepted methods of training and pruning. All are based on fruiting habits and tree response. It is wise to become

TABLE 9

Planting Distances of Fruit and Nut Trees, Approximate Time from Planting to Fruiting, and Approximate Annual Yield per Mature Tree

Kind of tree	Distance between trees (feet)	(meters)	Approximate time from planting to fruiting (years)	Approximate annual yield per mature tree
Deciduous fruits				
Apple				
dwarf	8–12	$2\frac{1}{2}$–4	2–3	$\frac{1}{2}$–1 bu.
semidwarf	15–25	5–8	4–6	2–5 bu.
standard	35–40	9–10	6–10	5–10 bu.
Cherry				
Sour	20–25	7–8	3–6	25–50 qt.
sweet	30	8–9	6–8	25–50 qt.
Peach	20–25	7–8	3–4	1–4 bu.
Pear				
dwarf	8–12	3–4	3–4	$\frac{1}{2}$–1 bu.
standard	20–25	7–8	6–8	2–4 bu.
Plum	20–25	7–8	5–8	1–5 bu.
Quince	15–20	5–7	4–6	2–5 bu.
Evergreen fruits				
Avocado	20–25	7–8	5–7	100–150 lb.
Grapefruit	18–25	6–8	4–6	300–600 lb.
Lemon	20–25	7–8	4–6	200–400 lb.
Lime	15–25	5–8	4–6	100–200 lb.
Orange	20–25	7–8	4–6	200–400 lb.
Tangerine	20	6–7	4–6	200–400 lb.
Nuts				
Almond	20–25	7–8	5–6	20–25 lb.
Filbert	15–20	5–7	5–6	20–25 lb.
Pecan	50–60	12–15	5–10	100–120 lb.
Walnut				
black	50–60	12–15	6–12	80–100 lb.
Carpathian	40–60	10–15	6–12	80–100 lb.
English	40–60	10–15	6–12	80–100 lb.

thoroughly acquainted with the principles and effects and then follow a systematic pruning method. Low-headed trees are preferred in home plantings. Experience in pruning is an excellent teacher provided it is accompanied by observation and follow-up. Trees in home plantings can be closely observed and results of pruning operations checked with yield and quality performance. The home gardener can be equally aware of achievement with well kept, productive fruit trees as he is with his vegetable garden and ornamental plantings.

Pest control in tree fruits around the home is often a difficult problem. Inadequate coverage of foliage, fruits, or twigs is the principal pitfall of most home fruit spraying or dusting programs. Timeliness of sprays is critical. Small trees are much easier to spray and have complete

FIGURE 20-6   **Plums are generally hardier than peaches and can be quite productive in the home grounds but do require some pest control. (Photograph by K. A. Denisen.)**

coverage with the usual low-pressure hand sprayers or dusters. Just this reason alone makes dwarfs and natural low growing types of trees become increasingly popular. Power sprayers with a tank capacity ranging from 5 to 15 gallons (20–60 liters) are available. However, their cost is not justified unless they can be used for a considerable number of trees or for vegetables, ornamentals, and other uses. Neighbors sometimes purchase adequate spraying equipment cooperatively to reduce the original cost and increase the rate of amortization. "All-purpose" or "one-pack" sprays are available, but they are considerably more costly if many trees need spraying several times during the season. Bulletins and pamphlets on home orchard spray schedules are usually available through the Extension Service and State and Federal Experiment stations. These will outline the specific pests of the area, give approximate time and frequency of spray application, and include the latest recommendations and regulations of newest, most effective pesticides.

Harvest for home use is similar to the commercial harvest with the possible exception that tree-ripeness is the guide rather than maturity. This implies the fruits are to be used for more immediate consumption

FIGURE 20-7   **The cherry harvest. In home plantings, all fruits can be tree-ripened.**

or processing. Dwarfs and low growing trees are convenient for the harvest and other operations. Careful handling helps retain quality of fruit, at fully ripe or storage maturity.

SMALL FRUITS

Grape Culture   A *cultivar* of grapes can be found for nearly every one of the many variable temperate and subtropical climates of the world. The American type, or labrusca grape, is native to the northern sections of North America. Some cultivars, e.g., Alpha and Beta, have nearly all the cold hardiness of their native forebears. However, their dessert quality leaves much to be desired so they are grown principally where other cultivars will not survive the severely cold winters. The Concord cultivar is typical of the labrusca grape and is very widely grown. It has become so well known that other newer cultivars are often described by being called of "Concord type." These are also termed "slip-skin" types. Fruit color ranges from red through purple, violet, blue, to light green, to nearly white. The most common color is blue. The European type, or vinifera grape, has skin which adheres to the fruit. It is a less hardy species of grape that produces exceedingly well through most of California and along the north coast of the Mediterranean Sea. Thompson Seedless, a

FIGURE 20-8 Crabapples are useful in the home for jams and jellies. (Photograph by K. A. Denisen.)

white grape, actually a faded green, and Tokay, a light purple, are popular cultivars. Hybrids of the European and American types, called French-American hybrids, are in increased demand for home gardens in milder sections of temperate climates. They are more cold hardy than European grapes, but generally less hardy than their labrusca parent. Muscadine grapes are less common than the other types but are grown considerably in Southeastern United States. They have fewer berries per cluster, but the berries are large and the plants quite productive.

Training systems vary with the type of grape, its function irrespective of fruit in the home grounds, and individual preferences. The Kniffin and fan systems are the most common forms for the American types. In extremely cold areas, the trunk may be trained nearly horizontal and close to the ground so the trunk and canes can be mounded and mulched for winter protection. Regardless of the training system used, best production is obtained when annual dormant pruning removes about 90 percent of the previous season's growth. The cordon system is frequently used in training the European grape. Many growers of muscadine grapes utilize the umbrella system.

Soil management and fertilizing of grapes in the home planting involves mulching, weed control, organic matter, and manure or commercial fertilizer. A constant mulch system of soil management replenished anually with straw, hay, corncobs, or other organic materials serves the various purposes of moisture retention, weed control, and organic matter maintenance. It is especially important in home grape growing that weeds are controlled by methods other than herbicides. The grape is especially sensitive to 2,4D and other hormonal weed killers. Extreme care should be exercised when these herbicides are used on the lawn and areas adjacent to grapevines. The vapors of 2,4D

FIGURE 20-9   The umbrella system of training grapes.

cause formative effects on grape leaves and shoots (Fig. 11–15, page 259) and may even delay ripening of fruits and accumulation of sugars.

Manure is a very satisfactory means of fertilizing grapes and maintaining organic matter content of the soil. The same effect can be obtained with commercial fertilizer and either a green manure crop or the mulch of organic matter. Nitrogen fertilizer seems to give a greater response than other nutrients. About $\frac{1}{2}$ lb. ($\frac{1}{4}$ kg) per vine of a 16 to 20 percent nitrogen fertilizer is generally sufficient to compensate for that used in mulch decomposition. If shoot growth is excessive and canes are unusually thick, it is better not to apply any nitrogen until a more balanced supply of nutrients is available to the plants. Excessive shoot growth is an indication of vegetativeness and will usually result in less fruit.

Pest control involves the use of sprays because dusts are generally ineffective against grape insects. Sanitation in the vineyard is helpful in destroying over-wintering stages of berry-moth and leafhoppers, two of the principal insect pests. Destruction of mummified berries helps control black-rot and mildew. A spray program utilizing ferbam and Sevin at about monthly intervals from pre-bloom until mid-summer will control most insect and disease pests of grapes.

Strawberry Culture

Photoperiod response and apparent lack of photoperiod response are the basic differences between spring-bearing and everbearing strawberries. Spring-bearing cultivars differentiate fruit buds during the short days of autumn while the plants are actively producing food reserves. After a rest period that varies in duration for different cultivars, the plants become fruitful. Long days are not needed for fruiting, but short days are needed for fruit bud differentiation. The everbearer is not influenced by day length insofar as fruit bud differentiation is concerned. Plants in the same bed may be in the various phases of differentiation, flowering, and fruiting all at the same time. Runner-producing cultivars of both everbearing and spring-bearing types will develop runners under a long photoperiod; consequently, both types are subject to photoperiod response with regard to runner initiation. Spring-bearing strawberries are by far the more common. The discussion here concerns primarily the spring-bearing types except as where otherwise indicated.

Cultivars of strawberries are numerous. One of the reasons for this is the limited adaptability of most cultivars. Occasionally cultivars like Surecrop, Midway, and Sparkle are quite well adapted to a wide range of climatic and soil conditions. However, for areas of specific climates, soils, or even cultural conditions, growers often find a cultivar that is very reliable, productive, high in quality, or capable of being shipped long distances. This becomes the standard cultivar for the area even though it may be rarely grown elsewhere. Since no cultivar is universally adapted, it behooves the strawberrry gardener to use those recognized as locally superior. He may, in addition, have small labelled plantings of new or different cultivars on a trial basis.

Planting strawberries is a skill that is easily acquired, but if incorrectly done will give poor growth, poor runner production, and poor yields. If

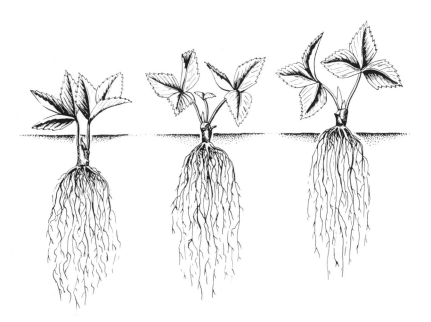

FIGURE 20-10   **Planting strawberries.** *Left,* **too deep;** *center,* **correct;** *right,* **too shallow.**

root tips appear darkened or dry, it is desirable to clip off the outer inch ($2\frac{1}{2}$ cm) of roots. In planting, roots should be fanned out from the crown, soil firmed about them, and the crown placed at soil level. When set too shallow, roots are exposed and drying will result. When set too deep, the growing point is covered and new leaves and runners are delayed in emergence or perhaps completely inhibited. If spring planted, blossoms are removed when they appear, to divert the food supply to plant growth and runner production. The early formed runner plants are most productive for the succeeding year's crop so early runner production is encouraged. No berries are harvested the spring they are set out. If fall planted, they should be set out early enough so they become well established by winter. These plants will be fruitful the next spring. Some growers have adopted the practice of purchasing vigorous dormant plants during the growing season, planting them at close spacing, fertilizing heavily and harvesting the crop the same season. This is an extremely intensive method of production.

Spacing systems will vary. The most common are the matted row and spaced matted row. Plants are set 12 to 30 in. ($\frac{1}{4}-\frac{3}{4}$ meter) apart in single rows 3 to 5 feet ($1-1\frac{1}{2}$ meters) apart. The broad range in spacing distances is due to the great variation in runner production between cultivars and if irrigation is available. Runners are trained to positions of least competition. If growing conditions continue to be favorable after the desired width of 20 to 30 in. ($\frac{1}{2}-\frac{3}{4}$ meter) is attained, it may be necessary to control runner growth. A range of 5 to 8 plants per square foot of matted row gives highest yields commensurate with large berry size. Overcrowding results in reduced yields, smaller berries, more foliage diseases and berry rots due to restricted aeration. The hill system which consists of runner pruning and resulting crown branching is used

FIGURE 20-11   Controlling weeds in strawberries with herbicides. Row on left is obscured by weeds.

extensively in some areas. Some gardeners prefer a narrow hedge-row and closer spacing of rows. This also involves runner pruning.

Weed control is a continuous problem in strawberries. Machine cultivation is difficult with most spacing and training systems. Fortunately herbicides have been used effectively in eliminating much of the hand weeding formerly done even in commercial plantings. The latest recommendations of local and state weed control specialists should be consulted in developing an herbicide program for any of the small fruits.

Fertilizing of strawberries may not be necessary the first year if there is abundant organic matter and available nitrogen in the soil. If low in general fertility, the addition of animal manure is highly beneficial. However, a complete commercial fertilizer is also satisfactory if applied at planting time or during the first season. A 1–2–1 ratio is generally satisfactory. Complete fertilizers for a producing bed should be applied after fruiting. If nitrogen is supplied before the fruiting season, i.e., in early spring, the immediate response may produce such heavy and dense vegetation that pollination is hindered and fungus diseases are more difficult to control. Soft berries may result from heavy spring applications of nitrogen. Phosphorus application in early spring may hasten ripening and increase berry size.

Mulching for winter protection is necessary in areas where winter temperatures fall below 15°F (−10°C). This protects against fruit bud injury in the crown and heaving of the soil. The mulch, which may consist of straw, hay, or other coarse materials, can also be used to delay spring blossoming by a few days. In the spring, the mulch is rearranged to leave a thin layer on the row while the remainder is placed between the rows. Berries are cleaner and weeds tend to be suppressed with this spring use of mulch.

Harvesting is usually completed within a 2 to 3 week period. Moisture supply is extremely critical at this time for the berries continue to enlarge even while they begin coloring for maturity. Picking is usually started in the morning as soon as dew has dried from the foliage. If irrigation is practiced, the water is applied after picking so it can be utilized in berry enlargement for the next picking, about two days later. Since strawberries are normally more than 90 percent water, their moisture requirements are high. Yields are nearly always increased by irrigation unless the harvest occurs in a season of extremely wet weather. Berries are picked when red and should be placed in the shade immediately to maintain quality.

Renovating the bed is done after harvest when it is retained for further production. Most strawberry beds are fruited for two years and

FIGURE 20-12   **Everbearing strawberries in the hill system with a summer mulch of chopped corncobs.**

some even longer. Renovation consists of cultivating and thinning plants. Reduction in stand permits the development of larger crowns. Many of the plants will produce runners again which may be kept or eliminated as required. If the matted row is narrowed to a band 6 or 8 in. (15–20 cm) wide, the runner plants are used to rebuild the matted row. If plants are selected for retention in the renovation process, new runner plants only act as competitors for space, nutrients, moisture, and light. Hand-thinning or pruning of excess runners is generally practical in home plantings.

Everbearing strawberries actually have cycles or waves of production rather than bearing continuously. They will produce two or three crops per year depending on length of growing season and other factors. Everbearers are best adapted to areas of relatively cool summers. Under such conditions they can be grown in a matted row with reasonably good success. By special methods of culture, everbearing cultivars can be grown successfully in quite warm climates. Close spacing of plants, continuous runner removal, and a summer mulch on the soil give the combined effect of favorable conditions for everbearing strawberry production. The plants are spaced 12 to 18 in. (30–45 cm) apart in rows which are similarly spaced. Runners are picked, pulled, or cut off as they appear. The summer mulch keeps the berries clean, stabilizes soil temperature, and helps to conserve moisture. Slowly decomposing materials like sawdust and chopped corncobs are superior to straw, hay, or leaves for this purpose. It is usually desirable to apply the fertilizer to the soil surface before mulching to counteract the utilization of nitrates by soil bacteria in mulch decomposition. During the year the planting is set out, it is best to remove the first series of blossoms. This operation, together with runner removal, conserves the food supply of the plant,

FIGURE 20-13   **Trellising of raspberries.** *Left*: **Red raspberries trained to stakes.** *Right*: **Black raspberries with wire supports.**

Horticulture for the Home

builds bigger crowns, and will result in much greater production later in the summer of the first year. As the plants become well established, blossoms are allowed to remain but runners are pruned. The manufactured food of the leaves thus yields berries instead of runner plants. The original plants may continue to be productive for several years if kept free of insect and disease troubles. Under this special treatment, everbearers can be very rewarding to the gardener and his family.

Insect and disease control is a much simpler operation than with the tree fruits. A knapsack or hand operated sprayer or a simple duster can give good coverage of insecticides and fungicides. Most fungus diseases of foliage, blossoms, and fruit can be controlled with such broadly effective materials as captan and ferbam. Sevin and methoxychlor control many insect pests of strawberries; Endrin controls cyclamen mite. Malathion is effective against spider mites. Materials poisonous to man are not used after fruits form. Rotenone can be used while fruits are present because of its non-toxicity to man and warm-blooded animals. Virus-free cultivars of strawberries are preferred over regular stock. It is helpful, especially in mild climates, to plant nematode-free plants in soil that has been treated for nematodes.

Bramble Fruits   *Raspberries* are the most widely cultivated of the bramble fruits. Red raspberries, because of their suckering habit, are readily established and then need periodic thinning and cutting of shoots between rows or clumps to keep within bounds. Black and purple raspberries do not sucker. They will spread by natural tip layerage which occurs easily in new plantings. When the planting becomes established, the pinching process in summer pruning tends to practically eliminate natural tip layering. All types of raspberries can be trellised. It is especially recommended if plantings are subject to strong winds. Canes heavily laden with fruit often need support to produce clean, attractive berries and to avoid cane breakage. New shoots arising from the crown in black and purple types are often quite weak in basal attachment, so they are greatly helped by trellises. Plantings of raspberries will be productive for 10 or 12 years with good care. Weed control is important, especially the control of perennial weeds. Various herbicides have been employed effectively on raspberries. Mulches are very helpful, but all decaying organic matter needs additional nitrogen fertilizer. Spring fertilizing with a complete plant food is a general practice for good raspberry plantings. The fertilizer can be spread between the rows or adjacent to the rows and worked in, or it may be incorporated into the decaying mulch. Appropriate dusts and sprays will control most insect and disease pests.

Blackberries, dewberries, Boysenberries, Loganberries, and Youngberries are similar in berry type and growth habit. They are distinct from raspberries because their berry remains attached to the torus or receptacle when mature. The torus is edible and is processed and eaten with the berry. The plants, while not generally as hardy as the hardier cultivars of raspberries, will survive cold climates with winter protection. Blackberries that are native to cold regions of Canada and Alaska are protected by a natural snow mulch. Because of their scandant growth, they are usually trained to trellises in the manner of grapes. Other cultural practices are similar to those of raspberries.

Blueberries   Highbush and lowbush blueberries are grown in home fruit plantings. There has been considerable hybridizing among the highbush types, and there are numerous cultivars which produce large sized berries. The low-bush types are native to bogs and swamps. They have been taken from their native habitat and grown in gardens, but no cultivars have become widespread in distribution. Blueberries require an acid soil. They represent an extreme among fruit plants for their acid-loving nature and prefer a soil pH of 4.0 to 5.0. The soil should be high in organic matter. Acidity of garden soils can be increased by applying and incorporating sulfur, aluminum sulfate, iron sulfate, or gypsum. About $\frac{3}{4}$ lb. ($\frac{1}{3}$ kg) of sulfur per 100 square feet will lower the pH of a sandy soil in the 4.5 to 6.5 range by 1.0. About $1\frac{1}{2}$ to 2 lb. ($\frac{3}{4}$ kg) are required for each 100 square feet on silt loam soils to reduce pH by 1.0. On heavy soils, large quantities of sawdust, acid peat, or other organic matter will increase the humus content of the soil and improve the structure, both essentail to good blueberry production. Nitrogen fertilizer should be supplied annually in the spring. Good air drainage is important and some areas need winter protection for plant survival in cold winters.

Gooseberries and Currants   The commercial fruit industry does not produce large quantities of these bush fruits. This is an important reason for having at least a small planting if the family likes the jams, jellies, pies and other products made from gooseberries or currants. They are relatively easy to grow, being less exacting in care than many other fruits in the home. Either clean cultivation or mulching is a satisfactory system of soil management. Commercial fertilizers at $\frac{1}{2}$ lb (200 g) per plant of a mixture containing about 10 percent nitrogen can be used in lieu of manure, compost, or peat. Adjacent trees or close proximity of grass and shrubs are extremely competitive to both currants and gooseberries. This can often be overcome by supplying additional nutrients and moisture. Full production is not attained until the bushes are about three or four years of age. Most fruit is borne on two- and three-year-old wood. Since older wood is less productive, renewal pruning is generally practiced. Low hanging, nearly horizontal branches are removed to encourage upright growth and produce clean berries. Spraying or dusting with rotenone is an effective means of controlling the currant worm, a principal insect pest of the fruits.

STUDY QUESTIONS

1. Why have home orchards almost totally disappeared? What is the current best alternative for a supply of fresh fruit for the home owner?

2. What role do dwarf fruit trees play in home fruit production?
3. What are the specific advantages of dwarf trees?
4. Why are small fruits considered better adapted to home grounds than tree fruits?
5. Distinguish between *Vitis labrusca* and *Vitis vinifera* grapes.
6. Why is it hazardous to use 2,4 D near grape plantings?
7. What are the advantages of everbearing strawberries for the home?
8. How can they be encouraged to be more productive and reliable?
9. Can raspberries or other brambles be used both for fruiting and ornamental purposes on the home grounds? Suggest how they might be used.
10. What is the principal limiting factor to more general use of blueberries as a fruit in the home grounds?
11. When is trellising considered essential for brambles and bush fruits?

SELECTED REFERENCES

1. Amerine, M. A., and M. V. Cruess. 1972. *Technology of Wine Making*. Avi Publishing Co., Westport, Conn.
2. Batchelor, L. D., and H. J. Webber (editors). 1948. *The Citrus Industry*, Vol. II: *The Production of the Crop*. University of California Press, Berkeley, Calif.
3. Childers, N. F. 1976. *Modern Fruit Culture*. Horticultural Publications, New Brunswick, N. J.
4. Darrow, G. M. 1966. *The Strawberry: History, Breeding, and Physiology*. Holt, Rinehart and Winston, New York.
5. Denisen, E. L., and H. E. Nichols. 1962. *Laboratory manual in Horticulture*. Iowa State University Press, Ames, Iowa.
6. Eck, P., and Childers, N. F. (editors). 1966. *Blueberry Culture*. Rutgers University Press, New Brunswick, N. J.
7. Upshall, W. H. (editor). 1970. *North American Apples: Varieties, Rootstocks, Outlook*. Michigan State University Press, East Lansing, Mich.
8. Weaver, R. J. 1976. *Grape Growing*. Wiley, New York.
9. Winkler, W. A. 1962. *General Viticulture*. University of California Press, Berkeley, Calif.
10. Woodruff, J. G. 1967. *Tree Nuts, Production, Processing, Products*. Avi Publishing Co., Westport, Conn.

Glossary

Abscission layer   A layer of thin-walled cells which are formed at the base of the petiole and facilitate leaf fall.

Absorption   The intake of water and other materials through root and leaf cells.

Achene   Fruit-type seed as found on the surface of strawberry fruits.

Adventitious (roots, shoots)   Arising at unexpected places.

Aerobic   Active in the presence of free oxygen.

After-ripening   Completion of maturation processes within seeds before germination will occur.

Air drainage   Air outlets and convection currents which prevent "dead" air and frost "pockets."

Anaerobic   Active in the absence of free oxygen.

Annual   Completion of life cycle within one year.

Annual rings   Growth rings visible in cross-section of woody plant stems.

Anther   The pollen-producing organ of the flower.

Antibiotic   Chemical capable of destroying disease organisms in tissues.

Anti-oxidant   Material used to prevent oxidation or discoloration of food.

Apical   The point or tip, e.g., of a stem.

Apomicts (apomixis)   Development of buds inside seed organs without fertilization.

Arboretum   Area devoted to specimen plantings of trees and shrubs.

Asexual propagation (clonal, vegetative propagation)   Increase in plant numbers by methods other than seedage.

Asphyxiant   Gas for killing animal pests.

Assimilation   Building of protoplasm from digested food.

Asymmetric balance   Balance without formality.

Auxin   Natural hormone produced in apical regions. It inhibits growth of lateral buds.

Axillary   In the angle formed by the leaf and stem.

Bactericide   A material for destroying or protecting from bacteria.

Berry   Fleshy fruit of cane and bush fruits (also strawberry).

Biennial   Life span of more than one year but not more than two years.

Biennial bearing   Producing fruit in alternate years.

Blanching   Shutting out light to prevent chlorophyll development or to cause its disintegration. Also treating vegetables with scalding water to stop enzyme activity.

Bonsai   Japanese art of growing miniature trees and shrubs by extreme dwarfing.

Bound water   Water held by the cell against freezing (see *free water*).

Bramble   Cane bush with spines. Fruit is a berry (e.g., raspberry, blackberry).

Branch   A subsidiary stem arising from the main stem or another branch.

Broadleafed (broadleaved)   Having leaves of great breadth in relation to length and thickness.

Bud-break   Resting buds resume growth.

Budding   A special type of grafting using a single bud as a scion.

Budstick   A shoot or twig used as a source of buds for budding.

Bulb   A plant storage structure which is generally globe shaped and formed by concentric layers of fleshy leaves attached to a disk-shaped stem plate at the base.

Bulblets   Tiny bulbs produced around the base of the mother bulb.

Callus   Wound tissue which develops from cambium or other exposed meristem.

Calyx   The sepals of a flower as a group.

Cambium   The cylinder of meristematic tissue between the wood and bark.

Canes   The externally woody, internally pithy stems of brambles and vines.

Capillary water   That water held in soil on and between soil particles. It is available to plants.

Carbohydrate   Energy containing product of plants consisting of carbon, hydrogen, and oxygen with the hydrogen and oxygen in the ratio of water (H_2O).

Carbohydrate accumulation   Storage of reserves in tissues and cells.

Carbohydrate utilization   Sugars are used in further vegetative growth.

Causal agent   The organism or condition directly responsible for causing a disease.

Cell   The unit of structure in plants.

Centigrade   Metric scale of temperature measurement (0°, freezing; 100°, boiling).

Chelates   Metal containing organic compounds useful in supplying deficient minerals to plants.

Chimera   Tissue in a plant of different genetic origin than other parts to cause differences in pericarp or their changed appearance.

Chlorophyll   Green coloring matter in leaves.

Chromosome   Gene-carrying threads or rods in the nucleus of cells.

Citrus   Evergreen fruits with thick rind or peel encasing sections of juicy flesh (e.g., orange, lemon, grapefruit).

Clone (clonal)   Derived from parent by asexual means.

Cold frame   An enclosed bed for propagating or protecting plants. The source of heat is solar energy (see *Hotbed*).

Coleoptile   Sheath-like structure over the shoot of the grasses.

Colloidal   Gelatinous with cementing or adhesive action as in clay soils or humus.

Companion crops Two non-competitive crops grown in the same area at the same time. One is short term and the other of long duration.

Compatible Different kinds or varieties that will set fruit when cross-pollinated *or* make a successful graft union when intergrafted.

Complete flower Has all essential organs for seed production.

Compost A soil and humus mixture high in nutrient content.

Conifers Cone-bearing trees. Most kinds are evergreen.

Conservatory Greenhouse with interesting and unusual plants for public viewing and study.

Corm Structures similar to bulbs in appearance but are enlarged modified stems rather than compacted fleshy leaves.

Cormels Tiny corms produced around the base of the mother corm.

Corolla The petals of a flower as a group.

Crop rotation Growing a sequence of several different crops to avoid continuous cropping with a single crop.

Cross-pollinated Pollen from the anther of one plant is transferred to a stigma of another plant.

Crown The perennial stem structure of a perennial herbaceous plant.

Cultivar A group of closely related plants of common origin which have similar characteristics. This term is replacing "variety."

Cuticle Layer of cutin or wax on the outer surface of leaves and fruits.

Cuttage Propagation from parts severed from the parent plant prior to rooting.

Day-neutral Without floral or vegetative response to length of day.

Deciduous No living leaves during dormant season.

Defoliation Loss of leaves.

Desiccation Drying out.

Dicot (dicotyledonous) Plant with two cotyledons or seed leaves in its seed, e.g., bean, oak.

Differentiation Changes in composition, structure, and functions of cells and tissues during growth.

Digestion Breakdown of foods into simple foods by enzymes.

Dioecious Male and female flowers on separate plants.

Disbudding Removal of flower or shoot buds.

Dominant Overriding factor of inheritance in a gene.

Dormant period Time during which no growth occurs because of unfavorable environment.

Double working Grafted twice. A variety is grafted to an intermediate stock.

Drupe Stone fruit. Seed is encased in a hard shell inside fleshy tissue (e.g., peach, plum, apricot, cherry).

Dwarfed Restricted plant size without loss of health and vigor.

Emasculation Removal of the anthers of a flower.

Embryo culture Growing seedlings from excised embryos on nutrient media.

Enzyme A biological catalyst which aids in conversion of food from one form to another.

Erosion Wearing away of soil by wind and moving water.

Evergreen   Retains functional leaves throughout the year.

Excising   Removing or extracting, as an embryo from a seed or ovule.

Extensive production   Use of large areas of land with a minimum of labor and capital.

F_1 The first generation progeny of a cross.

Fahrenheit   English scale of temperature measurement (32°, freezing, 212°, boiling).

Feeder roots   Fine roots and root branches with a large absorbing area (root hairs).

Fertilization (in sexual propagation)   Fusion of male and female germ cells following pollination.

Fertilizer analysis   Percent of available nutrients, N-P-K, in a fertilizer.

Fertilizer ratio   The lowest common denominator of a fertilizer analysis.

Fibrous root   A much-branched root system with lack of dominance.

Field capacity   Soil's ability to provide water for growing plants (see *Capillary water*).

Filament   The stalk supporting the anther in a flower.

Flat   A shallow rectangular container used for starting seedlings in a greenhouse or propagating frame.

Flower   Structure for sexual reproduction in plants. Also the attractive, colorful bloom for aesthetic purposes.

Focal point   Central object of interest.

Foliage   The leaves of a plant.

Foliar feeding   Applying liquid nutrients to the leaves.

Forcing crops   Those grown out of season as in greenhouses.

Formal balance   Precise and methodical treatment of an area to produce geometric exactness.

Free water   Water released by the cell when freezing occurs in intercellular spaces (see *bound water*).

Fruit   The edible product of a woody or perennial plant, which, in its development, is closely associated with a flower (botanically, a fruit is a ripened ovary).

Fruiting habit   The location and manner in which fruit is borne on woody plants.

Fungicide   A material for destroying or protecting from fungi.

Gamete   Sexually active reproductive cell.

Gene   Factor of inheritance carried in sex cells.

Genus   A unit of classification (see *Species*).

Germination   Resumption of growth of the seed embryo.

Girdling   Destroying an area of bark around the periphery of a stem resulting in death of the plant.

Grading   Separating or sorting produce according to size or quality.

Graftage   Method of inserting buds, twigs, or shoots on other stems or roots for fusion of tissues.

Gravitational water   Water in excess of a soil's capacity removed by drainage.

"Green thumb"   A term credited a person with the ability to grow plants successfully.

Ground cover   Plants used in lieu of grass for holding soil and giving foliage texture.

Growing season   Period between beginning of growth in the spring and cessation of growth in the fall.

Growth regulator (plant regulator)   A hormonal substance capable of altering growth characteristics of plants.

Hardening, hardening-off   Subjecting plants to adverse conditions to hasten tissue maturation for increasing hardiness.

Hardiness   That quality which causes plants to resist injury from unfavorable temperatures.

Hardpan   An impervious layer of soil or rock which prevents downward drainage of water.

Heading back   Pruning off the terminal portion of a twig or branch.

Herbaceous   Nonwoody. Dies back to the crown during the dormant season.

Heartwood   Central cylinder of xylem tissue in a woody stem.

Heat units   Number of degrees above a certain base temperature for a designated number of hours to serve as a guide to maturity.

Herbicide   A material for killing weeds.

Heterosis   Hybrid vigor resulting from a cross.

Heterozygous   "Mixed" hereditary factors. Not a pure line.

Hobby   An avocation or pastime.

Homozygous   Purity of type. A pure line.

Hormone   A naturally occurring or synthetic compound which stimulates plants in a specific manner.

Horticulture   Art and science treating of fruits, vegetables, and ornamental plants.

Hotbed   An enclosed bed for propagating or protecting plants. It has a source of heat to supplement solar energy (see *Cold frame*).

Humus   Partially decomposed organic matter.

Hybrid   The first-generation progeny from a cross of different cultivar, strains, or inbred lines.

Imperfect flower   Some of the floral parts are missing.

Incompatible   Different kinds or varieties do not successfully cross-pollinate *or* intergraft.

Inflorescence   The flowering structure of a plant.

Intensive production   A method of increasing the productivity per unit of land by increasing the expenditure of labor and captal.

Insecticide   A material for destroying or protecting against insects.

Insect vector   An insect which transmits a disease.

Intercalary meristem   A cambium type of meristem without a distinct separation layer between xylem and phloem (e.g., as in palms).

Intercropping   Growing short-term crops between rows of trees in a young orchard for full utilization of land.

Internode   The portion of a stem between nodes.

Juvenile stage   Early or vegetative phase of growth characterized by carbohydrate utilization.

Latent bud   An inactive bud not held back by rest or dormant period but which may start growth if stimulated.

Latent heat   Stored heat.

Lateral   From the side as of a stem.

Layerage   Propagation of plants from parts that remain attached to the parent plant while rooting.

Leaching   Washing of soluble nutrients down through the soil.

Liana   Vine and vining type of woody plant.

Lister   A plow used for ridging.

Location   A climatic or geographic area.

Long-day   More than 14 continuous hours of daylight.

Low-headed trees   Trees that have primary branches low on the trunk.

Marketing   Buying and selling.

Maturation   Accumulation of solids in tissues.

Maturity   The state of ripeness. Usually that stage of development which results in maximum quality.

Meristematic   Capable of cell division.

Metamorphosis   Life cycle (e.g., egg, larva, pupa, adult).

Monocot (monocotyledonous)   Plant with one cotyledon in its seed, e.g., sweet corn, coconut.

Monoecious   Separate male and female flowers on the same plant.

Mulch   Material placed on the surface of the soil for moisture retention, temperature control, weed control, clean produce, or other reasons.

Mutation (bud sport)   A genetic change within an organism or its parts which changes its characteristics.

Narrowleafed   Having leaves of much greater length than width (e.g., grasses and conifers).

Nematocide   A material for destroying or protecting against nematodes.

Node   Structure on a stem from which leaves arise.

Nursery   Establishment for growing, handling, or retailing woody and herbaceous plants.

Nutrients   Elements available through soil, air, and water, which the plant utilizes in growth.

Organic matter   Plant and animal residues and remains.

Ornamental plants   Those grown for beautification, screening, accent, specimen, color, attraction, and other aesthetic reasons.

Osmosis   Passage of materials through a membrane from areas of high concentration to areas of lower concentration.

Ovary   The flask-shaped female organ of the flower which contains ovules. The undeveloped fruit.

Ovicide   A material for killing insect eggs.
Ovule   The undeveloped seed.

Parenchyma   Unspecialized large cells in storage areas and leaves.
Parthenocarpic   Seedless as in some fruits.
Parthenogenisis   Reproduction without fertilization.
Perennial   Life span of more than two years.
Perfect flower   Contains male and female organs in the same flower.
Pesticide tolerance   Established quantity of a pesticide that can legally remain on harvested products in interstate commerce.
Petals   The usually showy and attractive structures around the essential organs of the flower.
Petiole   Leaf stem.
Phloem   A complex tissue of sieve tubes and parenchyma. Responsible for translocation of food in solution. The inner bark of woody plants.
pH *scale*   A means of expressing acidity or alkalinity. A reading of 7 is neutral (e.g., distilled water), below 7 is acid, and above 7 is alkaline.
Photoperiodism   Response of plants to the daily duration of light.
Photosynthesis   The manufacture of carbohydrates in green plants from the raw materials carbon dioxide and water.
Phytochrome   Pigment in the plant associated with day-length response.
Pinching back   Pruning the soft succulent tip of a shoot to stimulate lateral growth.
Pistil   The female part of a flower consisting of ovary, ovules, style, and stigma.
Plant propagation   The perpetuation or increase in numbers of plants.
Plow sole   Zone of compaction caused by plowing at the same depth continuously.
Pollen   The dust-like spores produced in the anther and containing the male germ cell.
Pollination   Transfer of pollen from anther to stigma.
Pome   Fruit with an embedded core like apple, pear, and quince.
Pomology   Science of fruit growing.
Primordia   Initials or beginnings of structures (e.g., root primordia).
Processing   Preparing for future use as in canning, freezing, preserving, and dehydrating.
Procumbent   Nearly prostrate, spreading.
Productive stage   The period of fruitfulness, productivity, or accumulation of carbohydrates.
Progeny   The young or seedlings of a plant.
Pubescent   Hairy or downy.
Pruning   The judicious removal of leaves, shoots, twigs, branches, or roots of a plant to increase its usefulness.

Quality   Various factors including flavor, texture, appearance, odor, and food value which influence judgment of produce.
Quarantine   Regulation forbidding sale or shipment of plants, usually to prevent disease or insect invasion of an area.

Receptacle   Basal structure of floral parts.

Recessive   A gene whose effect is masked by a paired dominant gene.

Rejuvenation   Stimulation of new growth on old plants usually accomplished by pruning.

Renewal (replacement) *spurs*   Grape canes near the trunk cut back to two buds to provide new fruiting wood in a favorable location.

Renovation   Reinvigoration or rejuvenation to thin plants, remove weeds, and form new plants as in a strawberry bed.

Reproductive   Ability to increase in numbers. Development of storage organs especially those giving rise to new plants.

Respiration   Oxidation of foods by plants and animals for energy release. Carbon dioxide and water are liberated.

Rest period (resting bud)   Period of non-visible growth. Controlled by internal factors.

Rhizome   Horizontal underground stem which forms both roots and shoots at its nodes.

Ringing   Removing a narrow strip of bark around the periphery of the stem to prevent downward translocation of food beyond that point.

Ripeness   State of complete maturation prior to breakdown (see *Maturity*).

Rodenticide   A rodent poison.

Roguing   Removing and destroying undesired plants such as those infected with systemic diseases.

Root hairs   Outgrowths from epidermal cell walls of the root specialized for water and nutrient absorption.

Rootstock   The flesy root of a herbaceous perennial plant with buds or eyes in the upper regions.

Runner   (see *Stolon*).

Sapwood   Outer portion of xylem tissue and the area of upward movement of water and nutrients in a woody stem.

Scaffold branches   The primary branches of a tree arising from the trunk.

Scandent   Vining or climbing.

Scarification (scarifying)   Injuring or scratching the seed coat to aid germination.

Scion (sion, cion)   Piece of twig or shoot inserted on another plant in grafting.

Seed   A ripened ovule consisting of a dormant plant in miniature that bridges the gap between generations.

Seed treatment   Application of fungicides to seeds for protection against damping-off and seed decay organisms.

Seedling   A plant grown from seed.

Self-pollinated   Pollen from the anther is transferred to a stigma of the same plant.

Sepals   The leaf-like structures encasing a flower bud and later subtending the open flower.

Sexual propagation   Increase in plant numbers by seeds.

Shoot   Current season's stem growth with leaves and buds.

Short-day   Less than 12 continuous hours of daylight.

Side-dressing Fertilizer application alongside row crops during growth.

Site A selected place of operation in a locality.

Slip A herbaceous or softwood cutting.

Slurry A thick suspension of an insoluble material in water.

Softwood (greenwood) Herbaceous. Prior to hardening of tissues.

Soil conditioners Chemicals which aggregate soil particles for improved structure.

Soil-less culture (hydroponics) Growing plants in nutrient solution without soil.

Solanaceous Of the nightshade family. Common vegetables include tomato, potato, eggplant, and pepper.

Species A group of plants of similar characteristics to have common use and response. Part of the classification scheme.

Spore Resting stage of a fungus capable of propagation. Usually one-celled.

Spotting board A device for marking a flat for spacing of plants or seeds.

Spur Short, woody stem (branch) which is the principal fruiting area of many tree fruits.

Stamen The male part of a flower consisting of anther and filament.

Starter solution Fertilizer dissolved in water and applied immediately following transplanting.

Sterilization (soil) Inactivation by steam or chemicals to prevent disease, insect, or weed problems.

Stigma The sticky upper surface of the pistil on which the pollen germinates.

Stock The root or stem on which a graft is made.

Stolon (runner) Horizontal stem near the ground surface which forms plants at nodes.

Stomate Pore in the epidermis of a leaf.

Strain A special type selected from a variety.

Stratification Subjection of seeds, usually in alternate layers with a well-aerated media, to complete after-ripening and break the rest period.

Stunted Unthrifty plant restricted in size and vigor because of unfavorable environment.

Style The "neck-like" support joining the ovary and stigma of the flower.

Subordination Severe heading back of all but one of multiple leaders to restore apical dominance.

Succession crop A crop following another crop on the same area during the same season.

Successive plantings Varied planting dates of the same crop to extend the harvest period.

Succulent Watery, as stems and leaves high in water content. Also a group of ornamental plants with thick, fleshy leaves and stems and high in water content.

Sucker Shoot arising from a root.

Sunscald Injured or killed bark or epidermal tissue from excess light intensity and other unfavorable conditions.

Tap root A vertical, central, and deeply penetrating root with few laterals.

Terminal The distal end as of a shoot, twig, or branch.

Terrarium A container, usually of glass, used to provide the total environment for growing plants.

Thatch Built-up layer of clippings and leaf debris on a lawn.

Thinning out Pruning off entire stems or branches to give remaining stems or branches more space and more reserve food.

Tissue tests Determination of plant food needs by chemical analyses of leaves or stems.

Topworking Changing the variety of a tree by inserting buds or grafts on its branches.

Torus The receptacle of bramble fruits.

Training Shaping or adapting plants to specific forms by pruning, tying, or bracing.

Transition stage The integration period of juvenile and productive stages of growth.

Translocation Movement of water, minerals, and food within the plant.

Transpiration Water loss by evaporation from internal leaf surfaces.

Transplanting Moving plants from one container or media to another.

Tree surgery A pruning art involving removal of large limbs, cleaning and treating wounds, and bracing weak trunks and crotches.

Tuber A foreshortened, thick, fleshy underground stem (e.g., white potato).

Turf Grass used for lawn.

Turgidity Plumpness as in cells full of cell sap and moisture.

Twig One-year-old branch or stem of a woody plant.

Undercutting A nursery practice in which a U-blade cuts through soil and roots under trees and shrubs.

Variety A group of closely related plants of common origin which have similar characteristics.

Vegetable The edible product of a herbaceous garden plant.

Vegetative Non-sexual. Also characterized by significant increase in size.

Viability Ability of seeds to germinate.

Virus A submicroscopic protein material capable of regeneration within the host to produce diseases or disease symptoms.

Water rights The legal privilege of water usage.

Waterlogged Without soil aeration due to poor drainage.

Watersprouts Rapid growing shoots that arise from adventitious or latent buds on branches or trunks.

Weed A plant growing where it is not wanted.

Wilting percentage Refers to water remaining in soil at time of permanent wilting in relation to the soil's capacity.

Woody Hard, tough, and fibrous. Non-herbaceous.

Xylem   A complex tissue of tough, fibrous, elongated cells forming vessels and woody tissue. Responsible for upward movement of water and minerals from roots to leaves.

Zygote   The new microscopic organism resulting from union of egg and sperm.

Index

Layerage, 158–162
Leaf, 38
Leaf-bud cuttings, 165, 167, 168
Leaf cuttings, 168, 169, 170
Legal aspects, water, 95
Light, 62–67, 111–113
Light intensity, 62–63
Light quality, 63–65
Lilac, 39, 372
Lily-of-the-valley, 173, 402
Liming, 84, 85, 344
Linkage, 303
Liquefied gas (NH$_3$), 84
Luxury items, 286

M

Margin, profit, 286
Marigold, 395, 407
Marketing, 284–288
Maturation, 121, 122
Maturity, characteristics, 265, 266
Maturity standards, 261–268
Mechanical harvesting, 267, 268, 269
Mendel, Gregor, 296, 297, 298, 299, 302
Meristem, 39
Metamorphosis, 241–242
Mid-rib, 38
Mimosa, 419, 420
Mint, 402
Mites, 242, 243, 257
Miticides, 257
Modified leader, 205, 206, 227–229
Moles, 360
Monocot, 36, 37
Monoecious, 41, 42
Mound layerage, 160, 161
Mountain ash, 369
Mowing, 352, 353, 358
Mulches, spring, 130, 131
Mulches, summer, 132, 133, 380
Mulches, winter, 129, 130
Mulching, 87, 88, 354–355, 443, 461

N

Narcissus, 399, 400
Natural enemies, 260
Nematocides, 257, 258
Nematodes, 243, 440
Nicotiana, 406
Nodes, 39
Nursery stock, 267, 268, 451
Nutrients, mineral, 67–68
Nutrients, non-mineral, 67
Nut trees, 449–453

O

Okra, 435, 440
Olericulture, 6
Onions, 433, 435
Open center, 206, 230
Orchards, 89, 92, 446–447
Organic matter, 78, 79
Organic, muck, 74, 75
Organic, peat, 74, 75
Ornamental crab, 368, 372
Ornamental herbaceous plants, 391–420
Ornamental horticulture, 6
Ornamentals, 31, 32
Ornamentals, marketing, 292, 293, 294
Ornamentals, woody, 364–390
Osmosis, 48
Out-door living, 328, 330–331
Ovary, 28, 41
Overhead irrigation, 104–107
Ovules, 41

P

Parents, 298
Parthenocarpic, 43
Patch budding, 191–192
Peaches, 227–229, 230, 292, 450, 453
Peas, 435, 441
Peony, 212, 213, 399
Pepper, 435, 442
Perennials, 27, 32, 398–405
Perishability, 286, 288–292
Pest control, 240–262, 415, 438–440, 453–455, 458, 463
Pesticide tolerances, 261
Pests, 359–361
Pests, roses, 382
Petiole, 38
Petunia, 407–409
pH, 79, 80
Phenotype, 303
Phloem, 35
Photoperiod, 65–67, 112–113, 458
Photosynthesis, 47, 62
Phytochrome, 111, 112, 113
Phytotron, 113
Piece-root graft, 182
Pinks, 410
Pistil, 41, 304–308
Pith, 34
Plant breeder, 296, 297
Plant breeding, techniques, 298
Plant, containers, 414–419
Plant patent, 312
Plant propagation, 136–198